D0906190

Game Theory in the Social Sciences

Concepts and Solutions

Game Theory in the Social Sciences

Concepts and Solutions

Martin Shubik

e MIT Press
ridge, Massachusetts
England

Second printing, May 1983
© 1982 by The Massachusetts Institute of Technology
All rights reserved. No part of this book may be reproduced in any form or by any means, electronic or mechanical, including photocopying, recording, or by any information storage and retrieval system, without permission in writing from the publisher.

This book was set in VIP Baskerville by DEKR Corporation
and printed and bound in the United States of America.

Library of Congress Cataloging in Publication Data

Shubik, Martin.
Game theory in the social sciences.

Bibliography: p.
Includes index.
1. Social sciences—Mathematical models.
2. Game theory. I. Title.
H61.25.S49 300'.1'5193 82-63
ISBN 0-262-19195-4 AACR2

To Claire and Julie. To Marian. And to Lloyd, who was the only person who understood what I was trying to say and was able to explain it to me.

Contents

Acknowledgments

This book is the outgrowth of many years of joint work with Lloyd Shapley. Our collaboration has been deep and fruitful to me, and it seems strange to be publishing this work under my own name. I feel that it does merit publication, though, and Lloyd has graciously consented to have it published in this form. I neither question his decision nor attempt to read motivations into it.

Preliminary versions of chapters 1–6 have appeared as jointly authored RAND Memoranda R-904/1, 904/2, 904/3, 904/4, and 904/6. Chapter 7 contains much material prepared jointly for a RAND Memorandum which was not finished. Much of the material in chapters 8–11, as well as some of the writing, comes from joint notes or files. Chapter 12 is primarily the product of this author, although even here I have benefited from Lloyd's wide knowledge and critical insights.

I am indebted to many colleagues who have taught me much over the years and corrected my many mathematical errors. These include especially Pradeep Dubey, Robert Weber, Herbert Scarf, Gerard Debreu, Andreu Mas-Colell, Robert Wilson, Robert Aumann, Michael Maschler, John Harsanyi, Reinhard Selten, Edward Paxson, Matthew Sobel, Richard Levitan, Charles Wilson, and David Schmeidler. My thanks also go to Bonnie Sue Boyd, Glena Ames, Karen Marini, and several others for excellent mathematical typing and support in the all too long gestation period for this book. I am grateful to John Nash and John Milnor who have kindly allowed me to reproduce some of their work. And finally, I wish to acknowledge both the Office of Naval Research and the National Science Foundation, under whose generous support much of the research leading to the writing of this work was done.

1 Models and Their Uses

1.1 Introduction

The objective of this work is to develop a fruitful application of the mathematical theory of games to the subject matter of general economic theory in particular and to suggest other applications to the behavioral sciences in general. After the general orientation and mathematical preparations that fill a major portion of this volume, the work will proceed through a series of detailed examinations of particular game-theoretic models representing basic economic processes.

Throughout I shall endeavor to maintain contact with the viewpoints, formulations, and conclusions of traditional economic analysis, as well as with the growing body of published work in economic game theory. Nevertheless, much of the material is new—in its viewpoint, formulation, or conclusions—and is offered as a contribution to a fundamental reshaping of the methodology of theoretical economics and other social sciences.

The economic models treated will cover a reasonably broad portion of static economics, and the game-oriented methodology employed will be reasonably systematic in its application, but I cannot pretend to completeness on either score. Indeed, I feel that a comprehensive, monolithic theory is probably not worth striving for, at least at present, and that the pluralistic approach is the only reasonable way to make progress toward a fully effective mathematical theory of economic activity.

Two arguments in support of this view merit discussion. The first has to do with the overall purpose of mathematical models in the social sciences. The usefulness of mathematical methods—game-theoretic or not—depends upon precision in modeling, and in economics as elsewhere, precise modeling implies a careful and critical selectivity. The nature of the particular question under investigation will determine both the proper scope of the mathematical model and the proper level of detail to be included. Compatibility between different models in the same subject area is desirable, perhaps, but it is a luxury not a necessity. A rigorously consistent superstructure, into which the separate models all nicely fit, is too much to expect perhaps even in principle. A patchwork theory is to be expected, even welcomed, when one is committed as in the social sciences to working with mathematical approximations to a nonmathematical reality.

The second and more remarkable argument for pluralism stems from the nature of multiperson interaction. The general n-person game postulates a separate "free will" for each of the contending parties and is therefore fundamentally indeterminate. To be sure, there are limiting cases, which game-theorists call "inessential games," in which the indeterminacy can be resolved satisfactorily by applying the familiar principle of self-seeking utility maximization or individual rationality. But there is no principle of societal rationality, of comparable resolving power, that can cope with the "essential" game, and none is in sight. Instead, deep-seated paradoxes, challenging our intuitive ideas of what kind of behavior should be called "rational," crop up on all sides as soon as we cross the line from "inessential" to "essential."

Three or four decades ago the "n-person problem"—as this conceptual puzzle will be called—received its first clear mathematical formulation at the hands of John von Neumann and Oskar Morgenstern (1944, esp. pp. 8–15). Beginning with their work, a surprisingly large number of ingenious and insightful solution concepts for n-person games have been proposed by many different authors. Each solution probes some particular aspect of societal rationality, that is, the possible, proposed, or predicted behavior of rational individuals in mutual interaction. But all of them have had to make serious compromises. Inevitably, it seems, sharp predictions or prescriptions can be had only at the expense of severely specialized assumptions about the customs or institutions of the society being modeled. The many intuitively desirable properties that a solution ought to have, taken together, prove to be logically incompatible. A completely satisfying solution is rarely obtained by a single method of attack on any multiperson competitive situation that is "essential" in the above sense.

Instead, we find that in any given application some of the available solution concepts may provide only halfway or ambiguous insights, while others may miss the point entirely. Since no one definition of solution is uniformly superior to all the rest, we shall repeatedly have to "solve" the same economic model from several different standpoints. Out of the various answers, whether they agree or disagree, we may come to a deeper understanding of the real economic situation and its social context.

This may seem an unsatisfactory state of affairs. But the class of phenomena with which game theory is concerned—the behavior of independent decision makers whose fortunes are linked in an interplay of collusion, conflict, and compromise—is one that has eluded all previous attempts at systematic mathematical treatment. Nor is game theory a wholly new departure for economics and the social sciences. On the contrary, we shall often find that one or another of the game solutions, *for a particular model,* corresponds to a standard, classical solution. What is new is the systematic use of the resources of game theory, inconclusive as they may be; this provides both the technique and the raison d'être of our investigations.

1.2 Plan of the Work

The present work is divided into several major parts, each essentially self-contained. This policy suits the subject matter as well as the specialized interests of some prospective readers. It also provides flexibility in exposition, such as the freedom to choose new mathematical notations convenient to the matter at hand. It inevitably entails some repetition of basic definitions and explanatory material, but even the straight-through reader may find this repetition enlightening since each reprise is fashioned to a new set of circumstances.

Much use will be made of simple examples. Many are highly artificial "toy" games, of no particular economic significance; if they are displayed prominently, it is because they will be used repeatedly in later chapters to illustrate new points about the theory as it unfolds.

Simple numerical examples will also be used to illustrate models. And even in their general forms, the models will often be "stripped down," by means of simplifying assumptions, to the bare essentials required to capture the particular phenomenon of interest. The key word here is "essentials." What is essential in a game-theoretic approach is all too often missed in the more traditional approaches, where gross or simplistic behavioral assumptions are found side by side with sophisticated and extremely general renditions of other aspects of the model.

Let me emphasize, however, that the reliance on simple models and examples is only an expository tactic and, to a lesser extent, a research strategy. The theories and methods themselves are deep and complex, and I have no basic quarrel with the quest for abstract mathematical generality, on either the economic or the game-theoretic side of the subject.

A small number of exercises will be found at the ends of the principal sections. Their primary purpose is to enable the reader to confirm his or her grasp of (usually) technical points raised in the text, and they are not intentionally difficult. Occasionally an exercise may also serve as an extension of the text, conveying perhaps an additional definition or result, or suggesting a fresh application.

This volume describes the main elements that go into the construction of a theory of multiperson games for application to the problems of political economy. Conceptual considerations are emphasized over formal machinery, and continuous attention is given to the interrelationships between the real situation, the mathematical model, and the abstract theory.

The first three chapters discuss the art of mathematical modeling in the social sciences, focusing on the questions that have special relevance in a game-theoretic analysis. What are the boundaries that define the model? Who are the decision makers, and how free are they to choose their courses of action? What are the rules of the game, and how do we reduce them to a systematic description? Model building in the social sciences is a substantive part of the theoretical investigation, not merely a routine preliminary step. Checking a model for realism and relevance requires not only knowledge of the subject matter but also awareness of the many technical pitfalls. Seemingly innocent modeling decisions or compromises can sometimes distort the results of subsequent mathematical analysis.

Chapters 4 and 5 are devoted to the motivational foundations: an overview of utility theory considering models of both individual and group preferences. Again, the special viewpoints of game theory (focusing, for example, on multipersonal, interpersonal, or strategic factors) are often called upon to good effect in developing these basic ideas in their most useful form.

Chapters 6 through 11 present the array of solution concepts offered by the theory of games for the exploration and analysis of the many facets of societal rationality. Stress is placed on the definitions of the solutions and the concepts behind them, with only enough simple theorems and examples to give a sense of their application. Throughout these chapters the relationships among the different solution concepts are stressed, and an "intersolutional" summary is laid out in tabular form at the end of chapter 11. The final chapter describes the applications of game theory that have been made, or are in prospect, in half a dozen different disciplines, both in and out of the social sciences.

A subsequent volume will be devoted to the application of this methodology to economic problems. Part I of that volume discusses the special role of money in economics. Part II studies one-sided, "open" models of oligopolistic competition, that is, market models in which the selling firms are regarded as active players with strategic discretion, while the customers are treated as a mechanism, represented in the model merely by a demand function. This approach leaves much to be desired, since many of the most interesting and vital problems concerning oligopolistic competition are undoubtedly dynamic in nature. We cannot even pretend to deal adequately with such problems at this stage in the development of economic theory. The main thesis here is that the theory of games provides a unifying framework for many apparently diverse approaches to the study of oligopolistic competition. Virtually all of the old results can be obtained within this framework, and some new results as well.

Setting up these static models serves the added purpose of formulating a critical problem that is investigated in part III: the relationship between oligopoly theory (or, more generally, the theory of the firm and the so-called partial equilibrium analysis) and the general equilibrium model of a completely closed economy. It is well known that microeconomic analysis in general and oligopoly theory in particular are concerned with the concept of industry and the existence of firms as separate institutions, whereas the firm or individual as a decision-making entity appears, at best, as a very indistinct shadow in current theories of general equilibrium. Our ultimate goal is to reconcile these two approaches in a game-theoretic framework.

Part IV deals with closed models of the economy formulated as games in strategic form and solved using the noncooperative equilibrium solution. In this part the key strategic role of money and financial institutions emerges as the linking mechanism between the implicitly dynamic, essentially institution- and money-based partial equilibrium analysis of the theory of the firm and oligopoly theory and the explicitly static, noninstitutional and nonmonetary general equilibrium theory.

Part V sketches other applications of the theory of games to political economy, in particular to the concept of ownership, to externalities and public goods, and to the assignment of joint costs.

1.3 Mathematical Models in Economics

Mathematical models are used to focus analytical attention on selected features of the economic world. Factors that seem to be relevant are abstracted from the real phenomena and fitted together, in a more or less simplified representation, while those felt to be irrelevant are ignored. Subjective judgments cannot be avoided in economic model building; but the validity of many simplifications and omissions can be tested objectively, at least in principle, by determining whether a larger model, with finer detail or with marginally excluded factors included, would yield substantially the same results.

A full discussion of the art of economic modeling is not intended here. I wish mainly to draw attention to places at which the game-theoretic approach entails special difficulties or dangers, and to post suitable warning signs.

1.3.1 The role of game theory

The questions one is forced to ask, and answer, in transforming an economic model into an economic game, serve to put the model through a series of salutary tests. Indeed, one important function of game theory in economics is to criticize—destructively if need be. It can be used on occasion to expose the inadequacies of old viewpoints and old solutions as well as to provide new viewpoints and new solutions. A case in point is a model discussed in the next volume in which a number of producers compete with each other in the presence of a passive demand function. A careful game-theoretic scrutiny reveals quite clearly the limitations to this one-sided approach. The criticisms are valid independently of any constructive contributions game theory is able to make. In short, game theory provides a new system of tests for rational economic models. (It is really the systemization that is new; the tests and criticisms themselves usually reduce to old-fashioned questions of "common sense" in the formulation or interpretation of the model.)

What are these tests? What distinguishes a game model? The major conceptual ingredients, such as "player," "coalition," "strategy," "payoff," and "solution," will be examined in later chapters. For the present, suffice to say that *multilateral decision making* is the essence of a game model. A theory of games is, among other things, a theory of organization. It deals not so much with feasibility as with negotiability and enforceability—with the power of individuals or groups to influence the distribution of goods and welfare, whether by threats and collusion or by unilateral action. A "solution" to an economic game will accordingly carry sociological and political as well as economic implications.

Perhaps the word "game" was an unfortunate choice for a technical term. Although many rich and interesting analogies can be made to Bridge, Poker, and other parlor games, the usual sense of the word has connotations of fun and amusement, and of removal from the mainstream and the major problems of life. These connotations should not be allowed to obscure the more serious role of game theory in providing a mathematical basis for the study of human interaction, from the viewpoint of the strategic potentialities of individuals and groups.

The construction of mathematical models of economic processes has been practiced at least since the days of Cournot. Most of the older models, however, have set out to portray the behavior either of the single individual (person or firm) or of aggregated masses of individuals. A few have dealt explicitly with the bargaining process or with other forms of specific interpersonal interaction (Edgeworth, 1881; Böhm-Bawerk, 1891; Zeuthen, 1930). But little has been done in the past by economists to explore the complex mazes of power relationships that arise when individuals are able to form groupings and coalitions. It is toward an understanding of the role of such coalitions in economic matters that this work will be especially directed.

1.3.2 Uncertainty

In order to carry out our analysis, it will be necessary to limit our scope and neglect a number of topics of great importance. For one thing, the discussion of the nature and role of uncertainty in economics will be kept to a minimum.

Virtually all of our game models will involve the special kind of uncertainty that is caused by not knowing what other players are going to do; this is called strategic uncertainty. Some will also allow random moves, by players or by Nature, thus bringing the factor of risk (uncertainty with known probabilities) into the players' calculations. But many other kinds of uncertainty, common enough in applications, can only be represented in our conceptual scheme as a lack of knowledge of the "rules of the game." (In our usage the rules of the game include not only the move and information structure and the physical consequences of all decisions, but also the preference systems of all the players.)

Note that we are distinguishing between knowing what other players are going to do and knowing their preferences. This distinction lies at the heart of game theory.

Historically game theory has operated for the most part under the assumption of complete information: all the players know all the rules and can make all necessary calculations (von Neumann and Morgenstern, 1944, pp. 29–30). Only recently has some progress been made in extending the theory to gamelike situations in which the rules are incompletely known (Scarf and Shapley, 1957; Harsanyi, 1962a, 1967, 1968a,b; Aumann and Maschler, 1967; Mertens and Zamir, 1971; Dubey and Shubik, 1977a; Levine and Ponssard, 1977). The present volume hardly ventures into this area and hence must neglect many problems involving uncertainty, such as inventory management, the forecasting of demand (i.e., unknown preferences), or the budgeting of research or advertising costs. This work should therefore be regarded as complementary to what others have done on decision making under risk and uncertainty (Knight, 1933; Savage, 1954; Raiffa and Schlaiffer, 1961; Zabel, 1967, 1969, 1970; Radner, 1968).

1.3.3 Dynamic vs. static models

The most significant relationships between economic and noneconomic factors can be studied adequately only in a dynamic theory. The analysis in this book is, however, almost entirely static. It can detect instabilities and disequilibria, but it is ill-equipped to deal with societies in motion—with the changes in laws and tastes, customs and institutions, that together with economic forces guide the long-run evolution of the socioeconomic and political system.

In human affairs it is desirable to consider situations with an uncertain, possibly infinite, life span. Legally the corporation, for example, is an artificial person with an indefinite life. Games such as Chess or Bridge have definite terminations, but although firms and individuals may be born and may eventually die, there is each day a high probability that the economic, political, and social activities in which they engage will continue tomorrow.

Dissatisfaction with the economic theory of the firm has led to several attempts to formulate a behavioral theory that accounts for indefinite time horizons as well as the possibilities of search, learning, and the changing of goals (Cyert and March, 1963). There is little doubt that in many situations conflicts or clashes of interest are resolved by individuals modifying their goals or desires. Persuasion and debate have their influence. [Rapoport (1960) has stressed the differences between fights, games, and debates; Boulding (1962) has considered the problems of conflict resolution between different types of organizations.]

In the present work preferences are taken as given and unchanging, so that persuasion and learning do not enter into our discussions. Because of this restriction, certain problems which apparently pertain to economics may in fact have no solution in the context of our discourse, even though society solves them daily. If further resolution is required in such instances, our set of basic considerations will have to be enlarged.

We shall be heavily concerned with static, one-period models. Multiperiod models are not beyond the reach of game theory, but they are generally avoided here, our preference being to push forward those applications that seem to promise the greatest immediate returns in new advances and insights.

It is not impossible to have a static theory express dynamic phenomena. One way is through the use of what, in game theory, are called *strategies.* A sequence of decisions, even a whole policy of action, is reformulated as a single, contingency-laden decision or "game plan." The multimove game can then be reduced to a game in which each player makes just one move—his choice of a "strategy." This reduction, first given a general mathematical formulation by Emile Borel (1921), has more conceptual than practical importance, but even a modest application of it may significantly extend the temporal scope of a one-move model. In the study of oligopoly, for example, we shall be able to treat the movement of firms into and out of an industry by combining the entry-exit decision with the production decision.

Another way to have a static model express a process over time is to reinterpret the static solutions as steady-state solutions of a corresponding model that is in continuous (or periodic) operation, so that quantities of goods, for example, become rates of flow. This is a familiar method in many kinds of equilibrium analysis. In a game-theoretic model there is an added hazard in the interpretation, in that some of the solution concepts may not make sense as steady-state equilibria or may require special justification for the intended application.

1.3.4 Noneconomic factors

Much of our analysis will be concerned with markets and the evolution of price systems. By using more than one concept of solution to an economic problem, however, we can consider the resolution of economic affairs by nonmarket means as well. The market form and private ownership are not natural to all economic societies. It has been suggested that the impersonal market with prices may have emerged first in Greece, after the time of Aristotle (Polanyi, Arensburg, and Pearson, 1957). Auctions are also apparently of western origin; the earliest references seem to be Roman (Talamanca, 1954), though Herodotus notes a Babylonian auction for wives. In many societies exchange rates were fixed by fiat, and traders were a class apart (Polanyi, Arensburg, and Pearson, 1957). In other societies land and even children have been regarded as communal property, and as such were not owned, exchanged, or traded but were used, employed, or cared for according to sets of rules, customs, usages, and taboos specifying the relationship between individuals, families, groups, and their environment.

I make these observations here because my basic approach to economics is through the construction of mathematical models in which the "rules of the game" derive not only from the economics and technology of the situation, but from the sociological, political, and legal structure of the society as well. Private ownership of land, the rights of various public or private groups to tax, and the existence of certain financial institutions are examples of legal and social features that may require delineation in a particular model. Similarly, rigid prices, a class structure that sets money lenders and merchants apart from aristocrats, priests, and peasants, and the redistribution of economic goods by appeals to social justice or time-honored custom are noneconomic factors that may be present to some degree and that should be reflected in any realistic theoretical model. The existing tools of game theory permit us to make a beginning in this direction (Shapley and Shubik, 1967b).

1.4 Boundaries of the Model

A model is defined by its boundaries. The suitability of a model depends as much on what is left out as on what is put in. Especially important are the "boundary conditions" that tie the modeled structure to its unmodeled environment. Boundary conditions often conceal tacit assumptions that influence the behavior of the system; the model builder has a duty to expose such assumptions to critical scrutiny.

In building a one-period model, we run a danger of overconstraining economic activity if we insist that trade, production, and budget conditions all balance out exactly. There are several ways in which such stringency can be relaxed, in the interest of greater realism, without introducing an undesirable amount of special institutional detail. The model may be left partially "open," for example by stipulating an exogenous foreign sector; this is often done in models of national economies, where the accounting can be out of balance by an amount attributable to foreign trade. Another way is to move toward a dynamic model, with credit or temporal transfers of assets permitted, so that budget conditions do not have to balance in each period, but only at the end. Care must be used here, however, since if the model is of indefinite duration, books may never balance and certain essential constraints will be nullified (as in check kiting or chain letter schemes).

One of the fundamental problems that have hampered the integration of oligopoly theory with general equilibrium analysis has been the construction of satisfactory models of oligopolistic processes in closed economies. Other important areas where clear analysis has been delayed, by a lack of care in specifying whether the economic models are meant to be open or closed, concern taxation, public finance, and ownership: difficulties are encountered, for example, in dealing with external economies and diseconomies, public goods, and joint ownership. I believe that the methods developed here can provide the correct approach to all these areas, though they will be applied intensively here only in the case of oligopoly theory.

1.4.1 Game-theoretic closure

The terms "open" and "closed" have been used so far to refer to economic factors of all kinds. We shall also have occasion to distinguish between open and closed game models. Roughly speaking, game-theoretic closure requires that all represented activity that involves human motivations result from free decisions by full-fledged players of the game.

Thus a model of a wheat market that allows freedom of choice to the suppliers of wheat but represents the buyers only through an aggregate demand function would have to be classified as open, since even with individual demand schedules the buyers would remain mechanisms and hence be merely part of the environment from the point of view of game theory.

A general trading and production game, on the other hand, can be formulated as a closed model, incorporating the utilities and strategies of all individuals in the economy. A frequent characteristic of such models is that the economic gains are extracted from the nonhuman environment, or "Nature." This is in contrast to the example of the wheat market, where the suppliers' gains would be primarily extracted from their customers.

Some intermediate cases will also be considered. It may be of interest to close an open model to the extent of taking into account the utilities of all affected individuals, but with one group of individuals—consumers, for example—still constrained to act as strategic dummies. An instance of this type of model might be an investigation of oligopolistic behavior in which we are interested in the effects of alternative social policies; in this case it is desirable to consider the welfare of the public while nevertheless constraining them to be price takers and hence strategically impotent.

1.4.2 End effects

The boundaries so far discussed might be termed spatial, in that they limit and define the scope and content of a model. The temporal boundaries are also important. I have alluded already to the time element in dynamic models, but even a static, one-period game has a beginning and an end.

Setting the initial conditions for a game model is very much like the corresponding task for a mechanistic model, but there is one novel consideration. It is necessary to find a way to include, among the initial conditions, any prior commitments or other inherited restraints on the players' freedom of choice. The question of coalitions, for example, can be very delicate. Assuming that one wants to avoid arbitrary or overly detailed sociological modeling, how can one take proper account of the effect of a coalition in existence at the start of the game? Since the different solution concepts for cooperative games take somewhat different views of the coalition process, there is no single answer to this question.

At the end of the game we may come up against similar intangibles. The outcome—the final state of the system—must be described in sufficient depth to allow for its accurate evaluation in the preference scales of the players. Hence future prospects must be considered, as well as immediate profits and losses. Again, coalitional effects can be bothersome. Can a transgressor really "pay his debt to society" and regain a "clean slate"? Or if a man double-crosses his partner during the game, without incurring any specific penalty, then are not his future prospects dimmed in a society based upon trust? If so, how can we express this quantitatively?

Ordinarily we shall try to build our models so as to avoid these effects, or at least to minimize them. Ideally we should strive for a complete dissociation of the game of the model from all other games in the environment, past or future.

1.4.3 Level of detail

Yet another kind of boundary can be discerned: the threshold below which detailed information is not sought or, if available, is deliberately suppressed. A sense of proportion is important in working out the fine structure of a model. It is obviously foolish to spell out some parts in microscopic detail if their effect on the solution is going to be overwhelmed by a gross or arbitrary assumption elsewhere in the model. Omitting inessential detail not only saves trouble, but also increases the generality of the result.

For example, we shall often exploit the fact that our major solution concepts are relatively insensitive to the intricacies of strategy and maneuver. (This is because our primary concern will generally lie with the coalitional rather than the tactical aspects of competitive behavior.) Accordingly, the rules of the game can often be stated in a few descriptive words or formulas, rather than in a detailed catalog of possible moves and countermoves.

Suppression of available but inessential detail is usually accomplished in one of three ways:

1. approximation (for example, the use of simple functional forms in place of less convenient but more accurate representations);

2. symmetrization (the assumption that similar individuals, factors, or relations are in fact exactly alike); or

3. aggregation.

These techniques are all common in nongame models, and we have only two special warnings to make here. First, nonsymmetric solutions often exist to symmetric games and should not be unwittingly assumed away. Second, there are somewhat deeper pitfalls than usual when aggregation is applied to people or firms rather than to inanimate things such as goods, technological factors, or time periods, since the very identity of the basic "player" unit is at stake. For example, if the members of an organization are engaged in a significant game for the control of the organization, then it may be unsound to treat the organization itself as a single player in some wider model (Shapley, 1964a, 1967c).

1.4.4 Sensitivity analysis

The use of simple functional forms or other modeling simplifications as approximations to a complex real situation always requires justification. The true measure of the success of an approximation—that is, its "closeness"—comes in the output of the model, not the input. Thus, to test the validity of one's conclusions, the model user should engage wherever possible in what, in operations research, is called sensitivity analysis (e.g., Quade, 1964, p. 172). That is, the modeler should make slight variations in the parameters or other assumptions of the model and observe their effect on the solution.

Unfortunately, game-theoretic models tend to have a great many extra assumptions that enter the structure in such discrete ways that it is not easy even to say what is meant by a "slight" variation. Examples of structural assumptions are (1) the status of a player (free agent or behavioral robot?); (2) the availability of information to the players; (3) the presence or absence of a monetary system; (4) the choice of a solution concept. As a rule of thumb, we might say that two structures are close approximations if they are both plausible renderings into formal language of the same verbal description. If we find that the solution of the model is sensitive to the difference between two such renditions, then our original verbal description was faulty or incomplete at some vital point, and we must start over again.

The systematic use of intersolutional analysis affords some defense against undetected sensitivities in our models—particularly those of type (4) above. Indeed, this is the primary argument for the concerted use of several solution concepts in economics and the social sciences. Another defense is more of a retreat: we deliberately restrict our attention to the more qualitative features of the results and attach little weight to precise numerical predictions or recommendations that come out of the mathematical solutions. (This is not to say that numerical results will be avoided, but that their purpose is almost invariably nonquantitative—to give tangible illustration of logical relationships, orders of magnitude, or trends.) Thus in the next volume virtually all of the formal analysis of oligopoly will be conducted on the basis of a simple, linear demand function and almost equally simple, linear cost curves. My position will be that without leaving this simple world we can obtain most of the significant qualitative features of oligopolistic competition that are accessible to a static, one-sided treatment. Only occasionally will the introduction of more general functional forms add significantly to the force of our conclusions.

2 Decision Makers

2.1 The Concept of a Player

A game begins with, and is centered around, a specified set of decision makers who are called the *players*. Each player has some array of resources at his disposal, some spectrum of alternative courses of action (including attempts to communicate and collaborate), and some inherent system of preferences or utility concerning the possible outcomes. Thus in an economic "game" the rules will include tastes and technology as well as laws, initial endowments, and distribution and information channels, and the rules must relate each of these features to the individual player.

Outside the game model the players are all alike—the theory refuses to distinguish among them. Any and all differentiating properties ought to be embedded in the description of the game. Admittedly it is easier to preach than to practice this principle of *external symmetry*. But even in difficult cases (such as a game whose "players" are several firms and a labor union, or a regulatory board and a group of public utilities, or a bank and its customers), the very attempt to achieve external symmetry will be beneficial and should at least lead to a classification of the limitations of the model.

2.1.1 Free will and behavioristic assumptions

In our game models we shall always assume that the identified players are rational, conscious decision makers having well-defined goals and exercising freedom of choice within prescribed limits. What they do with that freedom is a question for the solution theories to answer, not the rules of the game. The answers, as has already been stressed, will generally fall far short of a deterministic prescription of behavior.

Despite this essential element of free will, the rules of the game may nevertheless severely restrict a player's behavior. An extreme example is the consumer who is constrained to accept the prices named by other players and to maximize his welfare subject to that constraint. Thus he may retain no conscious control over his purchases in the market. Yet even such a "strategic dummy" is not barred a priori from negotiations or other forms of cooperative activity, as contemplated in some solution theories.[1]*

In a model that is not game-theoretically closed there will be individuals who are represented not as players but as automata, or as an aggregate mechanism, and their behavior may or may not be related to any conscious optimization process. For example, we might represent the demand for the commodity in a market by a function $q = f(p)$, where q denotes the quantity that will be purchased at the price p. If we look behind this function, we may find a purely behavioral statement, based on habit, tribal customs, or sociopsychological considerations. On the other hand, we may find a summation of many separate welfare-maximizing decisions, which in a larger, closed model we could account for on an individual basis.

Having adopted a pluralistic attitude toward the role of solutions in game theory, we must stress that we do not propose just one model of the individual. The utilitarian, rationalistic model of political and economic man may be valid for much of the professional behavior of those individuals who are primarily politicians or economic agents. But as Dahl (1961) has noted, few citizens can be considered as being professionally involved in politics, and the typical housewife or employee can scarcely be described as individually exerting economic power.

A cursory examination of decision making in economic and political life, conditioned as it is by deadlines, costs of information and computation, lack of knowledge, lack of understanding of one's own desires, lethargy, and a host of other "frictions," shows that subconscious or semirational processes and routines account for a large part of human behavior—especially "nonprofessional" behavior. Behavioristic theories, though we shun them, are not totally in conflict with the game-theoretic approach. Most of the above-named frictions are manifestations of limited information, either within the rules or about the rules. To the extent that they constrain or routinize behavior, they could be built into a gamelike model (though possibly with incomplete information) by formalizing the requisite feedback mechanisms for adjusting actions, aspirations, and so forth (Cyert and March, 1963, esp. chap. 6). As long as the behavioral constraints leave some element of free choice in the *interpersonal* activities of the players, though, the theory of games will have a vital role to play.

* Notes will be found at the end of the chapter.

2.1.2 Economic units, players, and people

The player is perforce the basic decision unit of the game. He is also the basic evaluation unit. In economic models he may be an individual consumer, a household, a firm or financial institution, a manager, a labor union, a government agency, or even a whole nation or its leader.

Specification of the players in a model carries with it tacit assumptions. For example, if the players are firms, then any internal organizational problems are assumed to be accounted for, so that the firm "speaks with one voice." In particular, the problem of defining the internal social welfare function, for the aggregate of individuals (or impersonal interests) that make up the firm, is assumed to have been solved. Similar remarks apply to other types of aggregated players.

The modeling of aggregates as a single player always presents difficulties, precisely because the group-decision and group-welfare problems are, in a certain fundamental sense, unsolvable. In international politics, to take a glaring example, the representation of a country as a single individual can be dangerously misleading. The analogies between national and individual psychology may be convenient and superficially persuasive, yet they are inherently unsound.[2]

2.1.3 Dummies

A *dummy* is a player that is not a player in some vital respect. He may appear in the model for the sake of formal completeness, or as a limiting case when parameters are varied. A *strategic dummy* is one who is so constricted by the rules that he has no choice in his actions. He may nevertheless have a stake in the outcome, and this may have an effect on the solution or on our interpretation of the solution.[3]

Sometimes a player, though not a strategic dummy under the rules, becomes equivalent to one after later analysis has eliminated all but one of his strategies. The model builder should beware anticipating such eliminations too far in advance—for example, replacing a potential player by a mechanism because "he has only one reasonable course of action." Domination arguments to reduce the number of alternatives, which are so useful in decision theory (and in two-person zero-sum game theory), must be used with great circumspection in n-person game theory. (The Shooting Match to be analyzed in section 2.2.2 is a good example.) Deliberately unreasonable behavior, especially in deterrence situations but in other contexts as well, can be a significant factor and should not be dismissed out of hand.

Two other kinds of dummy should be mentioned. The first, a formal fictitious player, is sometimes invoked as a device to balance the books in a non-zero-sum game (von Neumann and Morgenstern, 1944, p. 505ff.). He is a strategic and coalitional dummy but has nevertheless a well-defined payoff function, equal to the negative of the sum of the real players' payoffs. This formal device may be contrasted with yet another type of dummy, the statistician's opponent in a "game against Nature"; that opponent has strategic choice but no motivation or payoff (Blackwell and Girshick, 1954; Milnor, 1954). We shall have little occasion to refer to either of these devices in the present volume.

2.1.4 Coalitions

The term "coalition" will ordinarily be used in a neutral sense, without structural or institutional implications. A coalition is a mathematical figment; it is not the analog of a firm or a trade union, where there would be codified rules limiting the actions of the officers and members. In the formal theory a coalition can be any subset of players (even the all-player set), considered in the context of their *possible* collaboration. Thus we shall ordinarily not speak of coalitions "forming" during the game, or being forbidden, and so forth; indeed, it could be said that all coalitions are "in being" simultaneously.

Conversely, if a firm, cartel, union, or other entity is represented in the model in such detail that the members of the group are players in the game, then it will not suffice to say merely that these players constitute a coalition. The fact that they are organized in some way, and are perhaps not entirely free agents, should be represented by stated institutional costs, commitments, and constraints included among the rules of the game. Their corporate existence can sometimes, in fact, be modeled in the form of an additional player—a *manager*, let us say, who is narrowly restricted in his actions and whose preferences are directly tied to the fortunes of the organization (Shapley and Shubik, 1967b).

It should be pointed out, however, that the internal workings of organizations and institutions are difficult to treat adequately with the predominantly static tools of this book. Accordingly we shall generally try to avoid such questions in our models, despite their high game-theoretic content. [Two empirical studies of political coalition formation utilizing game-theoretic models are Riker (1959) and Leiserson (1968).]

2.2 Illustrating the *n*-Person Problem

In a game model the very number of players may have a decisive effect on the nature of the game, on the nature of the analysis to be employed, and on the nature of the conclusions to be expected. Von Neumann and Morgenstern (1944, pp. 339, 403) remarked on the repeated appearance of qualitatively new phenomena with every increase in the number of players. The inherent difficulties in extending ordinary rationality concepts into multiperson situations have already been emphasized. A wide gulf separates the well-regulated domain of "inessential" two-person games from the wilderness area of "essential" n-person games. Since the pervasive n-person problem exerts such an influence over the theory as a whole, I have chosen for a first, simple excursion into mathematical analysis a pair of similar games situated on opposite sides of that gulf.

In the first example there are two players whose interests point in directly opposite directions and so, in a useful sense of the word, are parallel. As a result, straightforward "rational" arguments lead us convincingly to a satisfactory solution of the game, using the "minimax" principle.[4] But in the second example, with three players competing for the prize, there is no parallelism, and the same kind of "rational" arguments lead only to a treacherous sort of equilibrium, which on closer inspection turns out to be surprisingly unreasonable.

2.2.1 A two-person example

DART DUEL. *Players A and B approach balloons marked B and A, respectively, each with a dart to throw when he wishes. The player whose balloon is first hit, loses.*[5]

To analyze this game, let the accuracy functions (hit probabilities) be $a(t)$ and $b(t)$, respectively, with both functions assumed to increase continuously from 0 to 1. If your opponent throws and misses, you should obviously hold your dart until you have a sure hit. But if he throws and hits, the game is over. The only question, then, is how long you should wait before throwing your dart. It is obviously bad to throw too soon, but it is also risky to wait too long.

Let us assume that the players have decided to throw at times $t = x$ and $t = y$, respectively. Then A's probability of winning is $a(x)$ if he gets to throw first, or $1 - b(y)$ if B throws first. If we suppose that simultaneous-hit and no-hit games are decided by lot, then the following *payoff function* describes A's prospects:

$$P_A(x, y) = \begin{cases} a(x) & \text{if } x < y, \\ 1 - b(y) & \text{if } x > y, \\ \frac{1}{2}[a(x) + 1 - b(y)] & \text{if } x = y. \end{cases}$$

Since someone always wins,

$$P_B(x, y) = 1 - P_A(x, y).$$

Now suppose that A assumes, cautiously, that his time decision x will be guessed in advance by B, or perhaps deduced on theoretical grounds. He can then hope for no better payoff than

$$Q_A(x) = \min[a(x), 1 - b(x)],$$

since B can choose either to wait until $t = 1$ or to throw just before $t = x$. To make the best of it, A must try to maximize $Q_A(x)$. But this means choosing $x = t_0$, where t_0 is defined by the equation $a(t_0) = 1 - b(t_0)$, or

$$a(t_0) + b(t_0) = 1. \tag{2.1}$$

(See figure 2.1.) By throwing his dart at this time (if B has not yet thrown), A can assure a win probability of at least $a(t_0)$ regardless of what B knows or does.

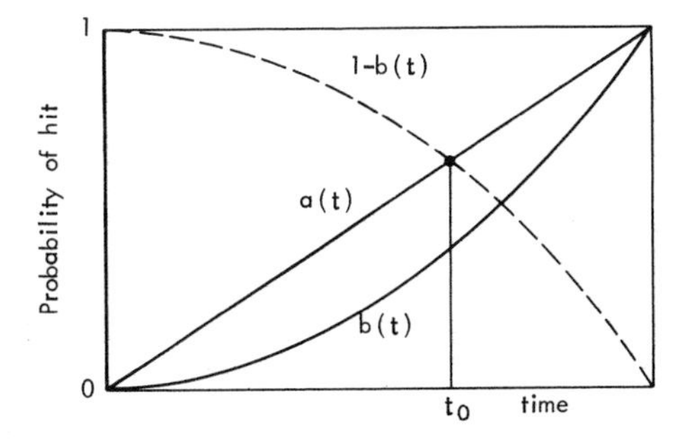

Figure 2.1
Dart Duel hit probabilities (1).

By the same token, a cautious B, by solving the same equation (2.1) and taking $y = t_0$, can assure a win probability of at least $b(t_0)$. But these two win probabilities add up to 1. There is no room for improvement over the "cautious" strategies. They are therefore *optimal strategies,* and the game is solved. The rule for correct play for either player can be stated very simply: "Throw your dart as soon as the sum of the accuracies reaches 1."[6]

For a numerical example, let $a(t) \equiv t$, $b(t) \equiv t^2$, as in figure 2.1, making A a uniformly better marksman. Then, by elementary algebra, both players should throw at time $t_0 = (\sqrt{5} - 1)/2 = 0.618$. A's probability of winning is 0.618; B's is 0.382.

2.2.2 A three-person example

SHOOTING MATCH. *Contestants A, B, and C fire at each other's balloon with pistols, from fixed positions. At the beginning, and after each shot, the players with unbroken balloons decide by lot who is to shoot next. The surviving balloon determines the winner.*[7]

This game does not correspond exactly to the Dart Duel. Here the problem is not when to fire, but at what. Also we do not know how long the game will last. As soon as one player is eliminated, however, the other two players become strategic dummies, and a direct calculation can be made.

Suppose that C is out and that A and B have fixed accuracies a and b, respectively. The next shot will result in one of the following events: an immediate win for A (probability $a/2$); an immediate win for B (probability $b/2$); or a miss and repetition of the status quo (probability $1 - a/2 - b/2$). Writing $P_{A,B}$ for the probability that A will ultimately win when pitted against B alone, we have

$$P_{A,B} = \frac{a}{2} + \left(1 - \frac{a}{2} - \frac{b}{2}\right) P_{A,B}. \tag{2.2}$$

Solving this equation gives $P_{A,B} = a/(a + b)$. In similar fashion we find that $P_{A,C} = a/(a + c)$, and so on.

Now suppose that the players are ranked in skill by $a > b > c$. Then it seems clear whose balloon A should attack in the first part of the match. His chances of an immediate hit are the same, but he would definitely prefer to shoot it out later with C rather than B since $P_{A,C} > P_{A,B}$. Therefore he aims at B's balloon. For exactly analogous reasons B and C both aim at A's balloon. (This is the "domination argument" mentioned in section 2.1.3.)

The strategic question settled, we can complete the calculations. Write P_A for A's initial chance of winning. Considering just the first round, we have

$$P_A = \left(\frac{a}{3}\right) P_{A,C} + \left(1 - \frac{a}{3} - \frac{b}{3} - \frac{c}{3}\right) P_A,$$

in analogy to (2.2). The equations for the other two players are similar in form. After a little pencil work, we obtain

$$\begin{aligned} P_A &= \frac{a^2}{(a+b+c)(a+c)}, \\ P_B &= \frac{b}{a+b+c}, \\ P_C &= \frac{c(2a+c)}{(a+b+c)(a+c)}. \end{aligned} \tag{2.3}$$

These add up to 1, since an infinitely protracted match has probability 0.

Apparently the game has been solved. We have deduced a rule for rational play: "Shoot first at the balloon of your stronger opponent." And we have found formulas for the resulting expectations. Everything seems in order. "What is all the shooting about?" the reader may wonder.

2.2.3 Critique of the solution

For reply, we insert some numerical values into our "solution" (2.3), say $a = 0.8$, $b = 0.6$, $c = 0.4$. Some further pencil work then reveals a paradox:

$$P_A = 0.296, \quad P_B = 0.333, \quad P_C = 0.370. \tag{2.4}$$

The order of skill has been reversed! The poorest shot has the best chance to win!

A little reflection shows what has happened. If A and B foolishly insist on being "rational," they end up cutting each other's throats, to the great benefit of C.[8] Another calculation can be made to show the extent of their folly. Had they somehow settled on the "irrational" policy of first shooting down C's balloon, then their winning chances would be 0.444 and 0.465, respectively—a striking improvement over (2.4). It is hard to see how they could be content with the solution given above if there were any way for them to agree, openly or covertly, on that "irrational" course of action.

There is another way in which A might make his superiority felt, which requires no collusion or "gentlemen's agreement" with B. Let A openly commit himself to shoot at C's balloon if ever C shoots at his (and misses!), but otherwise let him shoot at B's balloon as in his original "rational" strategy. Given this commitment on A's part, B will still find it advantageous to shoot at A's balloon. But C will now find it better to defer to the threat of retaliation and to attack B instead of A.[9] The resulting expectations come to 0.444, 0.200, and 0.356, respectively, with B definitely getting the worst of it. The key to this solution, which seems intuitively not unreasonable, is the presumption that A can commit himself to an "irrational" act of revenge, which will never actually be carried out—in short, that he can employ a strategy of deterrence.

We do not assert that the original solution (2.3) is necessarily wrong, or that it could not occur in practice among intelligent players. The moral of the example is, rather, that multiperson games cannot be properly analyzed or solved until adequate information is provided about the social climate—in particular, about the possibilities for communication, compensation, commitment, and trust. Given this sociological information, one can proceed to the selection of a suitable solution concept.

EXERCISES

2.1. *Suppose the accuracy functions in the Dart Duel have the form indicated in figure 2.2. What is the rule for optimal play?*

2.2. *In the Shooting Match, suppose that* $a > b = c > 0$. *What is the analog of the "rational" solution described in section 2.2.2?*

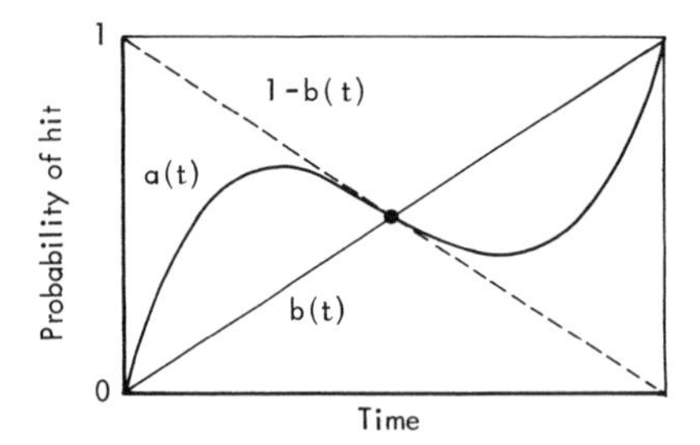

Figure 2.2
Dart Duel hit probabilities (2).

Suppose that instead of the three individuals shooting at balloons there were three bull moose or three elks willing to contest for control of the herd. Which two will fight? Or will there be a sequence of pairwise fights, possibly with the strongest wearing himself out fighting with the second strongest only to be beaten by the third contestant? In a private communication K. Lorenz has indicated that the type of fighting that takes place is species-specific, and it is not yet possible to generalize with any confidence about the overall nature of the competition that takes place within different species.

2.3 Increasing the Number of Players

In applications of game theory to economic, political, or social problems it is frequently important to be able to vary the number of players, and especially to consider the effect of letting the number of players increase without bound. The behavior manifested by mass markets or political systems with many millions of voters may differ considerably from the behavior of similar systems with only a handful of participants, and an elucidation of these differences is one of the major goals of game-theoretic analysis.

To illustrate the modeling techniques required, we begin with a market having just two traders, each with a large amount of goods. We may wish to compare the outcome from trade in this market with the outcomes in similar bilateral markets having four, six, or more traders. One way to do this is to replicate the market, that is, to add more traders identical to the existing traders in every way, including the same preferences, endowments, and strategic possibilities. This construction is equivalent to an expansion of the spatial boundaries of the market to encompass additional markets identical to the first (Edgeworth, 1881; Shubik, 1959a; Debreu and Scarf, 1963; Shapley, 1964d; Shapley and Shubik, 1966, 1969d).

Another approach is to fracture the original system by breaking the players in half, or into smaller pieces, and endowing the resulting new traders with suitably scaled-down versions of the attributes that described the original traders. This construction is equivalent to treating the original traders as aggregates who are now being disaggregated (Shapley and Shubik, 1972b).

For many purposes these two ways of generating games with larger numbers of players are mathematically equivalent, but this is not always the case. For one thing, if we pass to the limit, the replication method leads most naturally to a countable infinity of players, whereas the fracturing method leads most naturally to a game in which the players form an uncountable continuum (Shapley and Shapiro, 1960; Kannai, 1966). There are of course many other, less symmetrical ways of enlarging the player set while preserving other features of the model. For example, one or two large firms might hold fixed shares of the productive capacity of the industry while the others are fractured into progressively smaller units.

In economics, although many verbal accounts of market behavior, from Adam Smith on, pay lip service to the importance of the number of competitors, in most of the mathematical models of large systems the properties of mass behavior are implicitly or explicitly assumed a priori (through a demand function perhaps), rather than deduced as limiting properties of individual behavior in small markets. The game-theory approach, beginning in this case with Edgeworth (1881) but lying dormant for three-quarters of a century, allows us to make a careful distinction between assumptions and deduction in this regard. Indeed a major portion of current research in economic game theory is devoted to just this question: investigating the limiting behavior of various game-theoretic solutions and studying the extent to which these limits agree with other models of mass decision making.

2.3.1 Infinite player sets

Games with infinite sets of players have been attracting increasing attention as models for mass phenomena, both economic and political. Though unrealistic, or even absurd, on their face, such models are similar in purpose to the ideal constructs commonly found in the physical sciences, as for example when the large number of discrete particles that make up a fluid are represented by a continuous medium. The presumption is that the behavior of the infinite, continuous model, as analyzed by the methods of differential calculus, is indicative of the behavior that would be approached in the limit if a series of increasingly refined but increasingly complex finite models were considered instead.

Several different kinds of infinite-person games are found in the mathematical literature. In the *countable* case, the set of players can be identified with the sequence of natural numbers $\{1, 2, 3, \ldots\}$. Although each player can have a perceptible effect on the course of play, the rules are usually defined so that this effect dwindles to next-to-nothing for players who are far out of the sequence.

A simple illustration is the infinite-person voting game, in which the nth player casts a fraction $1/2^n$ of the total vote:

$$w_1 = \frac{1}{2}, \quad w_2 = \frac{1}{4}, \quad w_3 = \frac{1}{8}, \quad w_4 = \frac{1}{16}, \cdots.$$

If a two-thirds majority is needed to win, then every player can be shown to have some influence. This may be seen for player 4, for example, from the fact that $\{1, 3, 4\}$ is a winning combination, but $\{1, 3\}$ is not (Shapley, 1962c). (Curiously, if the majority required to win is set at five-eighths instead of two-thirds, then player 4 and those following him are essentially powerless.)

In the *uncountable* or *continuous* case, the players may be like the real numbers in an interval or the points in a multidimensional region. The possibility then arises of a "nonatomic" game in which none of the players have any influence as individuals or even in finite coalitions.[10] There is a strong tie here with measure theory and with the idea that finite sets of points, or other suitably sparse point sets, can have "measure zero" (zero weight, length, area, probability, etc.) and can therefore be disregarded. Nonatomic games have been pressed into service as models of large exchange or exchange-and-production economies.[11]

Continuous infinite games need not be nonatomic. Indeed among the first such games to be studied were the so-called oceanic games in which a favored few "major" players have individual voting power, which they wield amidst a continuum of "minor" players whose infinitesimal votes count only in the mass. The conceptual archetype of this mixed atomic-continuous model is a corporation with a small coterie of major stockholders, the other holdings being of insignificant size but very numerous.[12]

There are still other kinds of games with uncountably many players. One simple example is the game of unanimous consent, which, though a voting game, has no ocean of minor players since each individual is a "giant," capable of upsetting the applecart.[13]

2.4 Five Decision-Making Types

In light of our belief in the need for many models of economic, political, and social man, and in preparation for coming mathematical models of the individual, we shall attempt to delineate in this section five contrasting roles for the individual decision maker. Taxonomies can quickly become tedious, especially when the use for the various subdivisions is not clear. The present typology is meant only to be suggestive and is neither exhaustive nor exclusive; people and professions can readily be found in economic and political life that do not fit neatly into the categories suggested.

Our five types are: the citizen (as consumer, laborer, and voter), the industrialist or businessman, the financier, the politician, and the administrator or bureaucrat. Our eventual concern will be primarily with the first two.

2.4.1 The citizen

The citizen is considered in his roles as consumer, supplier of labor, small tradesman or producer, and participant in political life. For the most part he is (and regards himself as) strategically powerless as an individual. As a consumer he is almost always a price-taker, though in some markets he may bargain. If he is somewhat more sophisticated than most, he may do some price-forecasting and stockpiling. As a producer most individuals are wage earners, self-employed professionals, or small farmers or businessmen. In the ranks of the middle and upper executives and professionals there may be more leeway for bilateral bargaining, as there is for organized labor; otherwise most labor services are sold by price-takers.

Financially most citizens may be more correctly described as savers than as investors. Beyond their investment in their houses and consumer durables they buy financial claims rather than real assets. In this role they are again price-takers.

Few individuals are politically active, and even among those who are, few are active on more than one or two issues. This being the case, a politicoeconomic model describing voting in behavioristic terms could provide a good approximation of the political activity of most citizens in the game of politics.

2.4.2 The industrialist or businessman

The category of industrialists and businessmen is limited to senior executives of the larger business institutions of a society. Considering only the institutions appearing in the "*Fortune* 500," we estimate around five to fifteen thousand such individuals in the United States.[14] If we went to a 90 percent level of control of assets, this group might include several hundred thousand decision makers.

For the most part these businessmen are not direct owners but the trustees of the funds of others. Although money need not be their only measure of incentive, monetary measurements and profits play an important part in the shaping of their activities. In most instances in the United States they are constrained by law from overt collusion or joint action with competing institutions.

Many of the large corporations or business institutions dealing directly with the public are in an oligopolistic situation which can be best described as a noncooperative game among the businesses, with the customers represented passively as price-takers. Many businesses are also heavily engaged in intercorporate trade in which oligopolistic-oligopsonistic relations may be encountered; this situation would have to be modeled as a cooperative (or at least quasicooperative) game. And then there is the large sector of regulated industry, such as utilities and railroads, in which the situation may again be best represented as a cooperative negotiatory game, with the government's actions subject to a voting mechanism.

Finally, there is a considerable amount of economic traffic between private firms and governmental institutions. On the business side of this market noncooperative decision making is a reasonable approximation; on the government side policy coordination as well as voting considerations may influence strategy. An example is provided by the noncooperative sealed bidding conducted by government agencies after approval has been voted for a project.

The businessman will, in general, treat the individual citizen–customer–laborer–small stockholder as part of an aggregate whose behavior may be predicted.

2.4.3 The financier

The financial decision makers include the leaders of banking institutions of various sorts, insurance companies, savings and loan societies, and other financial institutions. At most they comprise a few tens of thousands of individuals who obtain funds mainly as oligopsonists or competitors. They are trustees of the funds of others. Profits and monetary measures of incentive are possibly even more important for this group than for other businessmen.

One of their major roles appears to be as outside evaluators and comparers of the monetary risk and monetary worth of enterprises. They are measurers of the levels of confidence, honesty, cooperation, and capability exhibited by other individuals and institutions.

2.4.4 The politician

The politician is a trustee of the "power," and to a certain extent the assets, of the mass of citizens. Though he may view them as an aggregate behavioral mechanism with no strategic initiative, he recognizes that this mechanism reflects the tastes, preferences, and habits of real individuals. Far more than the member of any other group, the political decision maker must consider interpersonal comparisons of welfare, both because of the needs of his constituents and because of their power to affect his career.

Almost all political decisions are best described in cooperative terms. Voting at any level (if the individuals are not treated mechanistically) involves discussion, debate, communication, and coordinated action—all characteristics of cooperative games. An exception is provided by the analysis of party politics during an election campaign, with the *parties* as the players. In a two-party system the game may become inessential and hence noncooperative (Shubik, 1968d). With more than two parties, though, cooperative possibilities reappear.

2.4.5 The administrator or bureaucrat

The decisions of the politician, businessman, or financier are implemented by bureaucracies of varying sizes. Most of them are sufficiently large that personal contacts between the top policy makers and the people in the implementing organization are relatively infrequent. There exist large cadres of high-level permanent civil servants, upper-middle executives, general managers, colonels, and so forth, who bear the main responsibility for implementation. As providers of continuity in organizations, they limit the strategic scope of the top decision makers.

A completely satisfactory theory of welfare economics would include the effect of bureaucracies upon the implementation of decisions. Possibly the appropriate model would be for each organization to be represented as a game, with constraints placed upon all members but with each member of the organization nevertheless considered a player (Shubik, 1962a; Shapley, 1972).

2.4.6 Comments

Much of microeconomic theory deals with the problems of optimal choice. The first two behavioral types we have distinguished are primarily "choosers." The next two types are more involved in the definition, generation, and presentation of alternatives, and the last in the implementation of decisions once the choice has been made.

Our major concern will be with the first two types of decision makers. In some instances we may wish to consider a single individual as playing two roles at the same time. The modeling of different roles for the individual decision maker is an ad hoc process; the special attributes of the real person and the special purposes of the model cannot be ignored. Even the most general economic theory should not be confused with a universal decision theory. This being the case, a certain amount of explicit modeling of the differentiated roles of the individual in an economy helps both to narrow the field of inquiry and to give the models more structure and relevance.

Notes to chapter 2

1. When we examine games with strategic dummies, we shall find that most of the relevant solution theories—not surprisingly—predict the failure of the dummies to enter into coalitions or to profit from negotiations. But this is a result of the analysis, not an assumption of the model.

2. How often, even in mature political commentary, one encounters dubious anthropomorphisms—"Egypt is friendly to . . . ," "France desires . . . ," "the Kremlin denies . . . ," "Peking believes . . ."—giving human shape, as in a cartoon, to nonhuman entities.

The problem of how to aggregate psychological attributes is certainly not outside the purview of game theory. Witness the progress made toward mathematization of concepts such as bluff, threat, toughness (in bargaining), and trust that were once thought to be ineluctably psychological in nature.

3. The effect is likely to be significant only in the case of very weak solution concepts, such as Pareto optimality, or in the branch of game theory in which side payments can be made outside the strategic structure of the game. For an example, imagine a three-person game with a $100 prize, in which *B* and *C* are strategic dummies while *A* must choose whether the prize goes to *B* or to *C*. [Compare von Neumann and Morgenstern's (1944, pp. 533–535) discussion of "removable" players.]

4. Since economic interests are rarely parallel in any sense, minimax theory has found little application in economics despite the occasional chapter given over to it in some modern textbooks.

5. This is an example of a "noisy duel," a class of games first treated by Blackwell and others in 1949. Burger (1959) has suggested the following economic scenario: Two merchants compete for a customer. Each makes his

pitch at a time of his choice. The customer (who is not a player) makes at most one purchase, and he makes it with a probability, $a(t)$ or $b(t)$, that depends on when he is approached and by whom.

6. An early throw is always incorrect, but a decision to throw later than t_0 will not matter, unless the other player, perhaps anticipating the mistake, also delays his throw but by a smaller amount.

7. Kinnard (1946, p. 246), Larsen (1948), Shubik (1954a), Gardner (1961), Boot (1967), and doubtless others. We have added the lottery feature in this "truel" to eliminate sequential effects. Again an economic setting can be created, if we try hard enough: Three merchants compete for a sale. Contacts with the buyer occur in random order, and at each contact the merchant attempts to persuade the buyer to *reject* one of the other products. The merchants have different "coefficients of persuasion." Etc.

8. See Shubik (1954a). For treatments of the "weakness is strength" theme in truels and other three-person games in the social-psychology literature, see Cole and Phillips (1967), Willis and Long (1967), Phillips and Nitz (1968), Cole (1969), Nitz and Phillips (1969), and Hartman (1971).

9. This is true unless B can also commit himself to retaliate. Curiously C would prefer most of all to shoot into the air, if the rules would permit it. This is because his only goal in the first stage of the game, so long as no one is attacking him, is to increase the chance of facing B rather than A in the second stage. If he could "pass," the expectations would be 0.381, 0.257, and 0.362. On the other hand, A could, perhaps, modify his threat to enjoin C from passing.

10. See Kannai (1966), Aumann and Shapley (1968, 1969, 1970a,b, 1971, 1974), Rosenmüller (1971b), Dubey, Neyman, and Weber (1981). The term "atom" denotes a point of positive measure; hence in the present discussion an atom is not a midget, but a giant.

11. A partial bibliography: Aumann (1964b, 1966, 1973), Vind (1964, 1965), Debreu (1967, 1969), Hildenbrand (1968, 1969, 1970a, 1974), Schmeidler (1969a), Cornwall (1969), Kannai (1970, 1971), Aumann and Shapley (1974), Richter (1971), Rosenmüller (1971a), and Dubey and Shapley (1976). Not all of the models considered in these works are games, strictly speaking.

12. See Milnor and Shapley (1961) and Shapley (1961a). More recent papers dealing with mixed models include Rosenmüller (1968), Gabszewicz and Dreze (1971), Gabszewicz and Mertens (1971), Hart (1971), and Shitovitz (1973).

13. See Aumann and Shapley (1968). For other work on infinite-person games, see Kalisch and Nering (1959), Davis (1962), Peleg (1963c), Schmeidler (1967), and Kannai (1969).

14. See *Poor's Register of Corporations, Directors and Executives* (New York: Standard and Poor's, published yearly) for a listing.

3 The Rules of the Game: Extensive and Strategic Forms

"Oh, what a tangled web we weave . . . ." Sir Walter Scott, *Marmion*

3.1 Introduction

The theoretical literature on bilateral monopoly, duopoly, and other situations that involve substantial bargaining rightly devotes much attention to the dynamics of individual moves. The details of process, be they price declarations, offers, counteroffers, threats, ploys, or bluffs, are often decisive in the resolution of the conflict. The information pattern is no less decisive as it unfolds over time, and the exact description of "who knows what when" can be extremely complex in a multilateral, multimove model. Game theory provides a framework and a formal language for dealing with these and all other details of the "rules of the game." This chapter will be devoted to an exposition and survey of this important area of descriptive game theory.

Our interests will, however, quickly lead us away from details of process. The question of tactical skill is usually not an issue in the kinds of economic models we shall study. Our point of view is that in the most important economic situations, the existing procedural rules are only of secondary significance; not only do they result from a long evolution, but they remain forever subject to modification or repeal when they stand in the way of underlying economic or political forces. This applies particularly to the procedures and traditions surrounding the workings of the marketplace.[1]

Thus, in most of our illustrative examples and larger models, the procedural rules of the game will be stated verbally and rather informally. Unless there is a particular point to be made, we shall forgo any direct use of the exact descriptive tools discussed in this chapter. To the reader uninterested in formalism for its own sake, this may come as a relief. The reader more inclined to mathematical rigor should in every case be able to construct at least one set of formal procedural rules that will satisfy the informal description given.

3.2 The Concept of a Strategy

Two descriptive forms are often contrasted in game theory: the *extensive form* and the *strategic form.*[2] In the former, we set forth each possible move and information state in detail throughout the course of the play. In the latter, we content ourselves with a tabulation of overall strategies, together with the outcomes or payoffs that they generate.

A *strategy,* in this technical sense, means a complete description of how a player intends to play a game, from beginning to end. The test of completeness of a strategy is whether it provides for all contingencies that can arise, so that a secretary or agent or programmed computer could play the game on behalf of the original player, without ever having to return for further instructions.[3]

For a source of simple illustrations we shall turn from the rich field of balloon popping to the still richer field of art dealing (Duveen, 1935).

ART AUCTION. *A painting is to be sold to the highest bidder. After each accepted bid any player may make a higher bid. If several bids are entered at once, the auctioneer accepts the one nearest the rostrum.*

Examples of strategies:

Player P_1 (*nearest to the rostrum*): "I will start bidding at $15 and increase my bid to $2 above anyone else's, provided that this is not higher than $24; otherwise I will bid $24 if I can and then stop."

Player P_2: "I will bid $17 initially unless someone bids more before I get in. I will enter only this one bid, or no bid at all."

Player P_3: "I will wait until someone else has bid, then raise in units of $1 up to a top of $19, unless I am bidding against P_2, in which case my top is $22."

If there are no other players, the following sequence of bids would result from this trio of strategies (parentheses indicate a simultaneous bid, not accepted):

1st round: P_1 bids $15 ($P_2$ bids $17),

2nd round: P_2 bids $17 ($P_3$ bids $16),

3rd round: P_1 bids $19 ($P_3$ bids $18).

If chance factors are present, a richer variety of lines of play can result, without necessarily increasing the complexity of the strategies themselves. To illustrate, we might modify the rules to require the auctioneer to choose at random among simultaneous bids. To describe the variety of plays that might result if the players again adopt the three strategies listed above, the "probability tree" in figure 3.1 is useful. The notation 17_2 means an accepted bid of $17 by player P_2, and the heavy dots separate equiprobable events. The probability of reaching a particular end is $(1/2)^k$, where k is the number of heavy dots on the path. Thus, for example, the probability that the leftmost path is followed, in which P_1 gets the painting for $21, is 1/8.

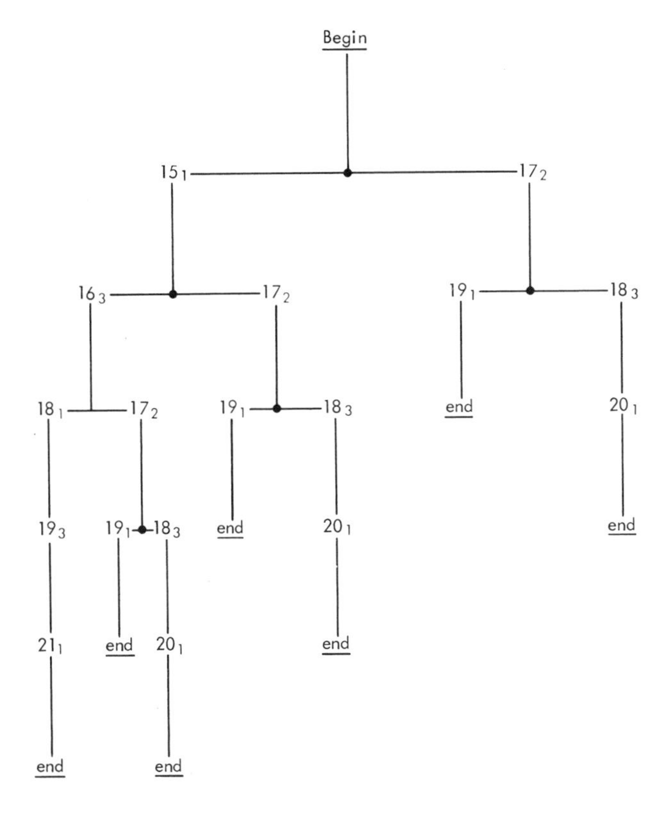

Figure 3.1
A probability tree.

3.2.1 Are strategies practical?

There is, in principle, no sacrifice of flexibility in adhering to a single, fixed strategy. Strategies that can be described in a few words, however, tend to produce a mechanical style of play. Highly adaptive behavior, if codified in strategic form, usually requires a lengthy and complex description. The reader will note that even the simple strategies we have been discussing contain provision for many contingencies that did not arise in the sequence of bids put forth above, or even in figure 3.1. In more realistic examples, a dismayingly large part of the planning that goes into formulating a strategy is wasted when the game is played. Since planning and computing are not costless, the strategic form is almost never used in practice in multimove games. Its value is primarily conceptual and theoretical.[4]

In multimove games the sheer number of strategies is usually such as to preclude any hope of a manageable enumeration, and other techniques must be found for practical work. Some calculations based on the game of Chess will help to bring home the astonishing scale of the problem.

Consider first a trivial game, which we might call One-Move Chess. Play is halted after one move on each side, and some arbitrary payoff is designated for each of the 400 possible end positions. In this game White has just 20 strategies, corresponding to his 20 possible opening moves. But Black already has 20^{20} strategies (about 10^{26}, or one hundred septillion!). We emphasize that these strategies are all operationally different, since any two of them will make different responses to at least one of White's moves.

Proceeding to Two-Move Chess, White has 7.886×10^{29} strategies, while for Black the number is in excess of 10^{52}. In Forty-Move Chess each side has more than 10^{1000} strategies. For the real game of Chess an estimate is more difficult to make, but hyperastronomical numbers like $10^{10,000}$ or $10^{20,000}$ seem to be indicated, since most strategies must include contingency-laden detailed instructions for playing an almost endless variety of meandering games, continuing sometimes for hundreds of moves until a fortuitous checkmate or stalemate, a thrice-repeated position, a lack of mating strength on both sides, or the "forty moves without progress" rule makes an end.

Note that the number of strategies is far greater than the number of possible courses of play. This is typical in multimove games.

3.2.2 Perfect information and mixed strategies

If at every point of time during the play of a game each player with a decision to make is completely apprised of the state of affairs, then we have a *game of perfect information* (von Neumann and Morgenstern, 1944). Of our examples thus far, the three-person Shooting Match was of this type, but the Dart Duel and Art Auction fell just short, since simultaneous actions were possible. (The latter are examples of "almost-perfect" information.) It is easy to think of examples in which significant information is longer deferred (the contents of an opponent's hand in a card game, or the effects of an advertising campaign) or even withheld completely from some of the players.

When perfect information is lacking in a game, a new strategic element assumes great importance in the analysis of optimum play: the deliberate randomization of decisions. In mathematical terms, one sets up a probability distribution, called a *mixed strategy,* over the original set of "pure" strategies, and lets it control the actual play of the game. It is perhaps intuitively clear that doing this is not helpful in a game of perfect information such as Chess in which one has no secrets to dissemble. [The first mathematical proof of this is due to Zermelo (1913).] On the other hand, in the Chess players' little preliminary game of Choosing Colors, in which one player conceals two pawns and the other picks a hand, it is clear that the only foolproof plan for either player is to randomize between right and left, with equal probabilities. Any other plan could be exploited by a knowledgeable opponent. (Note that the concealing player has perfect information but should randomize nevertheless.)

3.2.3 Perfect recall and behavior strategies

A further possibility is that a player at some point during the game may not be apprised of his own previous moves or of the information on which they were based. We are thinking not of schizophrenics or amnesiacs, but of teams. North, at Bridge, must bid without knowing the contents of South's hand, yet North and South can usefully be regarded in other respects as agents of a single player (Thompson, 1953a,b; Keeler, 1971). The technical term for this phenomenon is *imperfect recall* (Kuhn, 1950b).

In a game of imperfect recall it may be necessary for a player using a mixed strategy to make his random choice of strategy at the beginning of play, so that the actions of his agents will be properly correlated. In a game of perfect recall, on the other hand, the randomizations can be deferred until the actual decision points are reached, since there is no scope for profitable correlations.[5] Mixed strategies of this kind, with their on-the-spot randomizations, are called *behavior strategies* in the literature; they are far easier to work with, both in theory and practice, than general mixed strategies (Nash and Shapley, 1950; Kuhn, 1950a,b, 1953; Dalkey, 1953).

3.2.4 Comments

When we reduce a game description from extensive to strategic form, we lose sight of the actual pattern of information. The strategic form has, of course, its own peculiar information pattern: each player makes just one move (a strategy choice), in ignorance of what the others are doing. There is perfect recall, trivially, but not perfect information. Mixed strategies may therefore be worthwhile, and we cannot afford to dismiss them out of hand. But if we happen to know that the original model in extensive form had perfect information, then we can safely proceed to analyze the model in strategic form on the basis of pure strategies alone.

These informational concepts play only a minor role in the static economic models that will be our first concern, but their significance for economic analysis in general is far from negligible. To mention just two points: *imperfect information* provides a rational basis for explaining deliberately random behavior; *imperfect recall* pinpoints the fundamental informational obstacle to achieving optimality through decentralized decision making.

So far we have discussed strategies and information without reference to any exact descriptive model. In the next section we sketch a series of formal models that can be used to provide a rigorous basis for the study of the move and information structures of various classes of games in extensive form. (Some readers may prefer to go directly to section 3.4, reserving section 3.3 for reference or for later study.)

EXERCISES

3.1. *Show that the number of strategies for the first player in Tic-Tac-Toe, disregarding the symmetry, lies between* $9 \cdot 7^8 \cdot 5^{48}$ $(= 1.843 \times 10^{41})$ *and* $9 \cdot 7^8 \cdot 5^{48} \cdot 3^{192}$ $(= 7.461 \times 10^{132})$.

3.2. *In a two-player game,* P_1 *is an individual and* P_2 *is a team consisting of two people* P_{21} *and* P_{22}. *Everybody simultaneously calls out "Heads" or*

"Tails." If the three calls are the same, then P_2 wins; otherwise P_1 wins. Show that P_2 has a mixed strategy that wins with probability 1/2, but no behavior strategy that wins with probability greater than 1/4.

3.3 Modeling the Extensive Form

The earliest general-purpose formal systems for describing games in extensive form were those of von Neumann and Morgenstern (1944) and Kuhn (1950b, 1953). The one uses an approach through sets and partitions whereas the other leans more heavily on graphical or diagrammatic ideas; but the two approaches are nevertheless almost equivalent. Both are directly concerned only with *finite* games, that is, games in which the number of players, the number of choices at each decision point, the number of attainable intermediate and terminal positions, and the number of moves in each possible play-through of the game are all finite. Games such as Chess and Go (with suitable stop rules), Poker and Bridge (considering each deal as a separate game), and many other parlor games fall into this category. But most competitive situations found in the worlds of business or politics are only imperfectly modeled as finite games, since they usually have continuous strategic parameters, a continuous time frame, and an indefinite continuation into the future.

We shall first sketch the Kuhn model, which with minor variations has become the accepted general-purpose model for the extensive form. Then we shall note more briefly a number of other models that are better adapted to accommodate certain classes of infinite games.

3.3.1 The Kuhn game tree

Consider the following simple finite game:

FINGERS. *The first player holds up one or two fingers, and the second player holds up one, two, or three fingers. If the total displayed is even, then P_1 pays \$5 to P_2; if it is odd, then P_2 pays \$5 to P_1.*

Suppose the rules further state that P_1 moves first. Then a graphical representation of the game can be drawn as shown in figure 3.2. This is an example of the special kind of connected graph known as a "rooted tree."[6] The nodes (vertices) are labeled with symbols P_i or O_j, for player i or outcome j respectively. The starting node, or root of the tree, is distinguished by a circle with a dot in it.

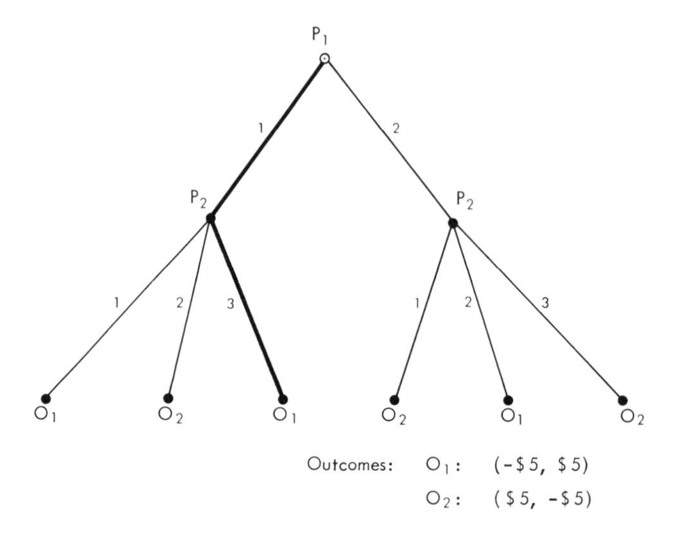

Figure 3.2
A game tree for Fingers.

Each node in the tree represents a *position* or *state* in which the game might be found by an observer. A node labeled P_i is a decision point for player i: he is called upon to select one of the branches of the tree leading out of that node, that is, away from the root. In our example P_1 has two alternatives, one finger or two fingers; accordingly we have labeled 1 and 2 the two edges leading away from the initial node. After P_1's move, the play progresses to one of the two nodes marked P_2; at either of these P_2 has three alternatives, which we have labeled 1, 2, 3. Finally a terminal position is reached, and an *outcome* O_j is designated. Thus any *path* through the tree, from the initial node to one of the terminals, corresponds to a possible *play* of the game. For example, if P_1 shows one finger and P_2 three, then the play shown by the heavy lines in figure 3.2 results, and, since $1 + 3$ is even, P_2 wins \$5 and P_1 loses \$5.

If random elements are present, another type of node will occur. For notational purposes we may pretend that chance, or Dame Fortune, is a player, and use the label P_0 for those nodes at which the tree branches due to a chance event. The edges leading out from such a node are labeled with the probabilities of the corresponding alternatives. Figure 3.3 illustrates this: the move structure is similar

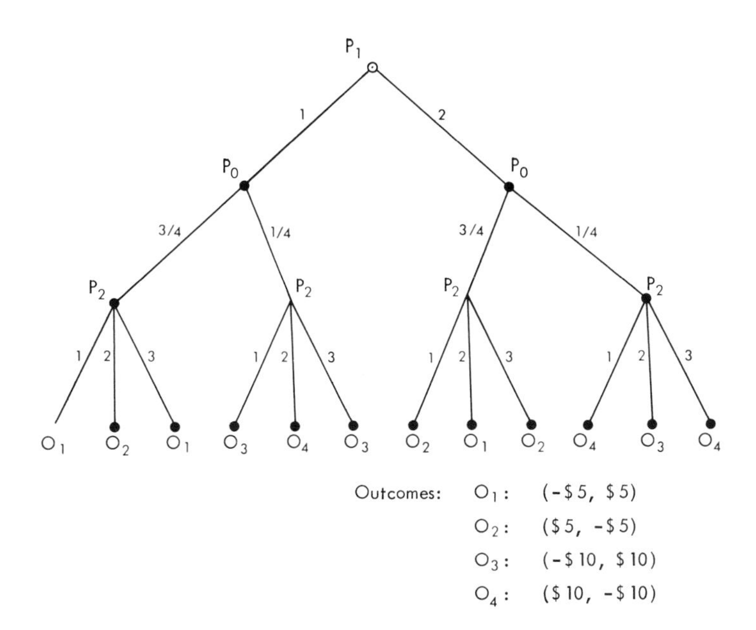

Figure 3.3
A game tree with uncertainty.

to that of figure 3.2, but a chance move has been inserted between the two personal moves that doubles the stakes with probability 1/4.

This notational apparatus equips us to handle any finite game of perfect information. The diagrams are virtually self-explanatory; but the apparent ease and concreteness of the representation is illusory, since in games of any complexity the tree will be unmanageably large and can be fully drawn only in the imagination.

EXERCISE
3.3. *Estimate the number of nodes in the Kuhn tree for Tic-Tac-Toe.*

3.3.2 Information sets

Let us now modify Fingers to require P_2 to move first, thus reversing the order of play. The tree of figure 3.4a then results. This is of course quite a different game—indeed the advantage is now entirely with the first player, who can win $5 with ease. The essence of the difference, however, is not in the timing, as such, but in the infor-

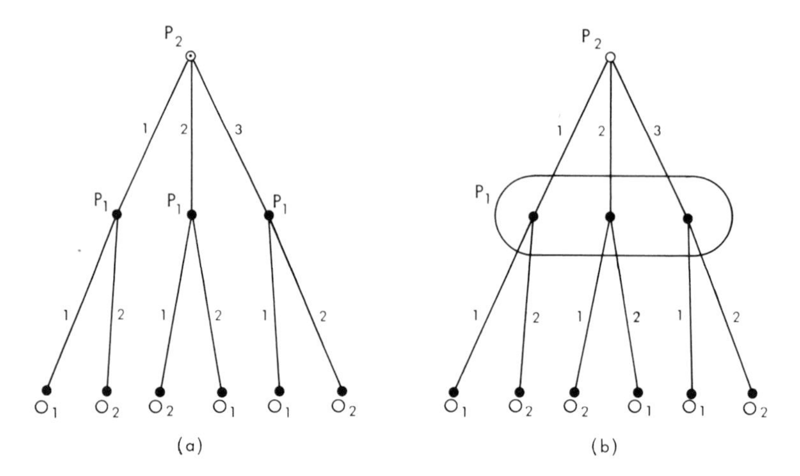

Figure 3.4
The information pattern.

mation pattern. It is not the fact of the opponent's prior move, but the knowledge of that fact, that conveys the advantage.[7]

To modify the example again, suppose that P_1 (still making the second move) is forbidden to look at P_2's fingers until his own are up. A method for indicating this rule is shown in figure 3.4b: a "balloon" is drawn, enclosing the three game positions among which P_1 is not allowed to distinguish. The set of points so enclosed is called an *information set.*[8] Of course, a more complicated example could have many information sets per player. Naturally no two information sets can have a node in common; nor can a single information set intersect any given path in more than one node. If an information set consists of a single node, the balloon is usually omitted.

Note that the player label is attached to the information set rather than to the nodes that comprise it. Note also that every node in the same information set must have the same number of edges issuing from it, and the way in which these edges correspond to each other must be made clear, either by their labels or by some definite convention such as always reading counterclockwise from the incoming edge.

For a more substantial example, let us describe the beginning of the Kuhn tree for a hand of Bridge. The root is a chance move (the deal), having about 5×10^{28} equiprobable branches. Each leads to a different node at which the Dealer must decide on a bid. The Dealer's nodes, however, are grouped into information sets (about 6×10^{11} of them)—one set for each possible hand he may hold. The multitude of points within each set represent the unknown-to-him contents of the other hands.

After the Dealer's selection of one of his 36 possible calls (35 bids or a pass), the next rank of nodes, belonging to player P_2 on the Dealer's left, is reached. These nodes are grouped in a different way to reflect the fact that P_2 is informed only of his own hand and the Dealer's call. (Any inferences he may make from the Dealer's bid are not "information" in our present, technical sense, no matter how "informative" the bid may have been in the practical sense.) There will be from 2 to 36 edges issuing from each of these nodes, depending on the Dealer's call, and each one leads to a node belonging to P_3, the Dealer's partner. (For example, "Seven No Trump" by the Dealer leaves only two alternatives: "Pass" and "Double.") The continuation of the tree should now be apparent.

The information set provides a basis for mathematical definitions of a number of significant informational concepts. *Perfect information,* for example, is simply characterized by the statement that every information set is a singleton. The definition of *perfect recall* is somewhat fuzzy and will not be given here.

Another important idea is that of a *position of complete information*; this is a node with the property that the branch of the tree that it defines—consisting of that node and all following nodes and edges—is "informationally independent" of the rest of the tree.[9] This means that there is no information set that ties any node of that branch to any node elsewhere in the tree. A position of complete information can be regarded as the starting point of a new game—a "subgame" of the original game—and it is often important to be able to recognize such positions.

EXERCISE

3.4. *Diagram Fingers with a probability p of an "information leak" revealing P_2's choice to P_1. Consider two cases: (a) P_2 is not told in advance that the leak will occur; (b) P_2 is told, with probability q, but P_1 is not told whether P_2 has been told.*

3.3.3 Strategies in the Kuhn tree

For a game in extensive form, the simplest definition of a strategy is *a function that associates with each of a player's information sets one of the alternatives issuing from that set.* When all players have fixed their strategies, it is as though in a toy railroad layout all the switches had been fixed, by the various players in control or by chance, with those belonging to a common information set fixed the same way. A train starting at the root of the tree runs through the network and arrives at a terminal, thereby tracing out the play of the game and disclosing the outcome.

A little reflection will reveal that this simple definition of strategy, though perfectly valid, may be very redundant. The decision points that can actually occur on a player's later moves are sharply reduced by his own earlier decisions. Yet a strategy, in the "toy train" sense, fixes *all* the switches—even ones that cannot be reached by the train. It would tell a Chess player, for example, what to do on his second move for every one of his possible first moves, even though the same strategy will have prescribed some particular first move.[10] To avoid the unnecessary tabulation of strategies that differ only in this vacuous way, a notationally more complicated but practically more efficient definition of strategy can be formulated, but we shall not pursue the matter here (see, e.g., Otter and Dunne, 1953; we were in fact already using such a definition implicitly in section 3.2.1 in our counts of Chess and Tic-Tac-Toe strategies).

The formal description of randomized strategies presents no special difficulties in the finite Kuhn tree. A *mixed strategy* is any probability distribution over strategies. We may think of a deck of cards being shuffled, with each card bearing a complete contingency plan for playing the game. Geometrically the set of all mixed strategies can be represented in the form of a *simplex* (figure 3.5), with one vertex for each pure strategy and with dimension equal to the number of such vertices minus one.

It is often more convenient and more intuitive to introduce strategic randomization by means of *behavior strategies,* wherein each player preselects, at each information set that belongs to him, a probability distribution over the alternatives available there (figure 3.6). Geometrically the set of all behavior strategies is also a convex polyhedron, with one vertex for each pure strategy. It is not ordinarily a simplex, however, but a Cartesian product of simplexes. Its dimension may be substantially less than the dimension of the corresponding mixed-strategy simplex. [For an instructive comparison of mixed and behavior strategies in certain Poker models see Kuhn (1950a) and Nash and Shapley (1950).]

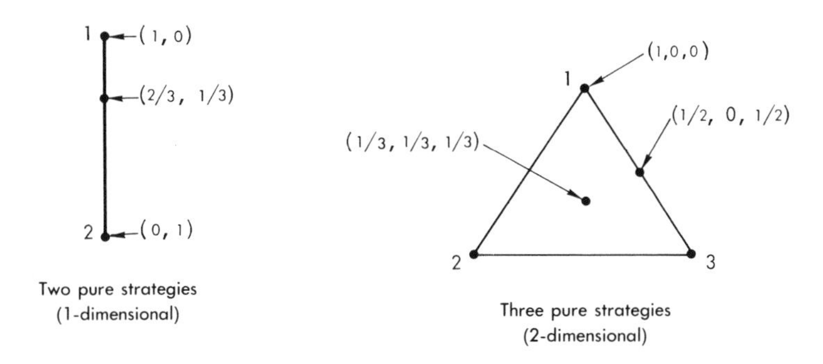

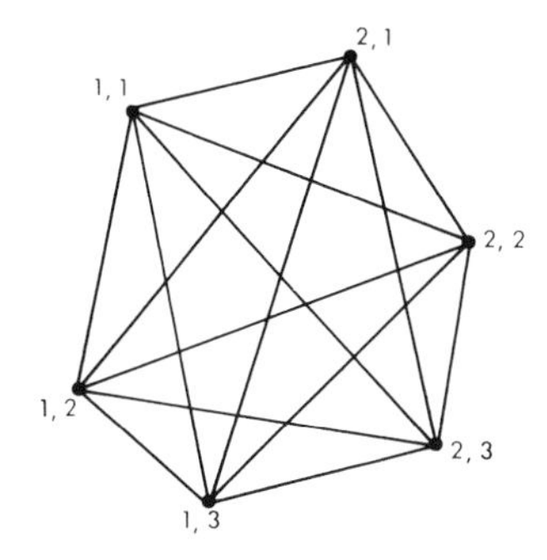

Figure 3.5
Examples of mixed-strategy simplexes.

Behavior strategies are more special than mixed strategies. That is, the effect of any behavior strategy can always be duplicated by some mixed strategy, but not conversely (Kuhn, 1950b, 1953; F. B. Thompson, 1952b; G. L. Thompson, 1953a). As already noted, however, in the important class of "games of perfect recall" the added possibilities for randomization afforded by mixed strategies are superfluous.

EXERCISES

3.5. *There are three firms in an industry. First they make production plans and announce them simultaneously; then they select their prices and announce*

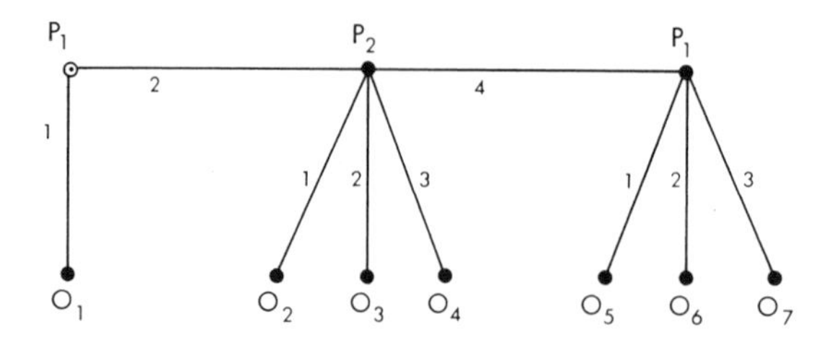

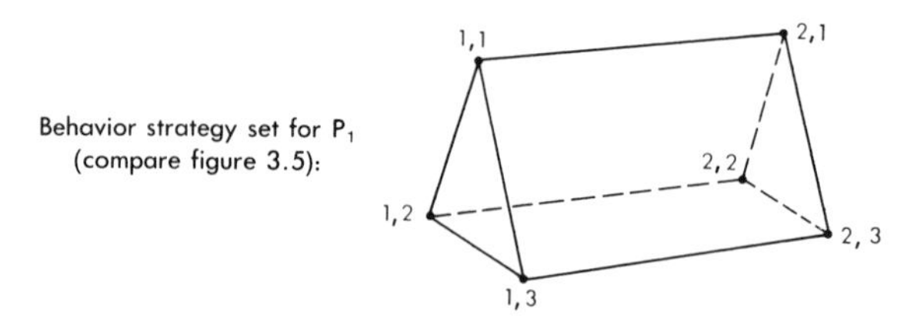

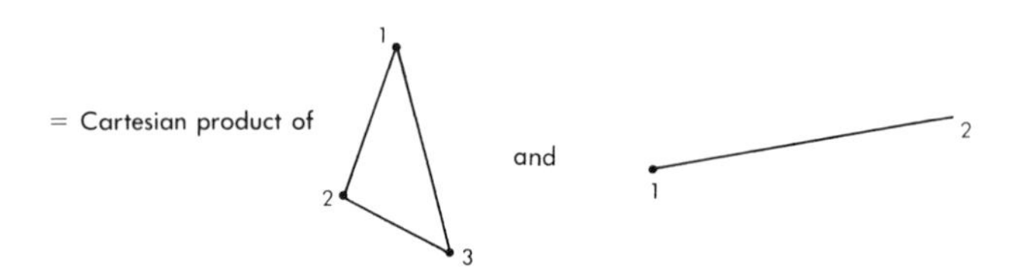

Figure 3.6
Examples of behavior strategies.

them simultaneously. Assuming two possible production levels and two possible price levels, show that each firm has 512 strategies, of which only 32 are essentially different.

3.6. *In the above example, what is the dimension of the space of behavior strategies?*

3.3.4 Simultaneous moves

In many game models the rules require the players to move simultaneously. The Kuhn tree does not provide a way to represent this; instead it gives us a choice among two or more equivalent representations in which the moves are depicted in different temporal order but with no information revealed.[11] Thus figure 3.4b showed moves for P_1 and P_2 that are effectively simultaneous; this could equally well have been indicated as in figure 3.7a. Since this is a common sort of situation, the more concise and symmetrical notation illustrated in figure 3.7b is often useful. The double labels indicate that two simultaneous decisions have been made.

In order to use this notation the players must be equally informed; usually, in fact, it is used only at a position where information is perfect except for the simultaneity. Indeed, in an important class of games called *simultaneous games* or *games of almost-perfect information,* the players are always fully informed of everything that has happened in the past and lack only information about the immediate present, that is, about what moves are being made concurrently with their own. With the aid of the notation of figure 3.7b, such games can be diagrammed without the use of balloons, and each node that actually appears in the diagram represents a position of complete information.

A game in strategic form is, trivially, a simultaneous game. Presently we shall consider other less trivial examples of this informational form.

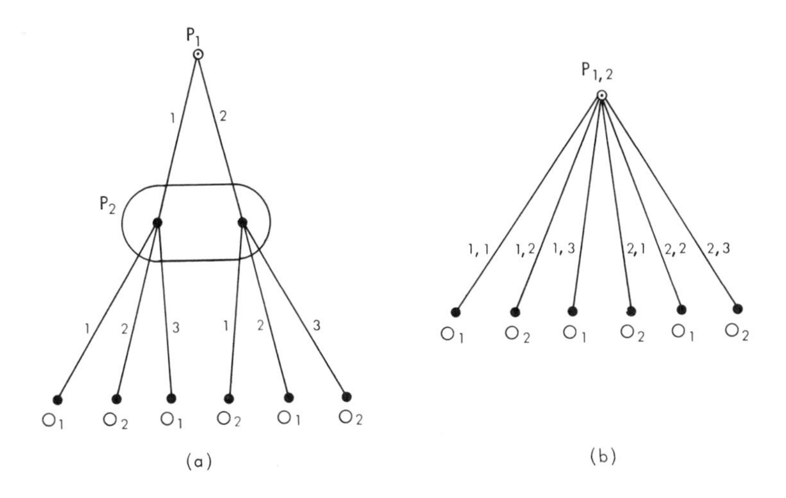

Figure 3.7
Simultaneous moves.

EXERCISE

3.7. *Using the simultaneous-move notation, make a diagram for the example of the three firms in exercise 3.5. Compare the number of nodes, edges, and paths with the number of nodes, edges, and paths in an equivalent Kuhn tree.*

3.3.5 Positional games

The game tree is an extremely useful device for didactic purposes, but one must often pay a high price for its use, in terms of redundancy. The rules of many games permit the same physical "position" to be reached through various different sequences of moves. Yet in a tree each sequence of moves must lead to a different node. The tree convention forces us to remember the history of the position, whether we want to or not.

In Chess, for example, different sequences often "transpose" into the same board position after a lapse of two or more moves by each player. In such a case it would evidently be wasteful in practice to conduct separate analyses of the continuation.[12]

The obvious answer is to abandon the tree in favor of a directed graph, thereby allowing different lines of play to merge. This leads to what we call a game in *positional form.*[13] This resembles in some respects a flow diagram for a computer program. The reader with a taste for such things may enjoy tracing through the diagram in figure 3.8, which depicts the three-person Shooting Match of chapter 2, laid out in positional form. (The light lines refer to misses.)

Although natural and useful for many applications, the positional form raises a series of new modeling problems that merit some discussion here.

Loops If there are loops (cycles) in the directed graph, then the same position can occur more than once, and a play of infinite length is possible. Several ways of treating this kind of situation have been developed. We shall not pursue the general case further, but a number of special structures with loops will be discussed in section 3.3.6.

If there are no loops, then the positional representation can be considered merely an abbreviation for a tree—namely the tree that results from "pulling apart" the merging lines of play and replacing each positional node by a "stack" of tree nodes, as many of them as that position had possible histories. Even when there are loops, it is sometimes useful to think in terms of the infinite tree that could be created by "pulling apart" the positional nodes in this way.

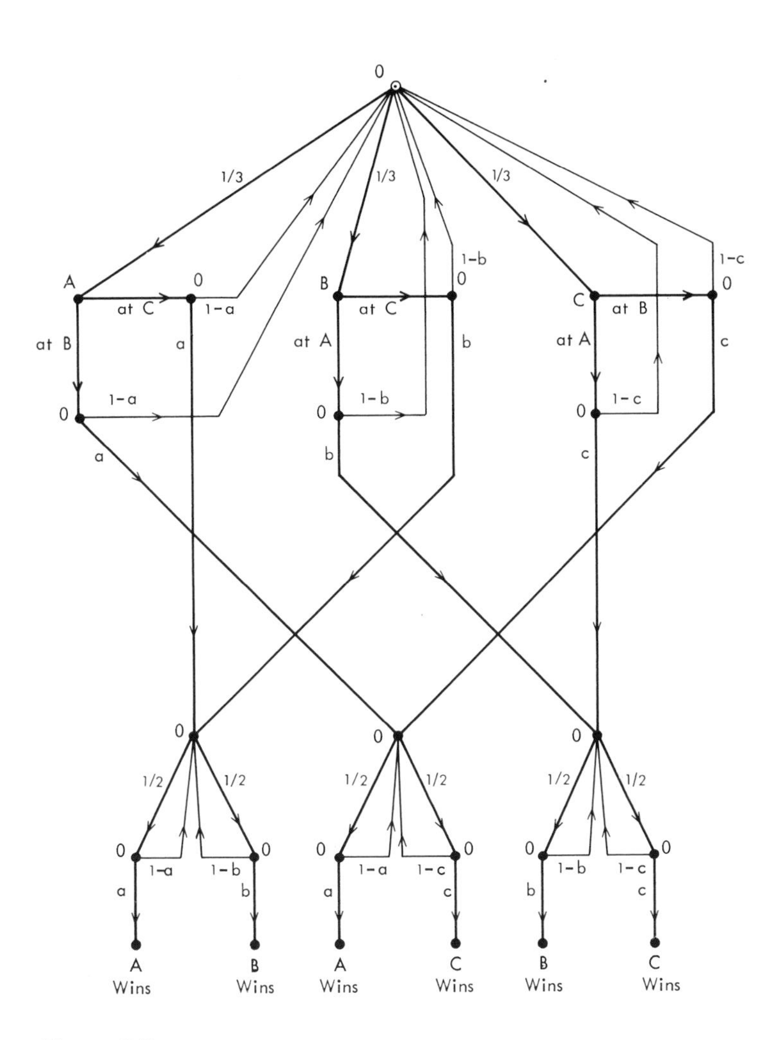

Figure 3.8
The Shooting Match.

Information Chance moves and simultaneous moves can be treated as before, but the use and interpretation of information sets become a good deal more complex. Usually it is assumed that all nodes having two or more incoming edges (more generally, positions having two or more possible histories) are positions of complete information. Indeed most positional games in the literature are games of perfect or almost-perfect information.

Strategies The proper way to define a strategy also becomes less clear. For one thing, the simple notion of a choice at each information set (the toy-train model) would presuppose that a player makes a fixed choice at each position, regardless of how he got there. True, we may be able to argue (for certain classes of games and certain concepts of solution) that there is no good reason for a player to make different choices at the same position. But that is hardly sufficient grounds for making such erratic behavior impossible or illegal.

For example, identical dispositions of goods and money in an economic model might arise either through a chance move, over which the players have no control, or through some action by one of the players which the other players might choose to regard as a "double-cross" or "breach of faith," although not a violation of the rules. In such a case some apparently "irrational" punitive measure might well be a reasonable response.[14]

In a game in positional form, strategies that prescribe the same action no matter how (or how often) a position is reached are called *stationary strategies.*[15] A stationary pure strategy prescribes a fixed choice for each position belonging to the player in question. A stationary mixed strategy is a special kind of behavior strategy in which the same probabilities are used at a position whenever it arises.

Spot payments A feature that is often found in games in positional form is that of "spot" payments, made during the course of play, which must be totaled to determine the final payoffs. More generally, the outcome may be some function of the cumulated "scores," as in the games of ruin or survival described in the next section. Sometimes, when a significant period of real time is covered by the model, the later payments are discounted in relation to the earlier ones before forming the sum.

The advantage of having a way of indicating spot payments in the diagram is that positions which differ only in the amounts that have accumulated can be represented by a single node. This is especially useful when we are representing the repeated play of a single basic game. For example, figure 3.9 represents the Fingers game played three times. If spot payments were not used, we would need two nodes at the second level, corresponding to the two money states $(-5, 5)$ and $(5, -5)$ possible at that time, and three nodes at the third level, corresponding to $(-10, 10)$, $(0, 0)$, and $(10, -10)$. If there were more variety in the payoffs of the basic game, the savings would of course be much greater.

Repeated play of the same game is a favorite tool in experimental gaming, and there is a considerable literature on the subject, both theoretical and empirical, with special emphasis on certain two-person non-zero-sum cases. For a bibliography see Shubik (1975d) or Rapoport, Guyer, and Gordon (1976).

EXERCISES

3.8. *Two players alternately name numbers from 1 to 10, and the player who makes the total 100 wins. Describe the positional form of this game, and then solve it by determining who should win in each position, working backwards from the end. [This game is attributed to Bachet de Meziriac (1612); see Vorobyev's account (1970b).]*

3.9. *An unusual game of Solitaire is diagrammed in figure 3.10. Show that any mixed stationary strategy is better than any pure stationary strategy. What definition of "strategy" for games like this, in which a path may have several nodes in the same information set, will ensure that the mixed strategies will include the behavior strategies (Shapley, 1953d; Isbell, 1957a)?*

3.10. SPELL-A-NUMBER.[16] *P_1 and P_2 alternately write down the digits of a real number between 0 and 1 in decimal form, with P_1 moving first. P_1 wins if the number thus produced lies strictly between 9/22 and 13/22; otherwise P_2 wins. Does either player have a winning strategy, and if so, what is it?*

3.3.6 Recursive, stochastic, and survival games

While it is not difficult to imagine the Kuhn tree model being extended, with great generality, to the case of games of infinite duration,[17] the infinite models that receive the greatest attention usually have some kind of special repetitive structure, based on an essentially finite foundation. In this section we shall briefly describe a few of these models.

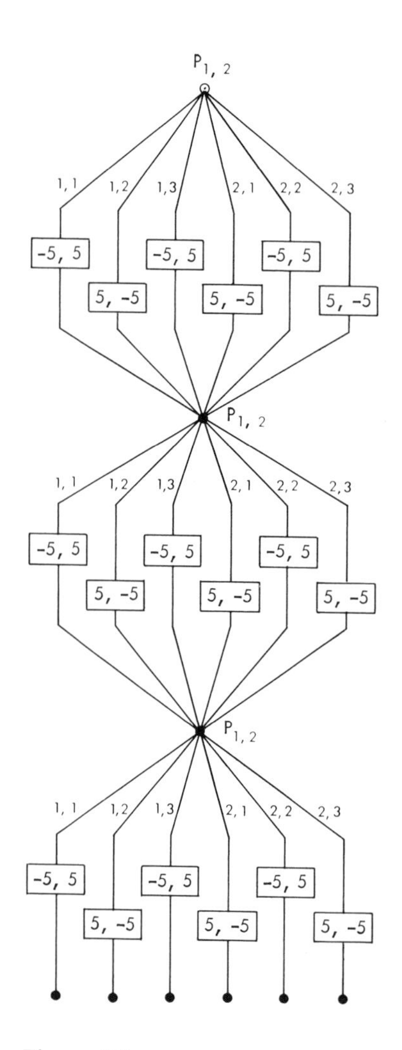

Figure 3.9
Spot payments in a repeated game of Fingers.

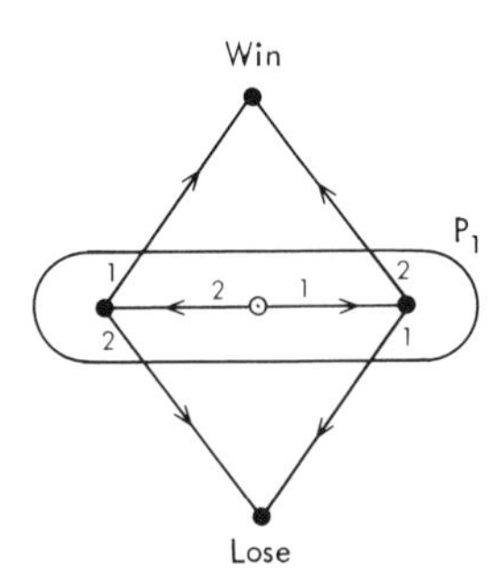

Figure 3.10
A game of Solitaire.

Recursive games Consider first a game consisting of a finite set of component games, tied together by a "supergame" that controls the transitions among the components. After playing one of the components, and depending on its outcome, the players are either sent back to play another component—possibly the same one over again—or they are paid off and the whole thing ends. A structure of this kind is called a *recursive game* (Everett, 1957).

Figure 3.11 shows two examples. The first has just one component, which is repeated if both players choose 2. The second begins in component B but may shift to A or C, or even revisit B, before the end is reached.

It is clear that a recursive game may go on forever. This can happen even when the players adopt entirely reasonable strategies. Thus, in the first example, P_2 would lose \$5 if he should ever choose his first alternative. On the other hand, player P_1 would very likely lose \$10 if he should choose his first alternative. Both players would therefore be well advised to persist in choosing their second alternatives.

In a recursive game the payoffs in the case of nonterminating play are defined to be zero. Thus a notion of status quo is implicit. In particular, we cannot add or subtract a constant to the terminal payoffs of a recursive game and expect its strategic properties to remain unchanged.

Stochastic games The recursive game might be a good format for the trials of Sisyphus or the occupant of some circle of Hell, but it lacks

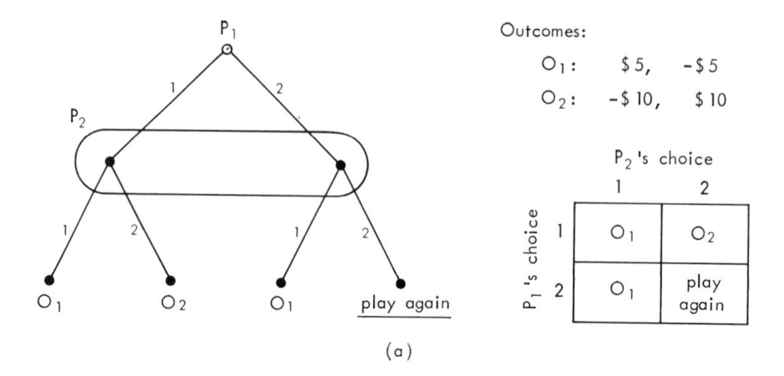

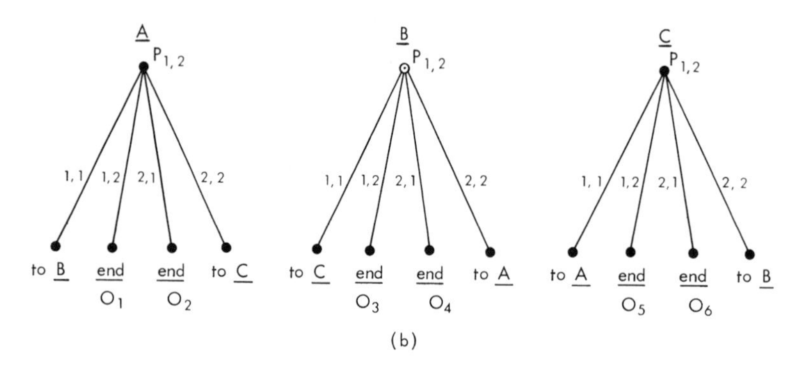

Figure 3.11
Recursive games.

several features that are needed for an adequate model of recurrent human affairs in general and economic affairs in particular. Spot payments and chance moves, which are not provided for, are important in many applications. Income may accrue during the game, and an individual's income as well as the options he faces may depend not only on previous decisions but also on "acts of God" such as storm and drought, fire, disease, or accident. The probabilities of these events may themselves depend on previous decisions. A driver may have an accident when completely sober, but he has the option of increasing the probability of an accident by drinking heavily prior to driving. Thus we are led to the idea of a controlled stochastic process—a chance-dependent sequence of events in which the actors can influence both the odds and the available future options as well as the gains or losses incurred.

Expressing this idea, we define a *stochastic game,* like a recursive game, as consisting of a supergame and a set of one or more component games, the latter sometimes called *states* as in the theory of stochastic processes.[18] The players start at whatever component has been designated the *initial state* and choose their first-round strategies for that component. Their choices determine both the spot payments (positive or negative) that are awarded at that time and the probabilities that are used to decide whether the game ends without further payoff or continues for another round in some specified state.

More formally, suppose that there are two players and n states and that in the kth state the first player has r_k alternatives and the second player s_k alternatives. Further suppose that upon reaching the kth state the players select alternatives i and j, where $1 \leq i \leq r_k$ and $1 \leq j \leq s_k$. Then the rules would specify a *payment,* a_{ij}^k, and a set of $n + 1$ *transition probabilities,* p_{ij}^{kl} $(l = 0, 1, 2, \ldots, n)$, which govern the passage from state k to state l. Here $l = 0$ means termination.

A recursive game is a special kind of stochastic game in which all probabilities are 0 or 1 and all payments are zero except at a set of terminal states associated with the different outcomes of the recursive game. The greater generality of the stochastic model has its price, however, in technical and conceptual difficulties and in weaker results. For one thing, the sum of the spot payments may become large without bound or may oscillate and fail to converge to a limit even though bounded.

The simplest device to cope with this problem is to stipulate that all the numbers p_{ij}^{k0} be positive, so that on every move there is a finite probability that the game will end (Shapley, 1953a). This has the effect of assuring that never-ending games have probability zero, no matter how the players behave, and that the total payments converge in expected value to a finite limit.

A second expedient is to apply a *time discount* to the payments. That is, a payment of a_{ij}^k received on the tth round is considered to be worth only $(1 - \beta)^t a_{ij}^k$, where β is a constant between 0 and 1.[19] This, too, has the effect of making the sum of the spot payments converge and setting a uniform upper bound on the total (discounted) payoffs.

A third approach is to set all $p_{ij}^{k0} = 0$ and allow infinite total payments, but to assume that it is the rate of income, not total income, that motivates the players. One sums the first t spot payments and divides by t. As t becomes large this expression, if it converges, approaches the *limiting average payment,* which is regarded as the true objective of the game.[20]

Generally speaking, the basic technique in analyzing both stochastic and recursive games is to postulate a *value* for each player in each state. Then each inner game can be studied by itself, with its payoff being the sum of the spot payment, if any, and the mathematical expectation of the postulated value (perhaps discounted) of the new state to which the action moves. If these inner games can now be solved, and their values expressed as functions of the postulated state values, then a set of simultaneous equations is obtained from which in principle the true state values can be determined. The value of the game itself is defined as the value of the initial-state game. We have already applied this method in the three-person Shooting Match (equation 2.2); it has something in common with functional-equation methods in dynamic programming, the theory of harmonic functions, and the calculus of variations.[21]

Survival games A distinction was just made between the spot payments, or their sum, and the players' "true" objectives. This distinction is carried further in the games known as survival or attrition games. In the prototypical *game of survival* two players, each with some specified number of "chips," agree to sit down and play a finite zero-sum game such as Two-Handed Poker and to continue to replay it until one of them has lost his entire stake. There are only two terminal states, but there is also the possibility that the game never terminates. The payoff then need not be zero, as in a recursive game, or even a constant. For example, it might depend on the limiting average size of the players' fortunes (in chips). As might be expected, best play in each round will generally depend on the players' current fortunes, sometimes in a rather complicated way (Hausner, 1952a,b; Bellman, 1954; Sakaguchi, 1956; Milnor and Shapley, 1957). These games are closely related to the classical problems of "gambler's ruin" and "random walk," and their analysis leans heavily on the theory of semimartingales [see, e.g., Snell (1952); for another kind of game based on martingale considerations see Dynkin (1969)].

A somewhat more general form of survival game is the *game of attrition,* sometimes studied in a military context. Each player has quantities of different types of weapons or other items. Each "engagement" causes certain numbers of these to be lost or captured, depending on the strategic choices and perhaps on chance. There may or may not be a replacement cycle, and play continues until one of the players is wiped out. The outcome may be win–lose, or the payoff may depend on the numbers of surviving items (Blackwell, 1954; Isbell and Marlow, 1956; Berkovitz and Dresher, 1959, 1960a,b; Dresher, 1961; Romanovski, 1961; Vorobyev, 1968; Vakriniene, 1968, 1970; Vrublevskaya, 1968, 1970).

Another extension of the simple survival model is the *game of economic survival,* in which the players are corporations or owners of corporations.[22] The players do not merely survive or go bankrupt; the stream of *dividend payments* also figures in the final evaluation. Moreover, an external interest rate is postulated, so that a firm may go out of business voluntarily, not because it fears bankruptcy but because its assets, if liquidated, will bring a higher rate of return in some other enterprise.

A game of economic survival is a species of stochastic game; it shares with the other survival and attrition models the property of playing virtually the same "inner" game over and over again. The key difference is in the role of the asset levels of the participants. To regard these games directly as stochastic games would entail introducing a very large number of states representing all the possible distributions of assets, and would not do full justice to their essentially repetitive nature (Shubik and Sobel, 1980).

EXERCISES

3.11. *The following table describes a one-position recursive game (two-person, zero-sum):*

	P_2's choices	
P_1's choices	*\$10, –\$10*	*–\$10, \$10*
	play again	*\$10, –\$10*

What must P_1 do to win?

3.12. *Suppose P_1 and P_2 have ten pennies between them and are committed to playing Matching Pennies until one player wins them all. Model this survival game as a recursive game, and determine the value of each position by the method of simultaneous equations.*

3.3.7 Information lag[23]

One other type of repetitive structure warrants mention here. The essential idea is illustrated in figure 3.12; here time flows from left to right, and P_1 and P_2 choose the *a*s and *b*s, respectively. The payoff is some function of the string $a_1b_1a_2b_2a_3$ . . . that is produced. The information pattern is described by saying that P_1 knows all the *b*s up to and including b_{t-r} when he selects a_t and that P_2 knows all the *a*s up to and including a_{t-s} when he selects b_t. The numbers r and s have an obvious interpretation as information delay times. In figure 3.12 $r = 1$ and $s = 2$, and P_1 has perfect information but P_2 does not.

Only the sum $r + s$ is really significant here. Thus, by renumbering moves as $a_i' = a_{i-1}$, $b_i' = b_{i+1}$ (and adjusting the beginning), the above pattern can be shifted into one that has $r = 3$ and $s = 0$ (figure 3.13). Here P_2 rather than P_1 has the perfect information; yet the two patterns we have diagrammed are clearly equivalent. This illustrates our previous point that the essential ordering of moves in a game is more a matter of information than of simple chronology.

The number $\lambda = r + s - 1$ has been called the *time lag* for this type of information pattern. If $\lambda = 0$, there is perfect information on both sides, and no shift as between figures 3.12 and 3.13 is possible. If $\lambda = 1$, there is almost-perfect information; that is, the moves are effectively simultaneous. (The only shift possible in this case is illustrated by the difference between figures 3.4b and 3.7a.) In both these cases an analysis by recursive methods is possible, since there are many positions of complete information. When $\lambda > 1$, however, the analysis becomes substantially more difficult, since the players may never be able to shake off the effects of their past uncertainties and make a fresh start.[24]

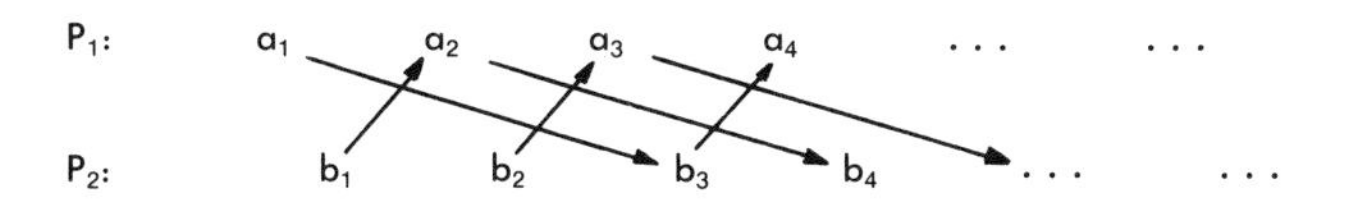

Figure 3.12
Information lag: first version.

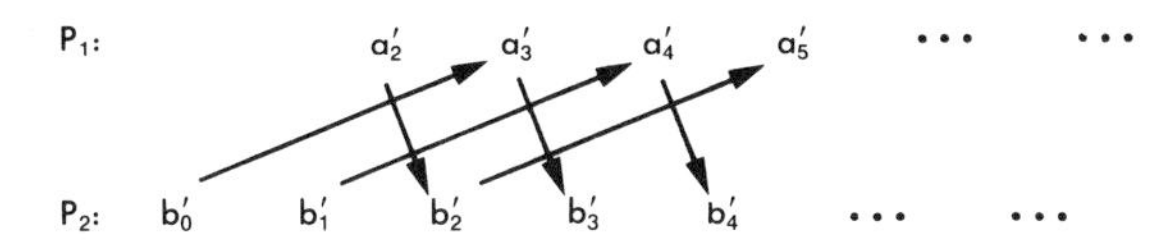

Figure 3.13
Information lag: an equivalent version.

There is a close connection between these games with information lag and "games of incomplete information," mentioned earlier, in which the players do not know all the rules when they start to play.[25]

3.3.8 Games of timing

The extensive-form models we have described thus far treat the game as a sequence of discrete events, and they are not especially well adapted to a continuous time frame. Although the principal ideas we have been discussing can be carried over in some form to continuous time, there are severe technical as well as conceptual difficulties. Satisfactory analysis in this area has usually depended on exploiting special properties of the model in question.

The Dart Duel of chapter 2 was an example of a game with continuous time. Each player, watching the other, had to decide when to throw his dart. We recall that the rule for optimal play was to throw one's dart as soon as the accuracies (hit probabilities) of the two players added up to 1. The game was analyzed in strategic form; this was easy to do despite the continuous infinity of decision points, because there were not really very many contingencies to consider.[26] One can even begin to sketch a game tree as in figure 3.14, but to be complete this diagram would require an infinity of branches, one issuing from every point of every vertical line.

Continuous-time games in which there are only a finite number of possible events are known as *games of timing* or (when the context permits) *duels*. The early literature stressed two diametrically opposed

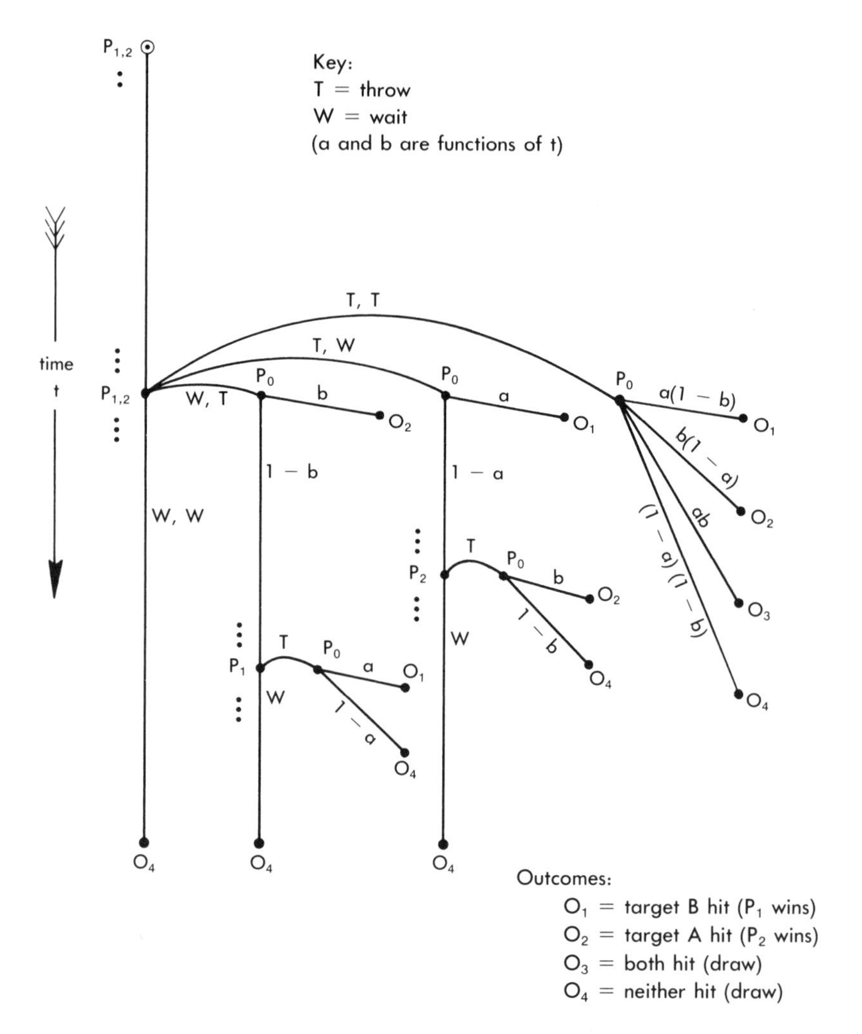

Figure 3.14
A game tree for the Dart Duel.

information conditions. In the *silent duel* the players learn nothing of what their opponents have done until the game is over; in the *noisy duel,* as above, they learn everything the instant it happens.[27]

In the silent case the extensive form quickly gives way to the strategic form. Time becomes like any other continuous strategic parameter and poses no unusual technical problems. Given the lack of information, solutions in mixed strategies are to be expected; typically these will consist of probability densities over portions of the time interval, combined perhaps with some "atoms" of positive probability. Figure 3.15 illustrates the solution to the symmetric silent duel in which each player has one bullet and the accuracy at time t is equal simply to t, for $0 \leq t \leq 1$.[28]

Returning to the noisy case, we remark that the players have not perfect information but almost-perfect information. It is the limiting case of a simultaneous game. Such games are usually best dealt with

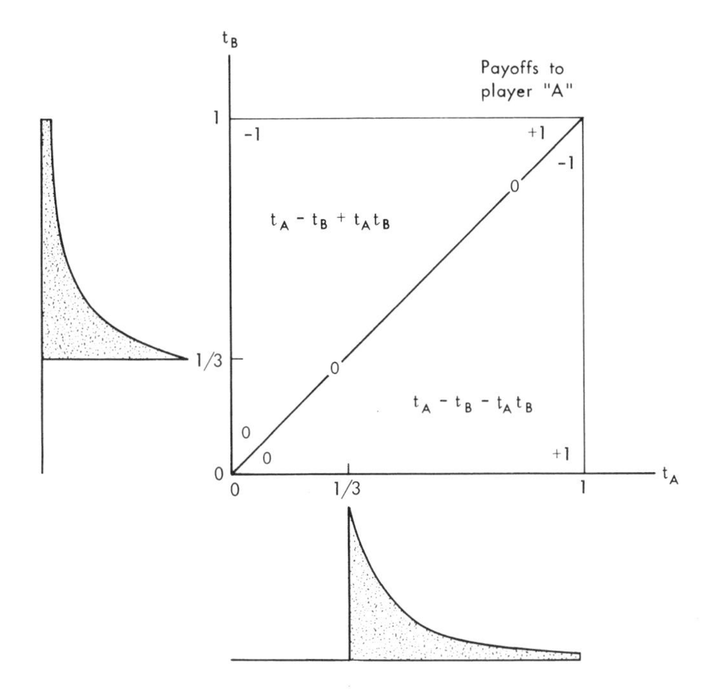

Figure 3.15
Probability densities for optimal firing in the symmetric silent duel.

in positional form, working backwards from the end to determine the value of each position, but the instantaneous information lag may give trouble. At critical times—for instance, at the beginning of the game—mixing over several different actions may be essential, just as in a discrete game, to keep oneself from being too predictable. Mixing over the *time* of some action, continuously in a small interval, may also be employed in order to cope with situations where exact simultaneity must be avoided (Dresher, 1961; Shapley, 1964b).

Another source of trouble lies in the nature of the continuous time variable itself. Consider the following example:

MUSICAL CHAIRS. *The first player to sit down after the music stops wins.*

In this game the best positional strategy is obvious: whenever you are standing and the music is not playing, sit down! But this positional strategy does not "solve" the game or even define a playable strategy; it does not give sufficient instructions to the user. In fact, this particular game has no strategies, pure or mixed, that are even approximately optimal.

EXERCISE

3.13. *Show that no pure strategy will suffice for the silent duel illustrated in figure 3.15. Verify the optimality of the mixed strategy given.*

3.3.9 Differential games

Games of timing involve only a discrete set of actions occurring within a continuous time framework. When actions or other positive decisions are taken continuously, as in the steering of a car or a ship, then we enter the realm of differential games. It would take us too far afield to attempt a full survey of this important and distinctive branch of game theory, which has a very extensive literature of its own.[29] Instead we shall make a few general observations on how some of the strategic and informational principles discussed in the preceding sections relate to differential games.

Mathematically the theory of differential games draws upon the fields of differential equations, dynamical systems, the calculus of variations, and especially control theory (which may be regarded as the theory of one-person differential games). This mix of technical disciplines has tended to attract a separate class of mathematical specialists from those attracted to the general body of game theory, and their work often reveals a lack of awareness of basic game-theoretic issues and ideas. Some, for example, have yet to discover how unsuited terms such as "optimal" and "rational" are in connection with noncooperative equilibrium strategies in non-zero-sum games. Even in the two-person zero-sum case, where differential games have been applied with notable success, the control-theoretic approach, by its nature, has tended to saddle the investigators with a narrow and sometimes inadequate view of the concept of strategy.

The cornerstone of control theory is a general proposition known as Pontryagin's Maximum Principle, which can be interpreted as asserting the equivalence of two forms of optimization.[30] In order to optimize "globally" over an entire multistage or continuous process, one must optimize "locally" at each stage or instant of the process. The idea of local optimization, however, presupposes the ability to set up a function that evaluates all possible intermediate stages or positions through which the process might pass, thereby enabling the decision maker to operate without memory or foresight. The values of the immediately following positions tell him what he needs to know about the future, while the positional parameters tell him what he needs to know about the past.

This intuitively appealing and analytically powerful concept retains at least some of its force when we pass to systems with two or more decision makers. An indispensable key to its successful use is the existence of a strong value concept, such as the minimax value for two-person constant-sum games. Nevertheless the underlying logic of Pontryagin's principle begins to crumble when there is more than one player.[31] Local optimality, though usually sufficient, is no longer inherently necessary for global optimality, since a strategy can be optimal in the game as a whole even if it does not provide for exploiting or "punishing" every error that the other player might commit.

One can also have local without global optimality in games (or even in one-person control problems) with no comprehensive stop rules, such as games of survival, because the "solving backwards from the end" technique does not work (for specific examples see Milnor and Shapley, 1957). It is quite possible in such games to have passive "waiting" strategies which are locally optimal, because they maintain the value of the position, but which postpone forever the winnings that more aggressive action could achieve.[32]

The following differential game—a sort of two-dimensional Achilles and the Tortoise—will serve to illustrate several of these points.

TAG (Isaacs, 1965). *A pursuer P chases his quarry Q across an open field. The former wishes to minimize, the latter to maximize, the time to capture.*

Suppose that P is twice as fast as Q and that there are no obstructions of any kind. Represent the players as moving points in the Euclidean plane, and avoid the question of infinitesimal "dodging" by defining capture to be an approach to within some preassigned small distance ϵ. Then, referring to figure 3.16, if both players play optimally they will both head for the point D, and "ϵ-capture" will occur just to the left of that point. These optimal trajectories are in fact unique; nevertheless there are many optimal strategies (contingency plans). Consider, for example, the following strategies for P:

1. Run directly toward Q at all times.

2. When on the closed segment AB, run toward C; otherwise run toward Q.

3. First run to C, then observe Q's position and run to that point, then observe his position again and run to *that* point, and so on. (The ϵ-capture rule fends off Zeno's paradox!)

Strategy 1 is the obvious Pontryagin strategy; it is both optimal and locally optimal. It exploits every situation to the full, except for anticipating future "dodges" by Q. Wherever the two contestants find themselves, this strategy gives P the earliest *assured* time of capture. To many differential game theorists this would be the only "optimal" or "rational" strategy for P.

Strategy 2 is optimal but not locally optimal, even though it is, like 1, a positional strategy.[33] If Q should run "north," for instance, then P, when he reaches B, would be further from his quarry at E than he

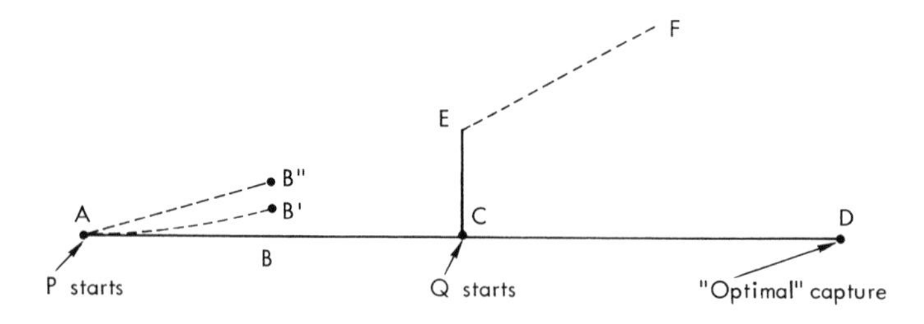

Figure 3.16
Tag.

would have been had he followed the pursuit curve AB', as dictated by strategy 1. He would nevertheless effect capture at worst at F, bettering the optimal capture time. (If AC is the unit, then CD is one unit in length and CEF is $1/4 + \sqrt{5}/4 = 0.809$ unit.) But Q's mistake in heading "north" would not have been exploited as fully by strategy 2 as by 1.

Strategy 3 is also optimal, since the distance between P and Q is at least halved at each observation. It is not locally optimal, however. Indeed it is not a positional strategy, even if we include time among the positional variables; P relies on his memory rather than on continuous observation of his target.

How does one choose among these (and many other) optimal strategies? A strategy like 3 may be easier to implement than 1, since so much less information must be acquired and processed. On the other hand, 1 and 2 require no memory capability, though 2 depends on a fixed landmark in the plane. The local optimality of 1 seems an attractive bonus. Yet if there is a real reason to believe that E will not or cannot play optimally, then our model of the situation is wrong, or at best incomplete, and probably none of the three is optimal. For example, if the maneuver CE could be predicted, then the straight path AB'' would lead to an earlier capture than either AB or AB'. In other words, if we adduce evidence to support the contention that local optimality is a desirable criterion in itself (and not just a technically attractive way to assure optimality), then in so doing we change the rules, and therewith the class of optimal strategies and (probably) the value of the game.[34]

An indication of the intrinsic technical difficulty of differential game analysis is provided by the fact that if we put a circular pond in the field, across which the players can see but not travel ("Obstacle Tag"), then no general solution is yet known, locally optimal or otherwise. This game was proposed many years ago by Isaacs (1965, pp. 135, 152). An idea of the trouble may be obtained from figure 3.17. If P runs directly toward A and Q directly away from B, then a position P' is reached at which P would rather be running toward the other bank of the pond.

Very serious conceptual difficulties arise when differential game theory is carried outside the domain of two-person zero-sum games (or other inessential games). The absence of a strong value concept means that the Pontryagin principle loses its grip on the situation. Functional equations can still sometimes be written down and solved,

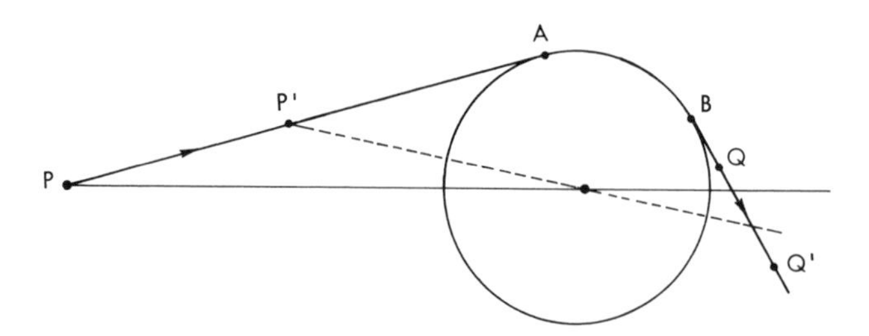

Figure 3.17
Obstacle Tag.

expressing a form of local optimality. But the "optimality" or "rationality" of the myopic and memoryless behavior they depict is seldom justifiable on heuristic grounds. The solution that is obtained is a form of noncooperative or "Nash" solution—an n-tuple of strategies with a weak equilibrium property (see chapter 9). But although this solution may uniquely satisfy the conditions for local "optimality," it is almost never the unique noncooperative equilibrium. Moreover, among the class of noncooperative equilibria it is usually far from the best or the most realistic.

EXERCISE

3.14. *Suppose that the payoff in Tag is 1 for capture, 0 for escape. Show that every positional strategy for P is locally optimal, regardless of the relative speeds of P and Q. Is the same true for Q?*

3.3.10 Models and context

This section has indicated how the modeler may use game-theoretic methods to incorporate phenomena that extend over time. Sometimes, and for some purposes, the relationship between the static and the time-dependent model of an economic or social process is reasonably good. It is not possible to make this conclusion a priori. It is our belief, however, that for much economic analysis the relationship is sufficiently good to make a study of static or steady-state models extremely fruitful. At many places in the preceding discussion we have had to motivate our treatment by intuitive or informal appeals to solution concepts. (Indeed most of the bibliographic references we have given in this survey are concerned as much with solutions as with problems of structure and modeling.) When we turn to the discussion in earnest of the various solution concepts in the later chapters of this volume, and then to actual economic models, it is desirable that the reader should try to consider the degree to which each argument might be made dynamic, even when no explicit dynamic analysis is presented.

Game-theoretic methods are a natural outgrowth of attempts to extend models of individual rational choice. Sometimes they are presented as though they are in complete opposition to other ways of modeling human behavior. This opposition is by no means true. The model of a stochastic game, for example, is not far removed from the type of flow diagram called for in simulations and behavioral models in modern psychology and social psychology. For another example, the controlled random walk represented by a game of survival is closely related to models used in learning theory. And the Kuhn tree itself is a generalization and elaboration of the familiar decision tree of modern decision theory (Raiffa, 1968). These and many other contacts of game theory with related fields will be surveyed in chapter 12.

3.4 The Strategic Form

In the strategic form of a game the whole move-and-information structure—what we ordinarily think of as the "rules"—drops out of sight. Strategies are now regarded not as complex sets of instructions but as abstract objects to be manipulated formally, without regard for their meaning. The outcomes or payoffs can now be given in tabular form or in mathematical formulas that may not reveal anything about the original extensive-form rules. It can be an interesting puzzle in cryptographic detection to try to infer the most likely information conditions, number of moves, and other elements of a game that is given to us in its strategic form. (It is somewhat like trying to guess the withholding tax laws from a study of paychecks.)

The strategic form is easily realized in vitro, that is, in the gaming laboratory. Occasionally a real game in vivo presents itself in strategic form: the players must really make simultaneous, independent, all-encompassing, initial decisions that determine the whole course of action. But this is quite unusual. It is therefore worth asking what is lost or forgotten when a game is "reduced" from extensive to strategic form. To put the question more sharply, suppose that two games have different extensive forms but come down to the same strategic form. To what degree may we regard them as analytically equivalent?

The answer, as so often occurs in these matters, depends on the purpose and context of the analysis. If we are just considering the problem of optimal behavior in, say, a two-person zero-sum game, then we are fairly safe in working only with the strategic form (but see Aumann and Maschler, 1972). Likewise, if we are only interested in solutions that depend on the "characteristic function" of a multiperson game (see chapter 6), then again the strategic form is enough.

On the other hand, there are at least three important areas of investigation in which the reduction from extensive to strategic form destroys essential information:

1. descriptive modeling,
2. study of nonoptimal behavior,
3. study of negotiations.

The first requires no further comment. The strategic form is deliberately nondescriptive—that is its virtue. The second heading refers to the problem of making explanatory models for such things as errors, trickery, surprise, skill, habit, stupidity, limited computing or communicating ability, complexity for its own sake (as, for example, when a Chess or Go player who has given his weaker opponent a handicap seeks to complicate the position), or simplicity for its own sake. Though seldom studied in a systematic, mathematical context, these are all proper topics for a general theory of games of strategy, and they all seem to require the apparatus of the extensive form.

The third area of investigation is the most treacherous, since the pitfalls are less apparent. Negotiations among the players are implicit, but only implicit, in most of the cooperative solution concepts we shall consider. Since these solutions generally do not look behind the strategic form, there is an implicit presumption that negotiations *during* the play are not allowed or are not relevant. It is a sad fact that we still lack a general theory of cooperative games in extensive form.[35] The standard solution theories tell us next to nothing about coalitional dynamics; they require us to assume, contrary to common experience, that all the "politicking"—coalition formation, deals, promises, threats, and so forth—takes place before the first move is actually played. Accordingly it may be misleading, when writing a verbal account of the meaning of a cooperative solution, to interweave negotiatory events with the moves and acts of the formal game, even though doing so may make the "story" sound more plausible.

What, then, saves our present endeavors from unrealism and futility? The best answer echoes the beginning of this chapter: we interest ourselves primarily in situations in which details of process, including details of negotiation, can be presumed irrelevant. To the extent that the Chicago Board of Trade or the Republican National Convention is a "game of skill" with myriad rules and special procedures (written and unwritten), present cooperative game theory can say virtually nothing about them. But to the extent that these institutions represent the battlegrounds for certain elemental economic or political forces, the present theory has at least some basis for making meaningful pronouncements.

Indeed many economic, political, and diplomatic competitions (in contrast to parlor games and military applications) are modeled most realistically not in the extensive form but rather in the strategic form or (more often) in the still more austere characteristic-function form (chapter 6). Inventing extensive-form rules for such games is often an exercise in artificiality. For example, political constitutions can be written—and can be game-theoretically analyzed—with little reference to the specific procedures for casting and counting ballots (Shapley and Shubik, 1954; Riker and Shapley, 1968; Banzhaf, 1968). In fact, voting procedures often change from election to election, and they may even be declared "unconstitutional" if they are found to violate the intent of the framers of the original constitution. In short, there is an underlying game, buried in the constitution, that is independent of any particular extensive-form realization.

As a second example we may consider a simple one-period duopoly situation with a large number of different ways of sequencing the price and production decisions and the flow of information (Shubik with Levitan, 1980). Many of these models might be plausible renderings of different actual duopolies, observed at particular points in time. Tactically they are different. But the differences dwindle in significance when we reflect that the "rules" are not rigidly fixed and may even be partially under the control of the players, who can, for example, buy more information (or more secrecy) or reschedule their decisions and announcements. To an operations researcher advising an actual firm, or to a Justice Department analyst evaluating a proposed merger, the absence of an adequate theory of negotiation in the extensive form might be regrettable. But a more detached observer, seeking only to understand the basic phenomenon of collusion in oligopoly, may be justified in hoping for some help from the existing theory of cooperative games.

To sum up, at the level of generality practiced in much of political science as well as in mathematical economics, the goal of accuracy in modeling may not only permit but even require us to avoid most of the detail with which the extensive form is concerned.

3.4.1 The outcome matrix

Thus far we have not taken much care to distinguish between outcomes and payoffs. Speaking generally, by an outcome we mean a possible state of affairs at the *end* of the game. But the statement of an outcome might have to include a description of the course of play that led to it, or even a probability distribution over several courses of play, as depicted in figure 3.1. Usually, however, the model builder will summarily discard anything that he regards as unimportant to the evaluation of the outcome by the players. In the Shooting Match we cared only about broken balloons, not about how or when they were broken, while in the Art Auction we cared only about the last bid; thus, from our initial sequence of bids (section 3.2) we may extract the "reduced" outcome "P_1 gets the painting for \$19," or, from figure 3.1, "P_1 gets the painting for \$19, \$20, or \$21 with probabilities $\frac{7}{16}$, $\frac{7}{16}$, and $\frac{1}{8}$, respectively."

If there are just two players, and not too many strategies, the outcomes (suitably reduced) can be displayed in matrix form, as suggested by figure 3.18. The outcome matrix is often too large to

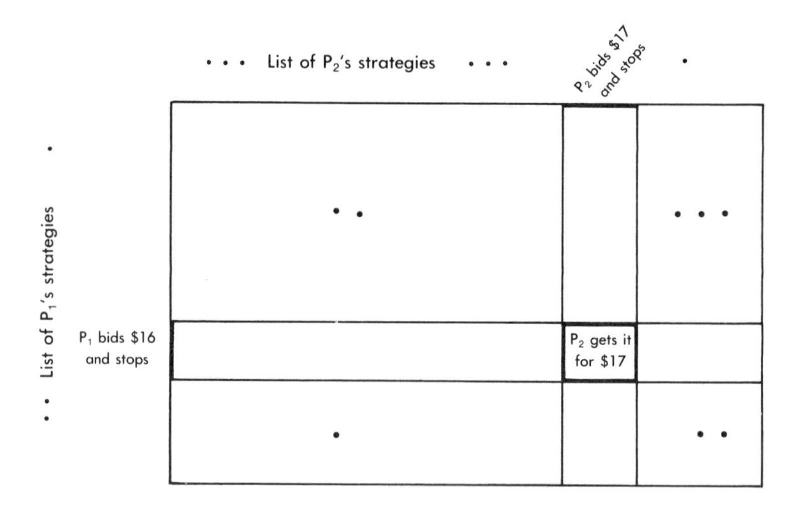

Figure 3.18
An outcome matrix.

write out in full, and with more players the array becomes multidimensional. Nevertheless it is a useful conceptual device. By its rectangular shape it reminds us that each strategy choice of each player must "work" against all possible strategy choices by the other players.

Figure 3.19a again uses the Art Auction as an example, assuming for simplicity that there are three players with two strategies apiece. Strategy 1 in each case is the initial one given in section 3.2, while strategy 2 is the strategy of never bidding at all. Thus, in the box marked by an asterisk, only P_1 and P_3 are bidding, and the sequence of bids leading to the stated outcome is 15_1, 16_3, 18_1, 19_3, 21_1.

3.4.2 Payoff vectors and matrices

Continuing with the example, suppose that the value of owning the painting is \$27, \$28, and \$20 to P_1, P_2, and P_3, respectively. If all three adopt the specimen strategies in section 3.2, then P_1 will obtain the item at a price of \$19, so his net gain is \$8. If we take zero to represent the value to each player of the status quo, then the outcome in question yields the *payoff vector* (8, 0, 0). Under another trio of strategies P_3 might win the painting at \$22; the payoff vector would then be (0, 0, −2).

If the outcome is probabilistic, the easiest thing to do is to average the payoffs. Thus in figure 3.1 P_1's average price is

$$\frac{7}{16}(\$19) + \frac{7}{16}(\$20) + \frac{1}{8}(\$21) = \$19.6875.$$

The payoff vector for this outcome is therefore (7.3125, 0, 0) in "expected dollars."[36]

Like the outcomes, the payoffs can be represented in matrix form. The coordinates are as before, but the entries are payoff vectors. Thus the number 12 with the asterisk in figure 3.19b refers to the gain of P_1 when he wins the auction with his first bid of \$15, as might occur if the players were to form a "ring" (Duveen, 1935).

It is common practice to display the payoffs to each player in a separate matrix; thus a (finite) two-person game is often described as a *bimatrix game.* A *zero-sum game* is one in which the payoffs to all the players add up to zero; similarly a *constant-sum game.* In such games one player's payoff matrix can be omitted since it can be determined from the others. In particular, a two-person zero-sum game in strategic form is almost always represented by just a single payoff matrix or payoff function, one player being designated the maximizer, the other the minimizer.

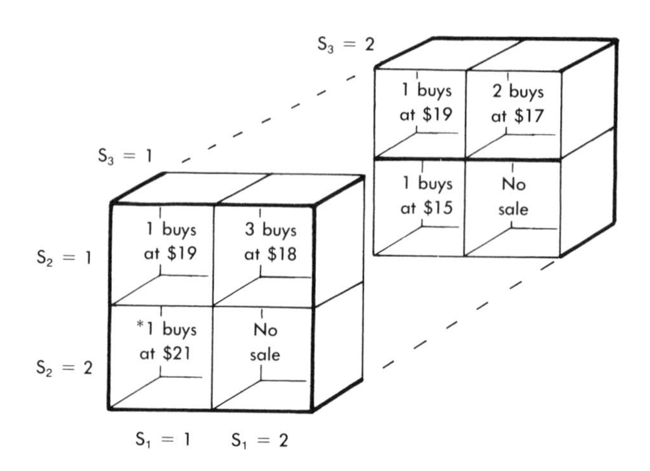

(a) Outcomes

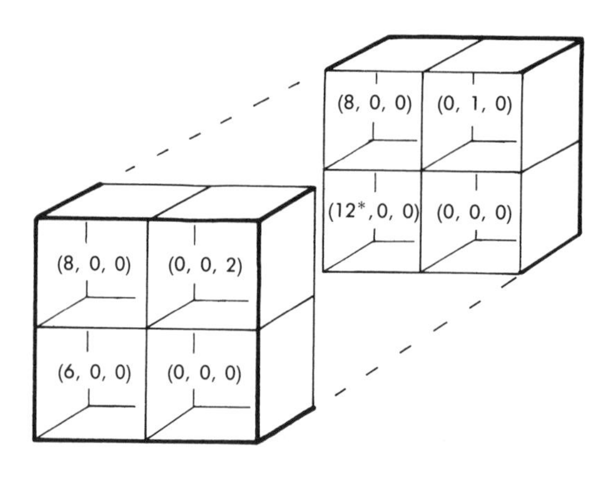

(b) Payoffs

Figure 3.19
Outcome (a) and payoff (b) matrices for the Art Auction with three players having two strategies apiece.

3.4.3 More general payoff functions

Speaking generally, an n-person game in strategic form is nothing but a set of n real-valued functions in n variables:

$$P_1(\sigma_1, \sigma_2, \ldots, \sigma_n), P_2(\sigma_1, \sigma_2, \ldots, \sigma_n), \ldots, P_n(\sigma_1, \sigma_2, \ldots, \sigma_n),$$

where $P_i(\sigma_1, \sigma_2, \ldots, \sigma_n)$ is the payoff to player i and where the domain Σ_i of each player's strategies σ_i may be set up in any convenient way. The matrix form, with its standardized numbering of the strategies, may needlessly obscure the patterns of interstrategic relationships, and another functional notation is often more expressive and more amenable to analysis as well. For example, in a given instance it might be more natural to represent a strategy for player i as a pair of numbers (i_1, i_2) than as a single number i.

When there are infinitely many strategies, the matrix form is of course no longer available, but a matrixlike representation is sometimes still handy. Thus in chapter 2 we considered the following payoff function for player A in a two-person constant-sum game:

$$P_A(x, y) = \begin{cases} a(x) & \text{if } x < y, \\ 1 - b(y) & \text{if } x > y, \\ \tfrac{1}{2}[a(x) + 1 - b(y)] & \text{if } x = y. \end{cases}$$

Here the functions a and b are assumed given, and the strategic variables x and y have the same domain, the closed interval $[t_1, t_2]$. This information is concisely displayed in figure 3.20a. (For another example see figure 3.15.) Much of the early work on infinite games was focused on such "games on the square," perhaps because of the strong visual analogy to matrix games.[37]

EXERCISES

3.15. *Write the payoff matrices associated with the game shown in figures 3.2, 3.4a, and 3.4b.*

3.16. *Analyze the two-person "game on the square" shown in figure 3.20b, where the strategies for both players either (a) range over the closed unit interval* $0 \leq s_i \leq 1$ *or (b) range over the open unit interval* $0 < s_i < 1$.

3.4.4 Other payoff indicators

Sometimes the passage from outcome to payoff is blocked by imponderables of valuation or interpretation that relate directly to the imponderables of the n-person problem we are trying to resolve. We may prefer not to dispose of these matters summarily, at the descrip-

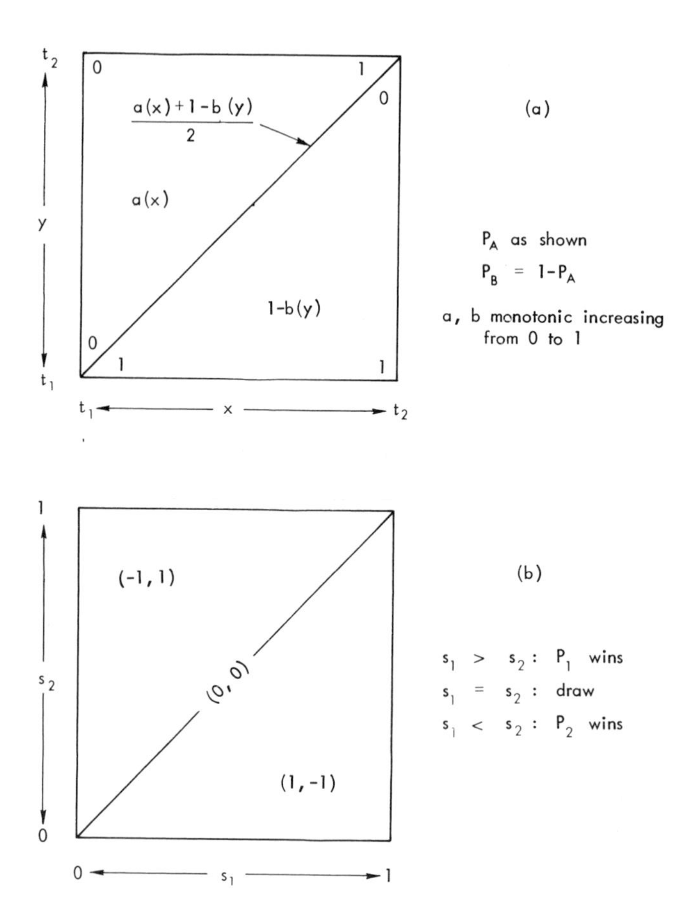

Figure 3.20
Games on the square.

tive stage of the analysis, but rather to carry them over into the solution stage. For this purpose more general forms of payoff indicator can be useful.

In simple cases it may suffice to introduce an undetermined coefficient or parameter into the expression for the payoff, and then adjust it or let it vary over a range as the solutions take shape. For example, the tradeoff ratio between two objectives, such as guns and butter, may be left undetermined.[38] More generally, it may be necessary to drop the idea that the payoff to each player is a number, whether determined or undetermined, or even an element of a completely ordered utility space.

The next chapter will discuss a variety of nonnumeric payoff indicators that have been used with some success in game-theoretic investigations. The price for so "loosening" the payoff-utility structure of the model must be paid, however, when the solution stage of the analysis is reached. Most solution concepts will require some adaptation or extension in order to cope with the new payoff scheme, and some may fall by the wayside.[39] The solutions that survive, for their part, are likely to be less determinate than usual, and less satisfactory in their interpretation.

In addition to the problem of setting up individual utility scales, there is the possibility that the stakes of the different players in the outcomes may not be cleanly separated. The end of the game, despite all our efforts at closing the model (see section 1.4.2), may fail to be "game-theoretically inert." The alliances and polarizations effected during the play of the game may have residual influences far into the future, and these influences, coalitional in nature, cannot be captured by merely adjusting the separate payoffs of individual players.

A suggestive approach to this modeling problem would be to use set functions in place of payoff vectors, thereby assigning a payoff number (or other indicator) to every *coalition.* These set functions would be tantamount to characteristic functions representing future games. An additive set function, in particular, would correspond to the inert case and would reduce to an ordinary payoff vector. Replacing the set functions by their value solutions (see chapter 7) would be one possible way of disposing of the problem summarily at the descriptive stage, but it would be interesting to try to carry the set-functional description through to the solution stage, after suitably generalizing the solution concepts to be employed.

In sum, weakening or generalizing the payoff concept does not really solve anything. It may, however, postpone a modeling problem to a later point where it can be shown to be irrelevant, or where it is swallowed by a bigger problem. Failing that, it may at least defer the problem to a point where the analyst's "customer"—the ultimate user of the model—can fairly exercise his own intuition in judging the matter.

Notes to chapter 3

1. For example, a commodity exchange might, by rule or custom, quote prices only in units of 1¢. Yet a mathematical model that ignored that rule and permitted a continuum of prices might actually be more realistic in the long run, since it is a fair assumption that the 1¢ rule would soon be revised if it were found that substantial economic values turned on the possibility of smaller increments. This point will be developed in more detail in section 3.4.

2. The strategic form is called "normalized form" or "normal form" by other authors; the present term seems more explanatory.

3. See Borel (1921). If negotiations or agreements are envisaged outside the formal rules, they must be assumed to take place before the selection of strategies, not during the actual course of play. Otherwise the strategic form loses some validity as a descriptive tool.

4. Rabin (1957) has shown using the theory of recursive functions that there are certain two-person win–lose games in which one player has a winning strategy but no "effectively computable" winning strategy, so that it would be impossible, for example, to write a computer program for him that would defeat any opponent. Rabin's game has infinitely many positions, but the rules, unlike the winning strategies, are effectively computable, and each play is finite in length.

5. The laws of Bridge appear to prohibit general mixed strategies, which would entail private understandings between partners. It seems likely, for this reason, that Bridge has no optimal strategies. See G. L. Thompson (1953b) and Keeler (1971).

6. A *graph* is a finite collection of *nodes* and *edges,* with a node attached to each end of each edge. The graph is *connected* if it is possible to go from any node to any other node along some path of connecting edges. In general, there may be several paths between two given nodes, so that closed loops or *cycles* exist. But if the connecting paths are all unique, the graph is said to be *acyclic* and is called a *tree.* A *rooted tree* is a tree with one node distinguished, called the *root.*

7. As von Neumann and Morgenstern put it, "preliminarity implies anteriority, but need not be implied by it" (1944, p. 51). In discussing information lag in section 3.3.7 we shall show how the timing of moves can sometimes be modified drastically without affecting the essential information pattern.

8. It might better be called a "lack-of-information" set, since it portrays a reduction of the information available to the player. The larger an information set, the less the player knows about the situation.

9. The term "complete information" (not to be confused with "perfect information") was used by von Neumann and Morgenstern (1944) to express an underlying assumption of their theory (as well as most subsequent theories), to wit, that the players at the beginning of the game are completely informed as to the precise state of affairs and can make all necessary calculations. The study of "games of incomplete information," which arise in many practical contexts, presents formidable conceptual and technical difficulties; for references to some recent work in this area see sections 1.3.2 and 3.3.7.

10. For another example we may look ahead to figure 3.6. Any strategy that instructs P_1 to play 1 at the beginning automatically makes the choice at his other node irrelevant. Of his six strategies, only four are operationally distinct.

11. The question of what makes two patterns of information equivalent has an extensive literature: see, e.g., Shapley (1950), Krentnel, McKinsey, and Quine (1951), F. B. Thompson (1952a), and Dalkey (1953).

12. In Chess the history of the board position is in fact sometimes relevant because of the rules for castling, for *en passant* captures, and for draws by repeated position. A rigorous definition of "position" in Chess is not so easy to formulate.

13. See note 6. In a *directed graph* each edge has a preassigned orientation, usually denoted by an arrow. In the case of a rooted tree the arrow is not needed since the rule "travel away from the root" gives an unambiguous orientation to each edge.

On positional form see Milnor (1953), Shapley (1953d), Berge (1957a,b), Holladay (1957), Isbell (1957a), Hanner (1959), Vorobyev (1963, 1970a), and several papers in the Russian collection *Positional Games* edited by Vorobyev and Vrublevskaya (1967).

14. The full implication of this point must await consideration of the role of "threats" in the various solution concepts to be defined.

15. See Shapley (1953c,d). Other terms sometimes used are *Markov strategies, positional strategies,* and (in the case of a loop) *steady-state strategies.*

16. Games of this type (where the winning set may of course be much more complicated than a simple interval) have been studied by Gale and Stewart

(1953), Wolfe (1955), Davis (1964), and Mycielski (1964b). They are related to the Banach–Mazur game of Pick an Interval and similar games: see Mycielski and Zieba (1955), Mycielski, Swierczkowski, and Zieba (1956), Oxtoby (1957), and Hanani (1960). Also see Mycielski and Steinhaus (1962), Mycielski (1964a, 1966), and Mycielski and Swierczkowski (1964), where such infinite games of perfect information are employed in investigating the axiomatic foundations of mathematics (the Axiom of Determinateness).

17. Aumann (1964c) considers infinite sets of alternatives at a node. Another kind of infinite generalization comes from the introduction of a continuous time scale.

18. See Shapley (1953c). For related work and extensions see Sakaguchi (1956), Gillette (1957), Takahashi (1962), Beniest (1963), Zachrisson (1964), Hoffman and Karp (1966), Charnes and Schroeder (1967), Blackwell and Ferguson (1968), Flerov (1969), Kifer (1969), Kushner and Chamberlain (1969), Pollatschek and Avi-Itzhak (1969), Rogers (1969), D. G. Smith (1969), Liggett and Lippman (1970), Maitra and Parthasarathy (1970), Schroeder (1970), Yoshikawa (1970), Parthasarathy (1971), Sobel (1971), and Kirman and Sobel (1974).

19. Equivalently we can suppose that the spot payments are banked and draw interest at the rate $\beta/(1 - \beta)$ per round.

20. This procedure turns out to be more or less equivalent to the limiting form of the discounted case as $\beta \rightarrow 0$: see Gillette (1957), Hoffman and Karp (1966), and Liggett and Lippman (1970).

21. See Bellman (1957), Scarf (1957), and Blackwell (1962, 1965). Little is known about stochastic or recursive games with more than two players, or about the two-person non-constant-sum case or the two-person constant-sum case with imperfect information in the supergame; see, however, Mertens and Zamir (1971), Basar and Ho (1974), Kirman and Sobel (1974), Levine and Ponssard (1977), and Shubik and Sobel (1980).

22. See Shubik (1959b), Shubik and Thompson (1959), and Gerber (1972). Unlike most of the other models discussed in this section, the main interest in games of economic survival lies in the *non*-zero-sum case.

23. The discussion in this section is based upon Scarf and Shapley (1957).

24. Nevertheless, using the flexibility afforded by shifting the pattern to give one or the other player perfect information, a certain kind of "generalized subgame" can be identified that aids in the solution (see Scarf and Shapley, 1957). A particular, deceptively simple game of this type (the "bomber–battleship duel") has received intensive study (see Isaacs, 1952, 1965; Isaacs and Karlin, 1954; Dubins, 1957; Karlin, 1957a; Scarf and Shapley, 1957; Ferguson, 1967; and Washburn, 1971). For some related games see Blackwell (1955), Matula (1966), Blackwell and Ferguson (1968), and Washburn (1971).

Information lag has also been considered in a continuous-time context (see, e.g., Jumarie, 1969, and Elliot and Kalton, 1972).

25. In addition to the references in section 1.3.2, see Krinskii and Ponomarev (1964), Rosenfeld (1964), Vorobyev (1966), Stearns (1967), Sweat (1968), Zamir (1969a,b, 1970), and Harsanyi and Selten (1972).

26. A strategy for player P_i is an ordered pair (t_i, f_i); the number t_i tells him when to act if the other does not act, while the function $f_i(t)$ tells him when to act if the other acts at $t < t_i$. In chapter 2 we started by eliminating all functions f_i but one, by a domination argument, and so were left with just the single strategic parameter t_i.

27. The study of duels was initiated in 1949 by Blackwell and Girshick and others in a series of Rand papers whose contents are summarized in Dresher (1961). Further references are Danskin and Gillman (1953), Shiffman (1953), Karlin (1953a,c, 1959), Blackwell and Girshick (1954), Caywood and Thomas (1955), Vogel (1956), Restrepo (1957), Burger (1959), Kisi (1961), Shapley (1964b), D. G. Smith (1967), Fox and Kimeldorf (1969, 1970), and Sweat (1969, 1971). The term "noisy" is unfortunately in conflict with engineering and information-theory usage, in which "noise" is the antithesis of information.

28. The optimal probability of firing in the infinitesimal interval $[t, t + dt]$ is equal to $dt/4t^3$ for $\frac{1}{3} \leq t \leq 1 - dt$. The mean of this distribution happens to be $\frac{1}{2}$, the optimal firing time of the associated noisy duel.

29. We shall not attempt a full bibliography either. Isaacs (1965) gives a highly readable account of the research through which he almost single-handedly opened the field. Also recommended are Avner Friedman's more recent book (1971), the volume of contributions edited by Kuhn and Szegö (1971), and, for current research, almost any issue of the *IEEE Transactions on Automatic Control*; for application to economics see Case (1979).

30. See, for example, Pontryagin (1957) or Rozenoer (1959). This is more or less equivalent to the "principle of optimality" of dynamic programming (see Bellman, 1957; Dreyfus, 1965; or Denardo, 1981).

31. Of course, a "principle" is not meant to be a rigorous theorem or a universal truth. It is, rather, like a signpost telling the specialist where to look for theorems or a vivid summing-up for the nonspecialist of the sense of a whole family of theorems. In any particular theorem expressing an instance of Pontryagin's principle there will be restrictive hypotheses or conditions of a more or less technical nature. The present critique, however, does not turn on technicalities.

32. There is a simple king-and-pawn end game in Kriegsspiel (double-blind Chess) in which none of White's locally optimal strategies wins, although the value of the initial position is a win for White. (White has a center pawn on

the sixth rank, blocked by the Black king.) The point is that White cannot force a win without running some risk of a draw, though he can make that risk arbitrarily small.

33. To see that strategy 2 is optimal, note that when P reaches B, then Q will still be to the right of B. The only way that Q might outwit his opponent would be to lead him back to the segment AB, whereupon strategy 2 would dictate a wrong turn. But Q will be caught before he can do this.

34. Locally optimal solutions yield another bonus: they solve a whole class of games at one blow, since they take no notice of the starting position. But how much would a real player pay for this aid to analysis, and in what coin?

35. The main obstacle seems to be that introducing explicit negotiations into the fabric of the extensive game makes the model too ad hoc and limits its generality. Perhaps the best hope for progress is with a recursive or stochastic model (see section 3.3.7), where one might insert a suitably vague, and hence general, renegotiation session between rounds.

36. It is clear that the problem of passing from outcome to payoffs is being grossly simplified. Indeed the whole of utility theory lies between these two concepts, as well as some considerations concerning the nature of money. Chapters 4 and 5 will be devoted to these subjects.

37. See Ville (1938), Wald (1945a,b), Dresher, Karlin, and Shapley (1950), Karlin (1950, 1953a,b, 1957b), Glicksberg and Gross (1952, 1953), Shiffman (1953), Gross (1956, 1967), Sion and Wolfe (1957), Gale and Gross (1958), and especially Karlin (1959, vol. 2), which contains a thorough coverage of previous work as well as many new results. For some subsequent work see Djubin (1968, 1969).

38. See Shapley (1959c). Discussing the need for a theory of games with vector payoffs (i.e., with several numerical components per player), Rigby (1959) described the following experience in a military application: "However, each player of these strategies . . . generated both a time delay and losses to the moving forces. . . . Since the values of these delays and losses would be realized in a subsequent battle . . . it was clear that some sort of exchange ratio between attrition and delay must exist. Efforts to obtain estimates of such a ratio failed completely. In fact, several rather arbitrary weightings . . . were assumed . . . , but this was a rather unsatisfactory expedient. . . . [The] occurrence of vector payoff represents a failure to resolve some of the questions whose answers are needed in order to construct a game model. This being the case, one should not be disappointed that the theory does not produce a clear cut, well defined solution concept."

39. The value solution, for example, fails when only ordinal utilities are assumed; see the Bargainers Paradox in chapter 4.

4 Individual Preferences and Utility

4.1 Introduction: The Domain of Alternatives

The theory of games concerns motivated behavior. It is continually involved with questions of preference and utility, and yet to a remarkable extent it can stand aloof from the controversial issues of utility theory. Most of its concepts and many of its techniques are compatible with a wide variety of underlying preference structures. Although special assumptions about utility are not uncommon in game models, they are made more often for reasons of technical convenience than out of any conceptual necessity. There is no single, narrow view of utility that is forced upon us by the game-theoretic approach.

In this chapter we shall review some of the preference structures that are used in the study of rational behavior. As in our discussion of the art of modeling, our main interest is in exploring the connections with game theory, and we can hardly pretend that the "guided tour" that follows is a judiciously balanced survey of the subject for its own sake (see Fishburn, 1968, 1970a). For one thing, the standard viewpoint of utility theory envisages only a "one-person game," focusing on the preferences of a single individual, whereas our present view is colored by the intended application to multiperson models. The really interesting preference systems, for our purpose, are those in which many preference structures coexist and interact.

A theory of *economic* choice is, of course, an extremely small part of the general philosophical problem of choice. It may help to explain the actions associated with Mrs. Jones's purchase of a pound of bacon but not her decision to raise her hemline or murder her husband. "Rational" economic choice implies action in obedience to some more or less consistent scheme of preferences and expectations, relative to some specified *domain of alternatives*. In an economic model this domain is likely to be one of a small set of characteristic types. Perhaps the leading example is the "commodity space" comprising the possible "bundles" of goods that an individual could own or consume. The set of possible time sequences of earnings of a firm is another example.

A distinction must be made here between *choice* and *preference*: one is an act, the other a state of mind. The former refers to decisions or strategies, the latter to outcomes or prospects. In the ordinary approach to utility theory the two are often not sharply distinguished; this is because the basic or "primitive" experiment of the theory is to present a subject with a menu of alternative outcomes and then simply to award him the one he selects. Nothing of significance is allowed to intervene between the decision and the event, and thus the notion has arisen that preferences are themselves directly observable.

In most social-science applications, however, one cannot afford to ignore the intervening machinery, be it deterministic, probabilistic, or game-theoretic in nature. Even in an ordinary election, in which the voter marks a ballot listing the possible outcomes, the apparently straightforward correspondence between choice and preference is illusory and, as is well known, may often be clouded by strategic considerations. A ballot is not merely an expression of preference.[1]

In general, we may not find even a superficial correspondence between the decision space and the outcome space. A supermarket manager, for example, would probably be motivated not by preferences among the actual shopkeeping decisions (the hours he might keep, the prices he might charge, or the inventories he might acquire) but rather by their consequences (the sales or profits or prestige that he expects to result from his decisions). Only by working through the intervening deterministic, probabilistic, or game-theoretic processes, including especially the actions of his customers, suppliers, and competitors, can he translate his preferences among *outcomes* into preferences among *strategies*.[2]

The choice of terminology hardly matters when we are discussing formal preference structures in the abstract, except to the extent that the connotations of the words used may predispose us, by a sort of semantic persuasion, to regard various axioms or assumptions as more plausible or less plausible.[3] Our position will be, nevertheless, that preference systems are most properly applied to *domains of alternative outcomes*, not to domains of choices or decisions. We shall use the symbol $\mathcal{D}$ to denote such a domain of outcomes.

4.2 The Structure of Preference

The word "preference" certainly connotes a relation that is transitive. (See appendix A.1 for a glossary of order-theoretic terms.) If x is "preferred" to y and y is "preferred" to z, then it is hard to imagine that x is not "preferred" to z. We shall begin our guided tour, however, with two or three very weak preference structures that lack this basic property but are nevertheless capable of supporting meaningful game-theoretic or decision-theoretic investigations.

The first rudimentary structure is not even a relation but a *choice operator*. This is a function C, acting on sets of outcomes, that picks from each set S of possible outcomes a nonempty subset $C(S) \subseteq S$ of "preferred" outcomes. For an illustration, take the overall domain $\mathcal{D}$ to be a finite set of real numbers. Then a timid or "wishy-washy" personality might reveal the following choice operator:

$$C(S) = \begin{cases} \text{the median element of } S \text{ if } |S| \text{ is odd,} \\ \text{the two near-medians of } S \text{ if } |S| \text{ is even.} \end{cases} \tag{4.1}$$

The choice operator is an extremely weak utility concept. It entails no form of transitivity, and it may not even yield clear-cut binary comparisons.[4] Thus in (4.1) we have $C(\{3, 4, 5\}) = \{4\}$ and $C(\{4, 5, 6\}) = \{5\}$, so that sometimes 4 is chosen over 5 and sometimes 5 over 4. Nevertheless, as John Nash showed as early as 1950, it is possible to base at least the noncooperative side of game theory on nothing more than choice operators (see appendix A.2). Nash required compactness, convexity, and continuity of the operators, but he made no order-theoretic assumptions. He went on to suggest that choice operators might be an appropriate utility concept for groups as well as for individuals.

Another, hardly less rudimentary preference structure is the *binary preference relation*. This is in effect a list of all those pairs x, y in $\mathcal{D}$ of which it can be said that "x is preferred to y," written $x \succ y$. No overall ordering of outcomes is implied; there may be cycles (but only of length greater than 2) and incomparable elements. For example, if $\mathcal{D}$ were the seats at King Arthur's Round Table, we might have (see figure 4.1):

$$x \succ y \Leftrightarrow x \text{ is adjacent to } y \text{ on the left.} \tag{4.2}$$

Using just binary preference relations, Robert Aumann (1964b) treated a class of economic models from a game-theoretic standpoint (see also Schmeidler, 1969a; Gale and Mas-Colell, 1974; Mas-Colell, 1974; Shafer and Sonnenschein, 1974). Bezalel Peleg (1966), imposing the condition of *acyclicity*, which is related to but much weaker than transitivity (e.g., seat the knights along one side of a bar), showed

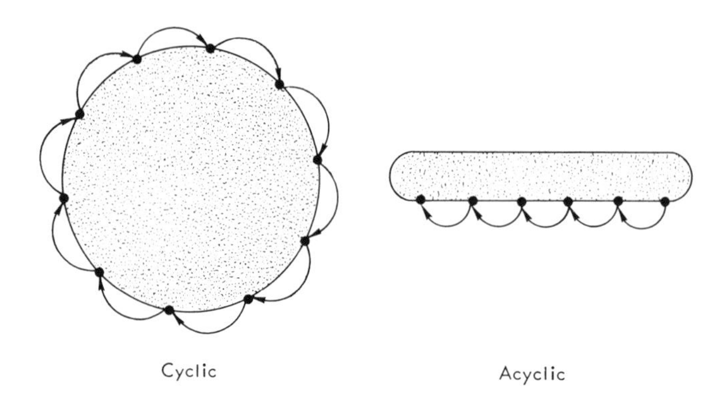

Figure 4.1
Some intransitive binary relations. The head of each arrow is preferred to the tail; there are no other preferences.

in a seminal note how to define several of the main cooperative solution concepts of game theory without transitive preference.[5]

A binary preference relation can always be extracted from a choice operator via the rule

$$x > y \Leftrightarrow C(\{x, y\}) = \{x\}, \tag{4.3}$$

but the converse does not hold, since a binary relation without transitivity tells us nothing about preferences among larger sets of alternatives. Indeed, since alternatives in real life do not usually present themselves in pairs, we would seriously question whether the *binary* preference relation, without transitivity, is a good primitive concept on which to base a robust theory of motivation.

4.2.1 Partially ordered preferences

A preference relation that is transitive but does not relate every two elements of $\mathscr{D}$ gives rise to an ordering that is weak or partial or both. (Here *weak* means that there are indifferent pairs, *partial* that there are incomparable pairs.) This permits us to distinguish between "indifference," for outcomes of equal desirability, and "incomparability," for outcomes that are never compared or, if compared, do not lead to an expression of relative desirability. This distinction is especially useful when we deal with group decision making, as in the case of an organization or coalition in which the internal decision process can lead to a "hung jury" in contrast to an affirmative declaration of indifference.[6]

Several special types of partially ordered preference systems have been studied with a view to application to the theory of games:

1. Duncan Luce (1956, 1959a) and others (Armstrong, 1939; Halphen, 1955; Tversky, 1969; Roberts, 1971; Jamison and Lau, 1973) have considered "semiordered" systems in which the inability to discriminate between nearby alternatives is the issue. For example, if $\mathscr{D}$ were the real line, there might be a number $d > 0$ such that

$$x \succ y \Leftrightarrow x \geq y + d \tag{4.4}$$

(see figure 4.2 and the exercises at the end of this section). Here d is called by psychologists the "just noticeable difference" or jnd.[7]

2. David Blackwell (1954, 1956) and others (Rigby, 1959; Shapley, 1959c; Aumann, 1964d; Peleg, 1970) have considered partially ordered preference systems in vector spaces. Here the inability to compare disparate factors in the outcome is the issue, as when both lives and money are at stake. The preference sets in these multiple-criteria models, that is, the sets $P(x) = \{y \in \mathscr{D}: y \succ x\}$, are "orthants" (rectangular cones) in form. For example, if $\mathscr{D}$ is the real plane and the coordinates of x and y are (x_1, x_2) and (y_1, y_2), respectively, we might have

$$x \succ y \Leftrightarrow x_1 \geq y_1, \quad x_2 \geq y_2, \quad \text{and } x \neq y, \tag{4.5}$$

as in figure 4.3. In another interpretation, the vector components are not "goods" but the utilities of the individuals that make up a group.

3. Robert Aumann and others have considered a similar mathe-

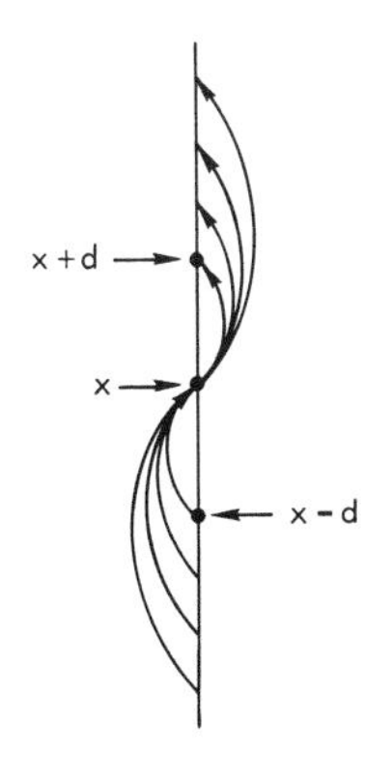

Figure 4.2
A semiorder. The pattern is repeated at each point x.

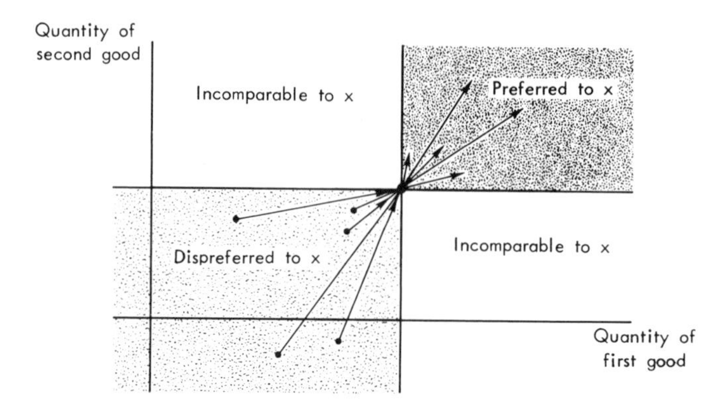

Figure 4.3
A partially ordered vector space.

matical structure but with yet another interpretation. It arises when $\mathscr{D}$ is a "mixture space," intuitively a set consisting of all probability combinations of risk neutrality that lead from a *complete* preference order to a linear or "cardinal" utility function (see section 4.4.2); such a space can take us from a *partial* preference order to a linear utility structure based on parallel convex cones (Aumann, 1962, 1964e; Sloss, 1971). For example, in figure 4.4 there are three "pure" outcomes, represented by the points X, Y, and Z, and an infinity of "mixed" outcomes, represented by the other points of the large triangle. As indicated, outcome Y is more desirable than any lottery between X and Z that gives less than a 25 percent chance of X, but it is less desirable than one that gives more than an 80 percent chance of X. Consistency with axioms concerning probability combinations, similar to those in appendix A.4, forces the preference sets throughout the triangle to be parallel cones, as indicated at the typical point W.

4.2.2 Extensions of relations*

In working with partial orderings or other incomplete relations, one often wants to extend or restrict a given preference system. Discriminating between previously incomparable outcomes would extend the system; imposing new preconditions for a preference judgment

*This technical section may be omitted on first reading.

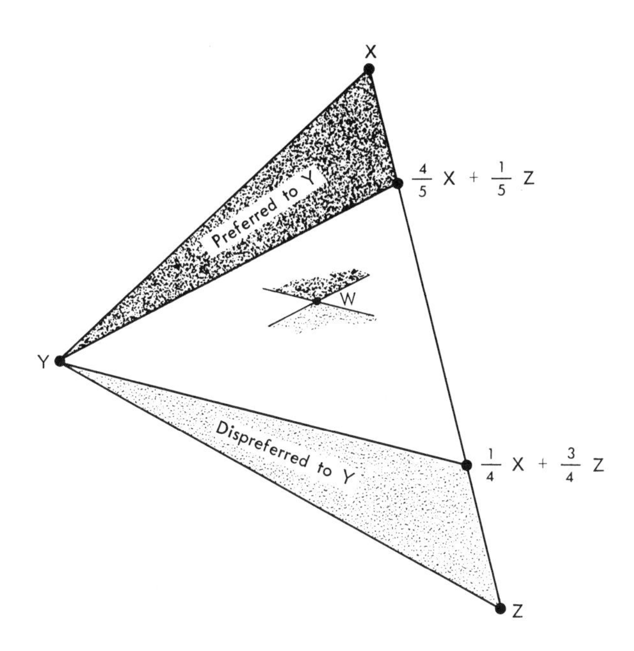

Figure 4.4
Partially ordered preferences in a mixture space.

would restrict it. A strong/weak terminology is often used here: we say that one preference relation is *stronger* than another if its preference sets are more inclusive, *weaker* if they are less inclusive. Thus, in the situation illustrated in figure 4.4, shrinking the preference cones to make them more pointed yields a weaker ordering, whereas making them larger and blunter strengthens the ordering. Making them half-spaces (but including only half of the common boundary) would produce a strong complete ordering.

The *intersection* of two or more relations $\succ_1, \succ_2, \ldots$ is defined by

$$x \succ y \Leftrightarrow x \succ_1 y, \quad x \succ_2 y, \ldots .$$

Evidently an intersection is weaker than, or at most equal to, each of the relations intersected. An example of the use of intersections is the unanimity or Pareto principle, which declares that the preference relation for a group of individuals should include the intersection of their individual preferences. Another example of the use of intersections is in the description of partial orderings, which are often conveniently represented as intersections of complete orders. In this connection, the *dimension* of a partial order has been defined as the smallest number of complete orders of which it is the intersection.[8] For example, the semiorder of the integers with jnd = 2 is two-dimensional because it can be expressed as the intersection of two complete orders:

. . . 1, 0, 3, 2, 5, 4, 7, 6, 9, . . .

and

. . . 2, 1, 4, 3, 6, 5, 8, 7, 10, ...

In extending a preference relation, the problem will often be how to incorporate some additional preference data with a minimum of disruption to the existing structure, or how to extend the relation so that some desirable new condition is fulfilled. For example, we might wish to adjoin the statement $x > y$ to a transitive relation that does not already relate x and y. If we are to preserve transitivity, we must also adjoin all other statements (unless already present) of the form $u > v$, where $u > x$ or $u = x$ and $v < y$ or $v = y$. Extending an order is thus inherently more complicated than merely extending a relation.

There are many types of conditions that one may wish to preserve, or to achieve, in an extension process. They include:

1. Order-theoretic conditions: acyclicity, transitivity, the semiorder property, completeness.

2. Topological conditions: continuity, openness or closedness of the preference sets.

3. Linear-space conditions: convexity, homogeneity, translation-invariance.

It is sometimes asserted that only the order-theoretic properties of preference have intuitive meaning in the context of the behavioral sciences. In our view this is a dogmatic overstatement. Nevertheless it is true that topological or other "technicalities," apparently innocent of any behavioral content, can sometimes have startling logical consequences and should be approached with great caution until their intuitive implications are understood.[9]

Two important extension processes merit special mention: (1) making an acyclic relation transitive and (2) making a weak ordering strong. For the first of these, let $>$ be an arbitrary binary relation, and let us define an extension $>^*$ of $>$ by

$x >^* y \Leftrightarrow$ there is a sequence $z_1, z_2, \ldots, z_p$ in $\mathscr{D}$, for some $p > 1$, such that $x = z_1 > z_2 > \ldots > z_p = y$.

The relation $>^*$ is sometimes called the *ancestral* of $>$.[10] The ancestral is clearly transitive; in fact, it is the weakest transitive extension. But if $>$ is a preference relation, that is, if it is irreflexive (see appendix A.1), then $>^*$ will also be a preference relation only if $>$ was acyclic to start with, since otherwise outcomes x in $\mathscr{D}$ will exist for which $x >^* x$.

For the second extension process mentioned—making weak orders strong—all we need is a tie-breaking rule for each indifference class. While this is not difficult to provide in principle (though the Axiom of Choice may be required), we may find that desirable topological or other regularities of the original preference system cannot be preserved. For example, if $>$ weakly orders the points of the plane according to the size of their abscissas, then $>$ cannot be extended continuously to a stronger order. Indeed any tie-breaking rule would necessarily produce situations in which points arbitrarily close together in the plane have distant points ranked between them.

There is clearly no unique way to extend a given incomplete ordering or relation; any given incomparable pair can be compared in either direction in the extension. In particular circumstances, though, some extensions may seem more natural than others, in the light of special properties of the given ordering or of the domain $\mathscr{D}$. Exercise 4.2 below provides an example (see also appendix A.1).

EXERCISES

4.1. *Show that the semiorder of the integers with jnd = 3 has dimension 3.*

4.2. *Let $\mathscr{D}$ be the real line, and consider the semiorder*

$x > y \Leftrightarrow x \geq y + 1.$

The "most natural" completion of $>$, based on the fact that the preference sets are completely ordered by inclusion, is simply

$x >' y \Leftrightarrow x > y.$

Exhibit another completion in which the point 1/3 is preferred to the point 2/3.

4.3 Utility Scales and Ordinal Utility

The emphasis thus far on "preference"—an undefined but presumably observable concept—has been a temporary concession to the behaviorist viewpoint. The decision maker has been a subject in an experiment—a "black box." His overt actions or manifestations of preference are observed and perhaps predicted, but his inward meditations are regarded as either inaccessible or irrelevant.[11]

The game-theoretic viewpoint, on the other hand, stresses free will and sophistication in strategic matters and is more congenial to the presumption that the decision maker can make all necessary computations and comparisons and then rate all the contemplated outcomes on a single scale of increasing utility to himself. This presumption does away with the possibility of incomparability, to say nothing of intransitivity. It also presents us with a new structural element to play with, namely the scale itself, as well as a new abstract quality, "utility," that the scale purports to measure, or at least to order.

Formally, to each individual i we associate an ordered set $\mathscr{U}^i$, called a *utility scale,* and a mapping u^i from $\mathscr{D}$ to $\mathscr{U}^i$ with the property

$$x > y \Leftrightarrow u^i(x) > u^i(y), \tag{4.6}$$

called a *utility function.*

It is debatable whether utility scales and functions have any basis in reality, whether they figure significantly in the way people actually think about their likes and dislikes. Everyday experience suggests that they do. People at least talk about their preferences as if they had personal "standards of judgment" existing apart from, but capable of being applied to, the particular domains of alternatives that present themselves from time to time. This is especially noticeable when absolute as well as relative judgments are expressed—"I like both candidates" or "He is the lesser of two evils"—or when changes are discussed—"So-and-so has gone up in my estimation" (for the word "estimation" one might read "utility scale").

In mathematical models $\mathscr{U}^i$ is usually given a numerical representation through identification with some subset of the real numbers. This is not always possible, however, since ordered sets exist that cannot be realized within the real number system. The well-known "lexicographic" order is an example:[12] if $\mathscr{D}$ is the real plane, then giving absolute priority to the first coordinate over the second yields the rule

$$x > y \Leftrightarrow \text{either } x_1 > y_1 \text{ or } (x_1 = y_1 \text{ and } x_2 > y_2). \tag{4.7}$$

It is well known that there is no mapping u^i of the plane into the line that satisfies both (4.6) and (4.7).[13] Another possibility is that there may simply be too many outcomes in $\mathscr{D}$, that is, more than the power of the continuum. For example, $\mathscr{D}$ might consist of all subsets of the unit interval [0, 1] or of all bounded real-valued functions on [0, 1], strictly ordered in some way. A real-valued utility function is therefore not as general as an arbitrary transitive relation. This loss of generality, however, is seldom felt as a serious restriction in economic modeling.

The possibility of a cooperative theory of games based on ordinal utilities was suggested by Shapley and Shubik as early as 1953, but it took the work of Aumann and Peleg (1960) to provide the conceptual clarifications and technical foundations needed before significant advances could be made.[14] In recent years applications of ordinal game theory have become increasingly numerous, particularly in regard to the "core" solution for economic models.

So long as only ordinal concepts are admitted, the utility scale may seem little more than a convenience—a concise way of organizing a large number of pairwise comparisons. It may seem surprising, then, that the very existence of such a scale can make any real difference to the theory. But this is indeed the case. In chapter 5 we shall see how utility scales, by providing a model for individual standards of judgment, can enable one to sidestep the paradoxes of collective choice that hinge on the notion of "irrelevant alternatives." At the present stage of our guided tour, however, the significant point is that when we shift the conceptual basis from an abstract preference relation to an abstract utility scale we obtain a new mathematical framework, allowing new kinds of conditions on the structure of preference to be formulated and investigated.

4.3.1 Invariance

Theories of economic (and other) behavior that rest upon a utility concept can be classified according to which transformations of the utility scales leave the propositions or predictions of the theory invariant. These transformations will in general form a group in the algebraic sense.[15] The so-called ordinal theories admit the group of all continuous order-preserving transformations:

$$T_f\colon u \to f(u), \tag{4.8}$$

where f is any invertible function mapping the real line (or some other representation of $\mathcal{U}$) onto itself in such a way that $f(x) > f(y)$ if and only if $x > y$. The so-called cardinal theories, on the other hand, admit only the positive linear transformations:

$$T_{ab}\colon u \to au + b, \tag{4.9}$$

where $a > 0$. Fixing any two points determines the whole scale.

Further structural assumptions on individual utilities are often made. Sometimes it seems appropriate to distinguish a particular zero level of utility, interpreted perhaps as the status quo or as the dividing line between "good" and "bad." Indeed experimentalists usually have to contend with the "status quo effect" in some form or other, whether or not they build a formal utility theory around it.[16] Such cardinal-utilities-with-zero are invariant only under the group of homogeneous linear transformations:

$$T_a\colon u \to au, \tag{4.10}$$

where again $a > 0$. Fixing one nonzero utility determines the whole scale.

An extreme case is the notion of *absolute utility,* in which there is both a lowest and a highest possible utility value, which may be labeled 0 and 1 (or, more vividly, "Hell" and "Heaven"). Such a scheme has been proposed by John Isbell (1959) as a basis for a theory of cooperative games and, in particular, for a multiperson bargaining theory. In one promising version the "1" plays no special role, so that the bargaining theory, but not the utility theory, is invariant under (4.10). Isbell's Hell, however, is not to be confused with the status quo.

The next section will provide a good illustration of the invariance principle at work.

4.3.2 Ordinal utility and bargaining

BARGAINERS PARADOX. *A and B have a continuous curve of possible agreements, increasing in value from 0 to a on A's utility scale while decreasing from b to 0 on B's scale, where $a > 0$, $b > 0$. Failure to agree is rated 0 by both A and B.*

This is the archetype of all bargaining games (Nash, 1950a; Harsanyi, 1956). We shall use it here to demonstrate an important limitation of ordinal utility (Shapley, 1969). The solution of such games and other aspects of bargaining theory will be considered in chapter 7.

To be specific (although our argument will be perfectly general), let $a = 1$ and $b = 1/2$, and let the possible agreements be worth x to A and y to B, where

$$y = \frac{1 - x}{2 - x} \quad \left(\text{hence} \quad x = \frac{1 - 2y}{1 - y}\right). \tag{4.11}$$

This equation yields the curve CC' in figure 4.5.

For reasons that will become clear, we cleverly introduce new utility indices as follows:

$$\tilde{x} = \frac{2x}{1 + x}, \; \tilde{y} = \frac{y}{2 - 2y} \quad \left(\text{hence } x = \frac{\tilde{x}}{2 - \tilde{x}}, \; y = \frac{2\tilde{y}}{1 + 2\tilde{y}}\right). \tag{4.12}$$

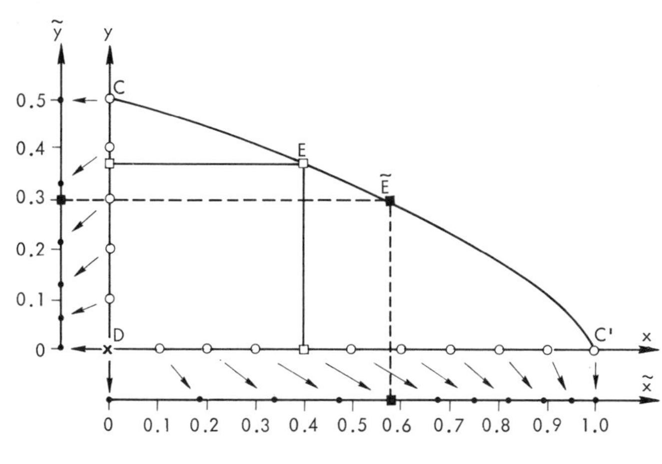

Figure 4.5
The Bargainers Paradox.

The transformations $x \to \tilde{x}$ and $y \to \tilde{y}$ are continuous and order-preserving within the ranges of interest; thus if we assume that only ordinal utilities are relevant, we have not altered the essential problem. The transformed scales are marked on the extra set of axes in figure 4.5; for example, the outcome E is worth $x = 0.4$ on A's original scale and $\tilde{x} = 0.5714$ on his new scale. The extreme outcomes C, C', D are not affected by (4.12); that is, their numerical utility values to each bargainer are unchanged.

Consider now a typical bargain, worth $\tilde{x}$ to A on his new scale. By (4.12) it was worth $x = \tilde{x}/(2 - \tilde{x})$ to him on his old scale. By (4.11) it must therefore have been worth

$$y = \frac{1 - x}{2 - x} = \frac{1 - \tilde{x}/(2 - \tilde{x})}{2 - \tilde{x}/(2 - \tilde{x})} = \frac{2 - 2\tilde{x}}{4 - 3\tilde{x}}$$

on B's old scale. This utility, finally, is equivalent to

$$\tilde{y} = \frac{y}{2 - 2y} = \frac{(2 - 2\tilde{x})/(4 - 3\tilde{x})}{2 - (4 - 4\tilde{x})/(4 - 3\tilde{x})} = \frac{1 - \tilde{x}}{2 - \tilde{x}} \qquad (4.13)$$

on B's new scale.

Now compare (4.13) with (4.11). Because of our cleverness at (4.12) the new utilities relate to each other exactly as the old ones did. This means that the bargaining curve CC' has merely been transformed into itself.[17] Since the disagreement point D is also unchanged, the whole problem appears in the transformed space exactly as it did in the original space. The utility diagrams are geometrically identical. The actual outcomes, however, yield different geometrical points. For example, we may calculate that the agreement at $E = (0.4, 0.375)$, as measured on the original utility scales, will yield the point $\tilde{E} = (0.5714, 0.3)$ on the new scales.

This argument shows that it is impossible to read off a solution to the real problem from just the ordinal information depicted in figure 4.5. It is not even possible to narrow the solution down to an interior subset of A. Since this impossibility is quite independent of the particular functions we chose at (4.11), we can draw a rather sweeping conclusion, to wit: *No resolution of the two-person bargaining problem can be made on the basis of ordinal utility alone.*[18]

Since bargaining lies at the heart of cooperative game theory, it is apparent from this example that an insistence on ordinal utility will place serious limitations on the ability of the theory to make explicit predictions. Indeed we shall find that those solution concepts which require only an ordinal preference structure tend to be rather vague in their predictions, even when there are more than two players.[19]

4.3.3 Bargaining and utility: Further discussion

This section covers three topics tangentially related to the preceding but of a more technical nature.

First, it should be noted that there are some *nonlinear* groups of utility-scale transformations that do not run afoul of the Bargainers Paradox. This raises the interesting possibility of intermediate utility types, between ordinal and cardinal.[20] One example is the group consisting of all transformations $T_{\alpha\beta}$, $\alpha > 0$, $\beta > 0$, where $\mathcal{U}$ is the positive real line and

$$T_{\alpha\beta}(u) = \beta u^{\alpha}. \tag{4.14}$$

This group, however, is really just the linear group in disguise. To see this, make the order-preserving change of scale $\tilde{u} = \log u$. Then the induced action of $T_{\alpha\beta}$ on the new scale is given by

$$\tilde{T}_{\alpha\beta}(\tilde{u}) = \alpha\tilde{u} + \log \beta, \tag{4.15}$$

which is linear after all.

Another example is the group consisting of all transformations $W_{\alpha\beta}$, $\alpha > 0$, $\beta > 0$, where $\mathcal{U}$ is now the whole real line and

$$W_{\alpha\beta}(u) = \begin{cases} \alpha u & \text{if } u \geq 0, \\ \beta u & \text{if } u < 0. \end{cases}$$

Unlike the preceding, this group is not linearizable. But it is separately linear for utilities above and below the special value 0, which might, for example, represent the status quo. Intensities of preference can be meaningfully compared when only favorable prospects, or only unfavorable prospects, are under consideration, but cross comparisons are meaningless.

The second tangential topic has to do with the effect of order-preserving transformations when there are three or more bargainers. We now have much less freedom to manipulate the bargaining surface or hypersurface than we did for the planar curve CC' in figure 4.5. In fact, bargaining surfaces in three or more dimensions have a certain rigidity, in that they cannot be mapped nontrivially onto themselves by transforming the individual utility scales.[21] Thus an ordinal theory for "pure bargaining" among three or more negotiators has some hope of success.

One suggestive way of picking out a "point of agreement" for the three-person case is shown in figure 4.6. Only ordinal comparisons are used. The idea is to exploit the fact that there is a unique point $u = (u_1, u_2, u_3)$ (not shown) above the bargaining surface, whose three projections P, Q, R are all on the surface. Similarly there is a unique v such that S, T, U are all on the surface, and so on. Note that v is beneath the surface but is strictly positive. Hence, on iteration, the sequence beginning $0, v, \ldots$ is strictly increasing and converges to a limit point on the surface. (A similar, decreasing sequence starting $u, w, \ldots$ converges to the same point from above.) We leave to the reader two problems: (1) to justify this limit point as a bargaining solution or fair-division scheme; and (2) to generalize the construction to more players.

A third, closely related question is the extent to which bargaining surfaces are "ordinally equivalent." When can they be mapped onto one another by order-preserving transformations of the separate utility scales? If all or most surfaces were ordinally equivalent, then we could map them onto some standard surface, say a hyperplane with positive normal. This is easily accomplished in the two-person case (see exercise 4.4), but it is not generally possible in higher dimensions. The proof is by counterexample; we shall give here the argument for three dimensions.[22] Consider six points:

$$(0, 2, 4),\quad (2, 4, 0),\quad (4, 0, 2),$$
$$(1, 5, 3),\quad (3, 1, 5),\quad (5, 3, 1).$$

As none of these majorizes any other, there is no difficulty in constructing a bargaining surface that contains them. Suppose that a trio of order-preserving transformations $T_1(u_1)$, $T_2(u_2)$, $T_3(u_3)$ did exist that mapped all six points onto some plane with positive normal. We might as well use the plane $x_1 + x_2 + x_3 = 1$, since any other coefficients could be absorbed into the definitions of the T_i. We would then have

$$T_1(0) + T_2(2) + T_3(4) = 1 = T_1(1) + T_2(5) + T_3(3),$$
$$T_1(2) + T_2(4) + T_3(0) = 1 = T_1(3) + T_2(1) + T_3(5),$$
$$T_1(4) + T_2(0) + T_3(2) = 1 = T_1(5) + T_2(3) + T_3(1).$$

If we add the three lines, we find that each of the nine terms on the left side can be paired with a just-larger term on the right side. Thus $T_1(0) < T_1(1)$, $T_2(2) < T_2(3)$, and so on. This means that the equations

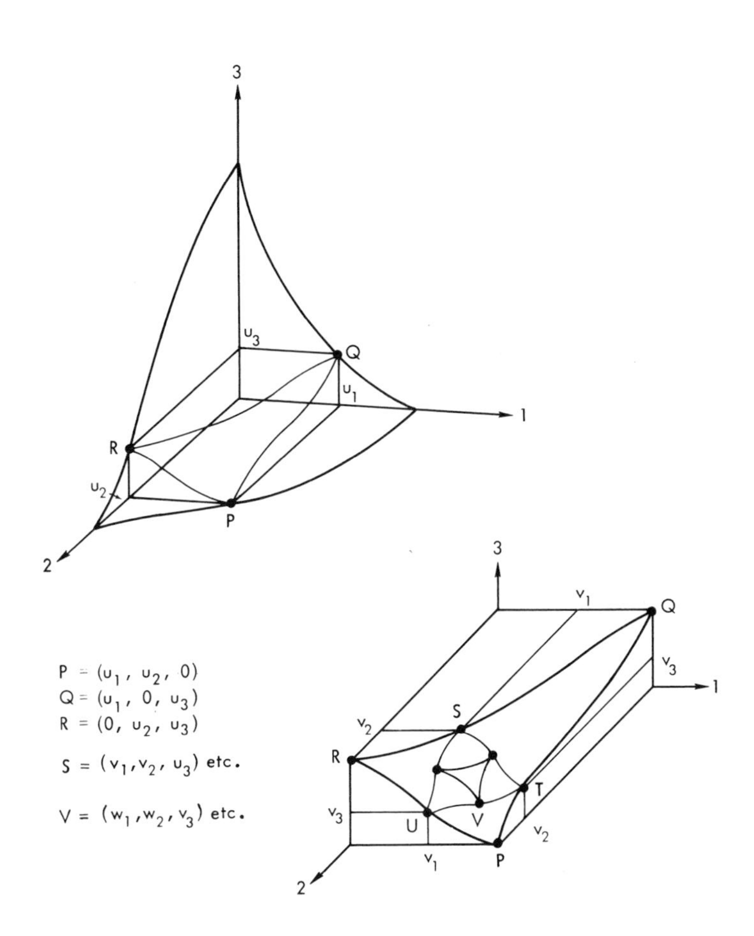

Figure 4.6
A three-person ordinal bargaining solution.

cannot all be true, so at least one of the six transformed points is off the plane. Hence no bargaining surface containing the original six points is ordinally equivalent to a plane.

EXERCISES

4.3. *Let T denote the transformation $(x, y) \rightarrow (\tilde{x}, \tilde{y})$ of (4.12) that takes the curve CC' into itself, and let T^n be the nth iterate of T. Show that any point on CC', other than C itself, can be moved as close as we please to C' by taking n large enough.*

4.4. *Find an order-preserving transformation of A's utility scale that maps the curve CC' of figure 4.5 into a straight line, leaving C and C' fixed. Is it unique?*

4.5. *Let f be a continuous, strictly increasing function of the real line onto itself whose fixed points do not form a convex set. [In other words, there are numbers $x_1 < x_2 < x_3$ such that $f(x_1) = x_1$, $f(x_2) \neq x_2$, and $f(x_3) = x_3$.] Let T denote the transformation $(x, y) \rightarrow (f(x), f^{-1}(y))$. Construct a bargaining curve CC' and a disagreement point D that are invariant under T, but with the property that $T(x, y) \neq (x, y)$ for all points (x, y) on the curve CC' except C and C'.*

4.6. *Find an ordinally invariant, player-symmetric way of selecting an agreement point in the three-person bargaining surface that is different from that described in the text.*

4.4 Cardinal Utility

The Bargainers Paradox was devised in order to show that a satisfactory theory of bilateral bargaining requires knowledge of something more than just an ordering of the bargainers' preferences. In some bargaining situations there will be material or institutional conditions to help determine the solution. For example, there might be a commodity that can be physically divided into equal shares, or a market mechanism that can generate prices. For general purposes, however, a theory that postulates more-than-ordinal preferences seems to be both necessary and in many ways more true to life.

It was once argued that nothing beyond ordinal utility is ever needed in economics or even has empirical validity.[23] The premise seems to have been that questions of utility enter economic analysis only through the theory of consumer behavior, where the "consumer" is narrowly construed as a price-taking automaton. But the game-theoretic approach, as was made clear in chapter 2, forces us to consider many other models of *Homo oeconomicus*.

4.4.1 Utility differences

The Bargainers Paradox hinges on the absence of any way of gauging the relative strengths of the contenders' desires. This is patently an artificial assumption; indeed it is difficult to imagine any real-life bargaining situation that might not be influenced decisively by the magnitudes or intensities of the anticipated utility gains or losses. It seems intuitively clear that an explanatory bargaining theory must accept comparisons between utility differences as well as between utility levels (Lange, 1934; Frisch, 1937; Allais, 1953).

We are therefore led to consider preference structures in which are ordered not only the individual elements of $\mathcal{D}$, but also (in a separate ordering) all *pairs* of elements of $\mathcal{D}$. Consistency between these two orderings, call them $\succsim_1$ and $\succsim_2$, would require that

$$a \succsim_1 b \Leftrightarrow (a, c) \succsim_2 (b, c), \tag{4.16}$$

where the notation (a, b) may be interpreted as representing the change from outcome b to outcome a. The idea that the pairs represent changes or differences is embodied in the following "crossover" property:

$$(a, b) \succsim_2 (c, d) \Leftrightarrow (a, c) \succsim_2 (b, d). \tag{4.17}$$

These are the essential axioms; if certain other, more technical conditions are met (see appendix A.3), then a utility index that is consistent with the two orderings exists and is unique up to an order-preserving linear transformation. In other words, a cardinal utility is determined.[24] There is nothing very mysterious about this result. The ordering $\succsim_2$ enables us to use any pair (a, b) of inequivalent outcomes as a measuring stick and so to calibrate the entire scale $\mathcal{U}$.

4.4.2 Other rationales

The resolution of bargaining problems is important to game theory, but it is only one of many reasons for basing theories of motivated behavior on cardinal utility. Indeed it is a rather unconventional reason, since most standard expositions of utility theory consider just one individual at a time. This section will review several "one-person" reasons for using cardinal utility, that is, arguments that consider a single preference or utility structure in isolation.

Intensities of preference We have already mentioned intensities in connection with bargaining. One can also argue that such intensities affect individual behavior, quite outside the context of give and take between individuals. This perfectly natural view does not lack either philosophical or empirical support (Stevens, 1957, 1961). But the penchant of many experimenters (especially armchair experimenters) for manufacturing situations in which intensity data are suppressed or ignored has led some to the careless conclusion that *only* ordinal questions can be answered by observation.[25]

Similarly, as noted earlier, some writers deny the empirical validity of cardinal utility in economics because certain sectors of microeconomic theory can be carried out without recourse to more-than-ordinal preference structures. General equilibrium theory is the most impressive example of this.[26] However, this objection in principle to cardinal utility has lost much of its force as the extent of the sectors that cannot be so treated has been more fully appreciated. The original psychological insight—that "how much more" does matter—has been rehabilitated to the level of a respectable hypothesis in economics.

Money and fiduciaries A second, more pragmatic support for the use of cardinal utility may be derived from the naive, layman's conception that the *monetary scale* can be used as a basis for evaluations and decisions and that it can be extrapolated well beyond the immediate business of the marketplace. This is illustrated by the attempt of businessmen to assign monetary values to intangible or "hidden" assets such as good will. In the history of economic theorizing both the use of monetary cardinal utility and the resistance to its use go back to the earliest applications of mathematical methods (perhaps to Daniel Bernoulli in 1738). The majority of applied models in economics and operations research come ready-equipped with dollars-and-cents data. It is often a path of least resistance to take the asset level of a person or firm to be an index of cardinal utility, after dealing with the more obvious distortions and intangibles on an ad hoc basis. The pitfalls inherent in this methodology are many, but it often works, and within limits it is theoretically sound.

Since real people tend to think in cardinal monetary terms whenever they can—so the argument might run—should we not endow the ideal people in our models with cardinal, money-based utilities whenever we can?[27] It must be added, however, that the monetary measure, so often used in practice, is more likely to be accompanied by an apology than by an attempt at theoretical justification.[28]

In assessing the role of money-based utility scales, it is well to remember that a significant fraction of all economic decisions are made by fiduciaries, that is, by people acting as agents or trustees for other people's interests (see section 2.4). The same pragmatic convenience that appeals to the model builder or operations researcher also makes itself felt in the trust agreements under which fiduciaries typically operate—be they presidents of corporations, managers of investment funds, purchasing agents for firms, executors of estates, or whatever. Maximizing the dollar value of the trust (under given constraints and methods of accounting) is a virtually universal criterion of performance. It follows that when we model decision makers who are acting as fiduciaries for assets owned by others, it is generally a fair approximation to ascribe to them monetary cardinal utilities.

Risk linearity In contrast to the preceding rough-and-ready approach, a justification of great theoretical elegance is provided by considerations of uncertainty in economic models. This relatively recent development has been largely responsible for the modern rehabilitation of cardinal utility. A formal discussion would be out of place here, but the highlights of the argument can be given in a few words.[29] The first step is to enlarge the outcome space $\mathscr{D}$ so that any "probability mix" of outcomes is itself an outcome.[30] The preference relation, extended to the enlarged space, is then assumed to admit a numerical utility scale and is further postulated to be compatible with the usual interpretation of probabilities, at least to the extent that the relative desirability of two outcomes is not changed if each is combined with an equal chance of a third outcome (see axiom P3 in appendix A.4; this compatibility can be expressed in other ways, having much the same logical consequences).

The effect of these postulates is to ensure the existence of a utility scale that is actuarially neutral with respect to probabilities, so that, for example, the utility of a fifty-fifty gamble between two alternatives is just the average of their separate utilities. This property is called *risk linearity* (or risk neutrality). It is not difficult to show that the only transformations that preserve risk linearity are the linear, order-preserving transformations (4.9); thus a cardinal utility scale is implied.[31]

Separability and additivity Yet another rationale for cardinal utility scales comes from the nature of the independent variables that go into the utility functions in typical economic applications. Goods are often assumed to be continuously divisible and homogeneous (or "fungible") in texture. This in itself does not decisively affect the nature of the utility scales, except perhaps to make more plausible the continuity or "Archimedean" axioms that help ensure that the scale can be embedded in the real-number continuum. But a further assumption is also often made, namely, that the "bundles" of different goods can be grouped into subbundles whose effects on the overall utility are felt separately.

For example, the utility function U might be such that a relation of the form

$$U(x_1, x_2, \ldots, x_m) \equiv f[u(x_1, x_2, \ldots, x_p), x_{p+1}, \ldots, x_m] \tag{4.18}$$

holds identically over some domain of interest. In that case we shall say that the goods represented by the subscripts 1 through p can be "separated out." Note that f here is a function of only $m - p + 1$ variables. Separability almost always simplifies the analytical work, and economists are often willing to assume it even when it only approximately holds in the real situation. Examples of potentially separable groupings of commodities include:[32]

1. categories of highly intersubstitutable consumption goods such as food, clothing, shelter, or entertainment;
2. inventories or assets attributed to separate operating divisions of a firm;
3. consumption of different members of a household (or other social unit), of which only the aggregate utility is being considered;
4. consumption of the same individual (or household or firm) in different time periods.

Separability does not of itself imply a cardinal utility. But when there are overlapping sets that can be separated out, then a special phenomenon called *additive separability,* which is a cardinal property, enters the picture.

For a simple illustration, let $U(x, y, z, w)$ be a utility function on four commodities, monotonically increasing in each variable. Suppose that U is separable in two overlapping ways, as follows:

$$U(x, y, z, w) \equiv \begin{cases} f[u(x, y), z, w], \\ g[x, v(y, z), w]. \end{cases} \tag{4.19}$$

Then, if some special cases are excluded (such as the case in which U is totally insensitive to one of the first three variables), it can be shown that monotonic functions α, β, γ, ϕ exist such that

$$U(x, y, z, w) \equiv \phi[\alpha(x) + \beta(y) + \gamma(z), w]. \tag{4.20}$$

Moreover, α, β, γ are uniquely determined up to positive linear transformations (with a common scale factor, to maintain additivity in (4.20)).[33]

In this example a cardinal utility scale exists for the "internal" evaluation of the first three goods, but the final, overall utility is still only ordinal insofar as we can tell from the information given. But if the overlapping separable sets link together all of the variables, then a fully additive (and hence cardinal) utility function can result. Suppose that the function U of (4.19) is separable in yet a third way:

$$U(x, y, z, w) \equiv h[x, y, t(z, w)].$$

Additional monotonic functions δ and ψ will then exist such that

$$U(x, y, z, w) \equiv \psi[\alpha(x) + \beta(y) + \gamma(z) + \delta(w)]. \qquad (4.21)$$

This at last makes it possible to represent the overall utility function in additive fashion, since if we define $U' = \psi^{-1}(U)$, we obtain

$$U'(x, y, z, w) \equiv \alpha(x) + \beta(y) + \gamma(z) + \delta(w).$$

Any other additive representation will be related to U' by a positive linear transformation; hence a cardinal utility scale has been distinguished.[34]

In these examples the variables x, y, z, w can be taken to be either scalars (single goods) or vectors (nonoverlapping sets of goods). In the latter case, of course, U' would not be fully additive, since nothing would be implied about the relation between the variables inside the functions α, β, γ, δ. However, additivity among two or more gross aggregations of goods is sufficient to determine a cardinal utility scale.

4.4.3 Summary

In sum, "one-person" arguments for cardinal utility can be based on a psychological principle concerning intensities, on the pervasiveness of money in human affairs, on a consistent attitude toward risk, or on the human or institutional penchant for compartmentalizing value judgments. Beyond that, the Bargainers Paradox shows that in a multiperson, interactive world, the need for solutions to bilateral bargaining situations provides a strong new argument for cardinal utility. None of these rationales, however, is context-free. In any given application, in economics or elsewhere, only a few, or none, may be persuasive. Hence we certainly do not propose that ordinal methods should be abandoned.

There may seem also to be a question of consistency. What if several of these rationales for cardinality are operative simultaneously but point toward different utility functions? One would hope that the competing rationales would reinforce each other so well "in the large" that minor differences would be ironed out in the give and take of social or economic interaction. Thus our habits in valuing things in monetary terms can both strengthen and modify our intuitive feelings about the intensities of our desires. Again, when monetary values are not directly risk-linear in life, opportunities for insurance, hedging, or speculation tend to appear. Again, repeated exposure to bargaining drives us to gauge the intensities as well as the directions of our desires. Again, a culture that habitually bargains (or gambles) will soon require some sort of money. Again, separable utility functions encourage the use of deputies or fiduciaries, which in turn invites the use of monetary criteria of performance.

Notes to chapter 4

1. See Farquharson's (1969) illuminating treatment of "sincere" versus "sophisticated" voting; also Gibbard (1973) and Satterthwaite (1975).

2. Rothenberg (1961), in his generally excellent survey, seems willing to equate orderings of states of the world to orderings of the "corresponding" public policies, calling them "logically equivalent"(!) (p. 5).

3. A synonymy: *Alternative* is a fairly neutral word, but suggests that a comparision or evaluation is to be made. An *outcome* is the end result of a process. The statistician's term *event* (from the Latin "out-come") suggests an occurrence at a particular point in time. Marschak (1950) uses *prospect*, a term that connotes a succession of future events or states of the world, extending perhaps up to some finite horizon; it also suggests a viewer for the prospect and hence a possibly subjective viewpoint. Arrow (1951) speaks instead of *social states*, implying the more objective viewpoint that is necessary for his multiperson context. Many variations on these terms are found in the literature, sometimes with insufficient care paid to the connotations.

4. Transitivity-like conditions could of course be added, such as $C(C(S)) = C(S)$ or $C(S) \subseteq T \subseteq S \Rightarrow C(T) \subseteq C(S)$. The operator (4.1) satisfies the first but not the second of these conditions. Unicity—$|C(S)| = 1$—would be another natural condition to add.

5. See also Adams (1965), Fishburn (1970b), Sen (1970), D. Brown (1974, 1975), and Blau and Brown (1981). Acyclic relations are also studied by von Neumann and Morgenstern (1944, chap. 12) in the context of what is essentially a group preference structure. An acyclic relation is sometimes called a *suborder* (see appendix A.1).

6. Some writers prefer the term "intransitive indifference" for the incomparability that arises, for example, in semiorders. In our usage "indifference" is always transitive.

7. The idea of "just noticeable difference" goes back to Fechner in 1850; see Luce and Edwards (1958), also von Neumann and Morgenstern (1944, pp. 608–616), Goodman and Markowitz (1952), Stevens (1957), and Luce (1959a,b).

8. See Dushnik and Miller (1941). In the case of a partial order based on a full-dimensional polyhedral cone in a linear space (as in figures 4.3 and 4.4), the dimension is the number of facets of the cone. However, the dimension concept has been mainly applied to finite or discrete domains $\mathscr{D}$ (see Ore, 1962; or Baker, Fishburn, and Roberts, 1972).

9. Schmeidler (1971) proves that if $\mathscr{D}$ is a connected topological space and $\succ$ is transitive and not vacuous, and if the preference sets are open and the preference-or-indifference sets are closed, then $\succ$ must actually be complete.

10. For the ancestry of "ancestral" see Quine (1944, p. 221) or van Heijenoort (1967, p. 4). Properly speaking, we have here defined the *proper ancestral*; the ancestral proper is the reflexive extension of our ancestral.

11. It may be argued that "to prefer" is itself an introspective construct that is neither more nor less observable than "to have greater utility for." Preferences can be observed, in a sense, by watching subjects in situations where they are making real choices among proffered outcomes. (Hypothetical or contrived choices are not enough, since the very form of a questionnaire may shape the subjects' thinking to fit the experimenter's formal structure.) But might not utility also be observed in this sense? With or without the aid of a choice situation, might one not hope to identify observable quantities that correlate with "level of satisfaction" or even "intensity of desire"? Luce and Raiffa (1957, pp. 23–32), however, regard statements about *preferences* to be statements about "manifest behavior" while relegating statements about *utility* to the status of operationally unverifiable abstractions. See also Suppes (1961).

12. See the papers by Hausner, Thrall, and Debreu in Thrall, Coombs, and Davis (1954); also Kannai (1963) and Fishburn (1971).

13. The stumbling block is not merely the fact that we are trying to make a two-dimensional set one-dimensional. Indeed the famous Peano curve maps a line segment continuously into a square in such a way that every point of the square is covered at least once, and at most four times (see, e.g., Hans Hahn's account in James Newman's anthology, *The World of Mathematics,* pp. 1965–1966). One can use the inverse of this mapping to construct a real-valued function $u(x, y)$ that is semicontinuous (upper or lower but not both) and that completely orders all the points (x, y) of the square without ties.

14. Though primarily aimed at eliminating the need for transferability, not

cardinality, from the von Neumann–Morgenstern theory, these works cleared the path for the purely ordinal approach as well. See also Aumann (1961b, 1964a), Peleg (1963a), Jentzsch (1964), Scarf (1967), and Shapley and Scarf (1974). For an investigation into some ordinal properties of two-person matrix and bimatrix games see Shapley (1964b); for an ordinal game theory applied to voting see Farquharson (1969); for an ordinal game theory applied to courtship and marriage see Gale and Shapley (1962).

15. A *group of transformations* is any set of transformations that contains the inverse of each transformation and also the product of each pair of transformations (that is, the transformation obtained by applying the pair in sequence).

16. This problem is of course related to the knotty problem of utility over time. The notion of status quo is meaningful only in a world whose main features are substantially constant, so that it makes sense to say that the "same" prospect can recur in successive epochs.

17. Applied separately, the transformations of x and y change the shape of the curve CC'; applied together, they do not. Many other joint transformations of x and y would have served as well as (4.12)—for example, the iterates $x \rightarrow \tilde{\tilde{x}}$, $y \rightarrow \tilde{\tilde{y}}$.

18. See Shapley (1969). By the principle of invariance any one-point ordinal solution would have to be a *fixed point* of the transformation T: $(x, y) \rightarrow (\tilde{x}, \tilde{y})$. Actually T does have four fixed points, namely, (0, 0), (0, 1/2), (1, 0), and (1, 1/2), but these all make unsatisfactory solutions for one reason or another. The first is too pessimistic, the next two are unacceptably one-sided, and the last is infeasible.

19. The "core" is a prime example (see chapter 6). In the extreme case of pure bilateral bargaining, as in figure 4.5, the core consists of the entire curve CC', since nothing less would be invariant under all order-preserving transformations.

The Walrasian competitive equilibrium is a sort of bargaining solution, and it is often unique even though it recognizes only ordinal preferences. The Bargainers Paradox is avoided by the presence of measurable commodities to which prices can be assigned (see, e.g., Eisenberg, 1961). These provide the necessary "metric" input into the solution concept. Only in exceptional cases (corresponding to a one-point core) can the competitive-equilibrium outcome of a Walrasian model be deduced from utility considerations alone.

20. The general condition for avoiding the paradox is that the fixed points of each element of the group form a convex set; such a group is called *unwavering* (see Shapley, 1969, and exercise 4.5).

21. For a precise definition, let a "bargaining surface" be characterized by the following two properties: (1) it is the intersection with the nonnegative

orthant of the boundary of some open set containing the origin; and (2) the vector difference of any two of its points lies outside the nonnegative orthant.

22. Bradley and Shubik (1974) show that six is the smallest number of points that will do the trick in three dimensions, while four points are enough in four or more dimensions.

23. "By the end of the nineteenth century many writers, notably Pareto, had come to the realization that it was an unnecessary and unwarranted assumption that there even exist utility as a *cardinal* magnitude. Since only more or less comparisons are needed for consumers' behavior and not comparisons of how much more or less, it is only necessary that there exist an ordinal preference field . . ." (Samuelson, 1948, p. 93; his emphasis).

Another quotation from a Nobel laureate will illustrate the depth of the disagreement: "It is actually possible to measure cardinal utility with a degree of approximation comparable to that with which we can measure the general run of demand elasticities. . . . And furthermore, there are many domains of economic theory where it is absolutely *necessary* to consider the concept of cardinal utility if we want to develop a sensible sort of analysis" (Frisch, 1964; his emphasis). See also Frisch (1926, 1932), Morishima (1965), Pollak (1965), and Camacho (1974).

24. Compare Suppes and Winet (1955). In their treatment, however, (a, b) is interpreted as an unsigned or absolute difference, so that (a, b) and (b, a) are equivalent. We find it difficult to give an intuitive meaning to the unsigned difference.

25. Cf. note 11. In a related discussion Koopmans (1972a, pp. 59–60) seems to maintain that by observing decisions we can determine the *direction* of preferences among the consequences of those decisions, but that the *strengths* of these preferences are by their nature not observable. In our view it takes an act of almost willful neglect on the part of the observer or experimenter to avoid gathering information on intensities of preference. A theory designed to explain how people make up their minds should pay attention to what people say about their thought processes, that is, to how they think they think. By this test the existence of preference intensities, as well as their widespread use in practical decision making, is practically indisputable.

26. Even in this sector, however, if uncertainty is included one must either work with a rather unsatisfactory and unrealistic extension of the Debreuvian model (see Debreu, 1959a; Radner, 1968) or make stronger utility assumptions. And even without exogenous uncertainty from "natural" causes, if general equilibrium theory is to be valid in an ordinal setting, it must be assumed that gambling, insurance, and other markets for risk have been ruled out (see Shubik, 1975a).

27. Kemeny and Thompson (1957) classify some of the modeling alternatives (p. 284) and explore their consequences in the context of a two-person game.

28. See Shapley and Shubik (1966, p. 808). A general theoretical justification may be given eventually, but there is needed first a theory of money that includes not only a consideration of the more obvious economic uses of money, but also an explication of its role as an efficient coding device in the reduction of economic information and the expression of economic strategies.

29. The axiomatization of a risk-linear utility was first carried out by von Neumann and Morgenstern in 1947, though in their formal derivation they do not make use of the notion of probability mixtures of outcomes. For other axiom systems in this popular field see Friedman and Savage (1948), Rubin (1949), Marschak (1950), Herstein and Milnor (1953), Blackwell and Girshick (1954), Hausner (1954), Savage (1954), Harsanyi (1955), Luce and Raiffa (1957), Aumann (1962, 1964e) (who develops a partially ordered risk-linear preference structure), Fellner (1965), and Dyckman and Smidt (1970). The Herstein–Milnor axioms are reproduced in appendix A.4.

30. This is formally analogous to the passage from pure strategies to mixed strategies described in chapter 3 (see also figure 4.4).

31. The better-known axioms for cardinal utility are based upon either risk linearity or the evaluation of utility differences. In many aspects of economic life both of these features appear to be reasonable; hence we could link both sets of axioms with a consistency condition to line up the scales produced by the different sufficient conditions.

32. It will be seen from this list that separability considerations come to the fore whenever the technique of aggregation is employed—whether of consumable goods or consuming individuals. The latter of course takes us back to multiperson utilities, though not in a game-theoretic context (cf. Camacho, 1974).

33. This is essentially the discovery of Leontief (1947a,b). [Readers of the "1947b" paper should be warned of a confusing misprint: the plus signs in Proposition III on page 364 ought to be commas.] Strengthenings and extensions of Leontief's work will be found in Debreu (1960), Gorman (1968a,b), Keeney (1971, 1972), and Koopmans (1972a,b), who gives a clear account of the whole question of additive separability and provides additional references. For some recent results on the *non*additive decomposition of utility functions see Farquhar (1974).

34. Functions such as the U of (4.21) have been called "ordinally additive" (see Pollak, 1967; Houthakker, 1960). Our point is that such "ordinal" additivity generally leads to a "cardinal" scale that is determined uniquely up to a positive linear transformation.

5 Group Preferences and Utility

5.1 Introduction

Much of the economic and social behavior in which we are interested is either group behavior or that of an individual acting for a group. The housewife, the breadwinner of a family, the corporate or union manager all act on behalf of groups. Group preferences may be regarded either as derived from individual preferences by some process of aggregation or as a direct attribute of the group itself. For example, it might be appropriate to consider the preferences of a family without considering the individual members if the family is to be treated as a decision unit, or "player," in some larger system.

On the other hand, game-theoretic methods provide an intriguing alternative to treating a group as though it were a sentient individual: we can cast the members of the group as players in an internal organizational subgame, vying for control of the group's actions in the larger game.[1] It may be neither profitable nor necessary to regard the group's corporate behavior as though it were governed by any kind of individualistic preference structure.

The following classic example, generally attributed to the Marquis de Condorcet (1785),[2] plays a central role in both the theory of group choice (voting) and the theory of group preference (utility aggregation):

VOTERS PARADOX. *Three individuals use simple majority rule to decide what to do as a group. Their personal preference orderings are*

$$A >_1 B >_1 C,$$

$$C >_2 A >_2 B,$$

$$B >_3 C >_3 A,$$

where A, B, C are three alternatives that they may have to face.

In this example, if the choice is between A and B, then the group will choose A by a 2:1 vote. Similarly, they will choose B over C, and

C over A. Hence the "revealed" preferences of the group are described by the binary relation

$$A >_{123} B, \quad B >_{123} C, \quad C >_{123} A.$$

This is not transitive; hence a group utility scale cannot be constructed.

When we move to larger groups, or a whole society, the concept of group utility takes on value overtones, becoming an expression of social welfare rather than being just a device to explain or predict the group's actions. Social welfare is (or ought to be) the optimand for planners, legislators, and other policy makers when they attempt to regulate or reform systems that, if left to themselves, would not necessarily seek the social optimum. For such users the revealed-preference approach, which infers a social utility function from observation of what actually happens, cannot suffice.[3] The prevailing method in welfare economics, as typified by the formulations of Bergson (1938), Samuelson (1948), and Arrow (1951, 1963), has been to make only minimal assumptions about individual preferences and behavior and then to investigate what can be deduced concerning group choice or group values (Mishan, 1960; Rothenberg, 1961; Fishburn, 1969; Sen, 1970; Plott, 1971). A weakness of this kind of approach is that the very sparseness of the assumptions may permit deceptively general-sounding conclusions to be drawn, by rigorously logical arguments, which become invalid when further assumptions are added to flesh out the bare bones of the model. A prime example of this will be examined in section 5.5.

It may be a worthwhile pastime to consider what assertions about group welfare can be established from minimal assumptions about individual preferences, thus avoiding the need for either empirical work or, alternatively, value judgments. In keeping with our pluralistic inclinations we neither agree nor disagree a priori with this kind of approach. We disagree, however, with the implication that by basing his assumptions on evidence drawn from some specified domain the economist has necessarily imposed a personal value judgment (Little, 1957). Accepting evidence and making value judgments are not equivalent.

5.2 The Utility Space

Utility spaces are important conceptual tools, both for our present discussion of group preferences and throughout game theory. We have already met them in connection with the Bargainers Paradox and in our discussion of outcomes and payoffs. If the n members of a society have utility scales $\mathcal{U}^1, \mathcal{U}^2, \ldots, \mathcal{U}^n$, then we denote the utility space for the society by $\mathcal{U}_N$, where N denotes the set $\{1, 2, \ldots, n\}$. Assuming that the $\mathcal{U}^i$ are real-number scales, $\mathcal{U}_N$ will be a subset of n-dimensional Euclidean space. Points in this space are variously called *utility vectors, payoff vectors, payoffs,* or *imputations,* as the occasion may demand.

Care must be exercised in distinguishing the utility space from a commodity or other space that describes the material outcomes. In the utility space all the "physical" features of the model (quantities, prices, even the numbers of different goods or time periods) have been stripped away, and analytical methods that depend on such features can no longer be brought to bear. For example, if a homogeneous cake is to be divided among n children, the equal-split solution will be conspicuous in the "cake space" because of its symmetrical location, but it may not stand out in any particular way in the utility space if the children's tastes or capacities are different.

Figures 4.5 and 4.6 illustrated the use of utility spaces to describe "pure bargaining" problems with two or three bargainers. Figure 5.1 gives some further examples with $n = 2$. We may think of each shaded region as the image in $\mathcal{U}_{12}$ of some unspecified domain $\mathcal{D}$ of material outcomes. In part a the image is not convex; in part b it is. Convexity is directly related to risk linearity. If both $\mathcal{U}^1$ and $\mathcal{U}^2$ are risk-linear (and hence cardinal) scales (see section 4.4.2), and if the domain $\mathcal{D}$ of alternatives for 2 is enlarged to include all lotteries over members of $\mathcal{D}$, then the achievable set in $\mathcal{U}_{12}$ will expand and exactly fill the "convex hull" of the original set, as indicated by the dotted lines. The straight-line boundary in part c might arise in similar

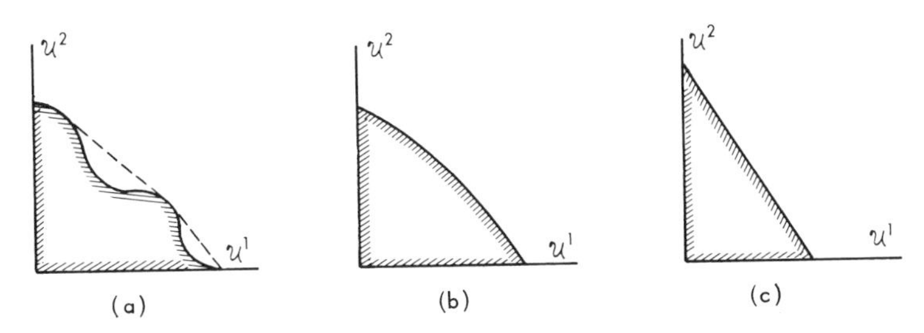

Figure 5.1
Outcome sets in the utility space for two bargainers.

fashion, from risk linearity applied to a small number of "pure" outcomes; three outcomes with utility vectors at the vertices of the triangle would suffice. It might also arise in quite a different way, through the availability of money or some other transferable commodity for which the bargainers have linear utility.

Unless the $\mathcal{U}^i$ represent absolute utilities in some sense (see section 4.3.1), their zero points, and hence the origin of $\mathcal{U}_N$, will have no special significance. Nevertheless in many applications there will be a most natural or convenient choice for the origin. In pure bargaining problems the "disagreement point" makes a natural point of reference. In other problems the origin of $\mathcal{U}_N$ could represent the status quo, the point of no trade, the break-even point, or even the bankruptcy level for traders with finite assets (the latter being a sort of absolute zero). In game theory one very often normalizes the utilities so that zero represents for each player his "individually rational" floor, below which he cannot be driven against his will.[4]

Similar remarks apply to the units of measurement in the scales $\mathcal{U}^i$ if all individuals have cardinal utility. For example, in figure 5.1c we could easily choose units so that the bargaining line has slope -1; this might simplify some calculations, but it would not have intrinsic significance unless we are prepared to assert that utility intensities are *interpersonally comparable* in the situation depicted. We do not flinch from this sort of assertion on doctrinaire grounds. At least two sources of interpersonal comparability can be important in economic game theory. First, the people or firms that inhabit our models will often "keep score" in dollars and cents or other monetary units (see section 4.4.2). Given that dollars can change hands, there will be a meaningful basis for interpersonal utility comparisons. The second source is the possibility that the game itself, or its solution, will induce an ex post facto comparison of utility intensities. As remarked earlier in this chapter, it is a natural and perhaps necessary feature of most bargaining processes for the bargainers to test and compare the intensities of each other's desires, and a similar remark applies to most "fair-division" schemes (Shapley, 1969; also chapters 7 and 10).

5.3 The Pareto Principle

When we set out to relate individual preferences to a preference structure of some kind for the group, we might wish to stipulate that group preferences come solely from the preferences of the individuals, or we might wish to permit other considerations to enter. In either case there is one simple condition that is nearly always assumed, called the *principle of unanimity* or, among economists, the *Pareto principle.* It comes in a "strong" form,

$$x >_i y \quad \text{for all } i \in N \quad \Rightarrow \quad x >_N y \quad \text{for all } x, y \text{ in } \mathscr{D}, \tag{5.1}$$

and a "weak" form,

$$x \gtrsim_i y \quad \text{for all } i \in N \quad \Rightarrow \quad x \gtrsim_N y \quad \text{for all } x, y \text{ in } \mathscr{D}. \tag{5.2}$$

The intuitive message is much the same: if every member of the group prefers x to y, then the group itself, in its corporate judgment, also prefers x to y.[5]

This principle, obvious though it may appear, has some far-reaching logical consequences. In some settings it may even lead to transitivity for the group preference relation, contrary to what the Voters Paradox would lead us to expect. In the rest of this section and the next we shall explore these formal consequences informally, putting to good use many of the concepts developed earlier here and in chapter 4. Without more than sketching the technical details, we hope to shed light both on the nature of the utility aggregation problem and on the power and limitations of axiomatic methods in general.

As a first step one can use (5.2) to prove that

$$x \sim_i y \quad \text{for all } i \in N \quad \Rightarrow \quad x \sim_N y \quad \text{for all } x, y \text{ in } \mathscr{D}. \tag{5.3}$$

In words, if two outcomes in $\mathscr{D}$ are "tied" in the eyes of every individual, then they must be tied in the group's judgment. Note the force of this condition. Although the relation $>_N$ may break many deadlocks arising from conflicting preferences in the group, it cannot include any tie-breaking rules that would resolve a case of universal indifference.

To proceed, let us assume henceforward that the *individual* preferences are transitive and, in fact, generate a (possibly weak) complete order. This means that we can set up a utility scale $\mathscr{U}^i$ for each individual i, together with a function u^i from $\mathscr{D}$ to $\mathscr{U}^i$, in such a way that

$$x >_i y \Leftrightarrow u^i(x) > u^i(y) \quad \text{for all } x, y \text{ in } \mathscr{D}. \tag{5.4}$$

Each outcome x in $\mathscr{D}$ can then be mapped into a utility vector $u(x) = (u^1(x), u^2(x), \ldots, u^n(x))$ in the group utility space $\mathscr{U}_N$. Since several outcomes may map into the same point in $\mathcal{U}_N$, it might seem that we lose the ability to discriminate outcomes among which the group as a whole might have strong preferences, in the sense of $>_N$. But this is not the case. Condition (5.3) guarantees that two outcomes having the same utility vector must in fact be group-indifferent. The binary relation $>_N$ on $\mathcal{D}$ therefore goes over unambiguously into a relation on $\mathcal{U}_N$ (for which we will use the same symbol), by the rule

$$\alpha >_N \beta \Leftrightarrow x >_N y \quad \text{whenever} \quad \alpha = u(x), \quad \beta = u(y). \tag{5.5}$$

Thus, with the aid of the Pareto principle we have moved the problem from $\mathcal{D}$ to $\mathcal{U}_N$, that is, from an outcome space with no particular structure to a utility space that is the Cartesian product of the n ordered sets $\mathcal{U}^i$.

For simplicity, let us now suppose that all the $\mathcal{U}^i$ are real lines, making $\mathcal{U}_N$ a real n-dimensional vector space. For further simplicity, let us suppose that all the vectors in $\mathcal{U}_N$ are achievable; that is, for each α in $\mathcal{U}_N$ there is an x in $\mathcal{D}$ such that $\alpha = u(x)$. The *preference sets* for N can now be defined: the "upper" preference set $P_N(\alpha)$ is the set of β such that $\beta >_N \alpha$, and the "lower" preference set $Q_N(\alpha)$ is the set of γ such that $\alpha >_N \gamma$. Figure 5.2 illustrates this for $n = 2$. The Pareto principle in the strong form (5.1) ensures that $P_N(\alpha)$ includes all points to the "northeast" of α, and $Q_N(\alpha)$ all points to the "southwest," as indicated by the hatched lines in part a. In general, the preference sets will also contain points outside these sectors, but $P_N(\alpha)$ and $Q_N(\alpha)$ cannot overlap, since $>_N$ is by assumption a preference relation and hence antisymmetric.

Figure 5.2b shows what happens if the group is assumed to be indifferent between two points: $\alpha \sim_N \gamma$. We must have $P_N(\alpha) = P_N(\gamma)$ and $Q_N(\alpha) = Q_N(\gamma)$, and so $P_N(\alpha)$ must include all points unanimously preferable to either α or γ (indicated again by hatched lines); similarly for $Q_N(\alpha)$.[6] The more such points we have that are indifferent to α, the more closely the preference sets will be "pinned down." The remaining open space, of course, comprises those points that are incomparable to α.

Let us now examine the consequences of assuming that the relation $>_N$ is complete, that is, that the group is "decisive" and can render a judgment of preference or indifference between any two outcomes. This means that for every α the sets $P_N(\alpha)$, $Q_N(\alpha)$, and $I_N(\alpha)$ completely fill up $\mathcal{U}_N$, where $I_N(\alpha)$ denotes the set of points indifferent to α. The strong Pareto principle (5.1) ensures that $I_N(\alpha)$ will have no "thick"

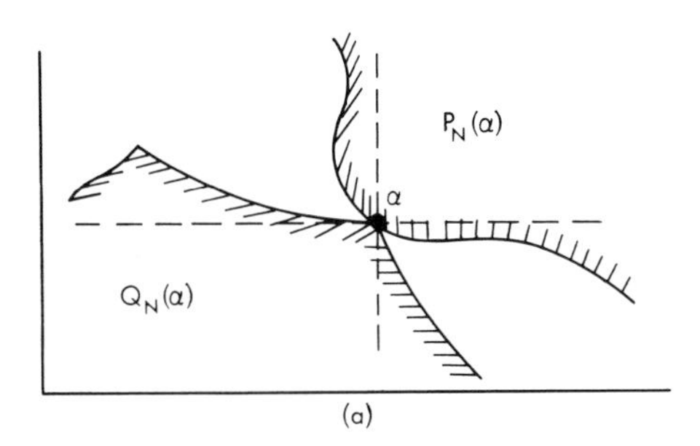

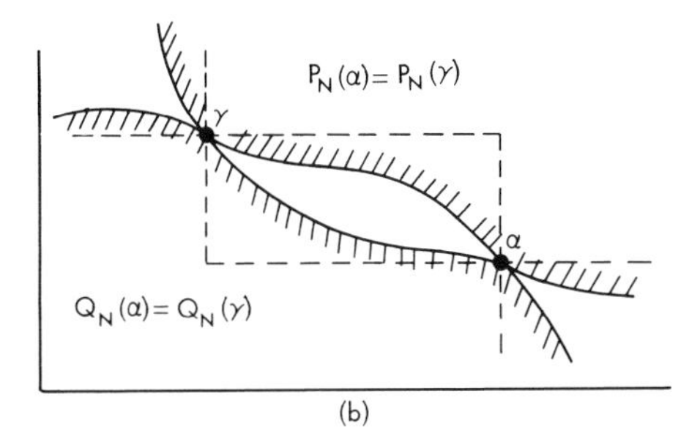

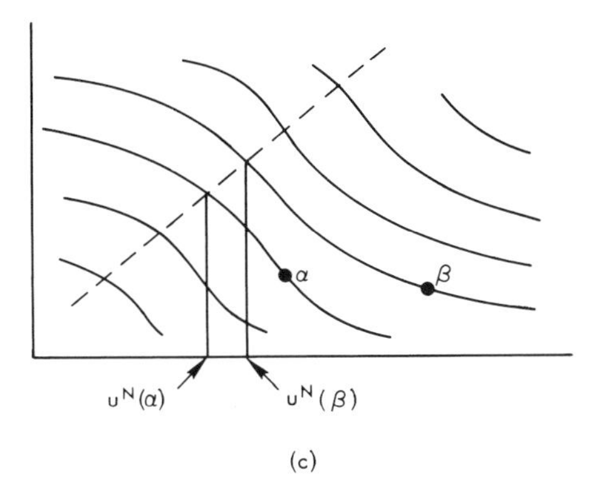

Figure 5.2
Preference sets in a two-dimensional utility space.

spots, that is, that the boundaries of $P_N(\alpha)$ and $Q_N(\alpha)$ coincide. It also ensures that this common boundary will be a continuous, monotonic curve (or, more generally, a hypersurface). An additional, somewhat technical assumption of continuity is required to ensure that this common boundary is in fact $I_N(\alpha)$,[7] but once we have indifference surfaces separating the upper and lower preference sets everything falls into place (figure 5.2c). Since two indifference surfaces cannot cross, the preference sets are properly "nested" or "stacked," so that

$$\beta >_N \alpha \Rightarrow P_N(\beta) \subset P_N(\alpha), \quad Q_N(\beta) \supset Q_N(\alpha).$$

Group preference is thus transitive after all. A group utility scale $\mathcal{U}^N$ and a group utility function u^N can therefore be defined. If we wish to be explicit, we can simply let $\mathcal{U}^N$ be any positively sloping straight line in $\mathcal{U}_N$, parametrized by the first coordinate u^1. This line cuts through every indifference surface $I_N(\alpha)$. To find the group utility of any x in $\mathcal{D}$, we first map x into the point $\alpha = u(x)$ in $\mathcal{U}_N$, then find the point β where $I_N(\alpha)$ intersects $\mathcal{U}^N$, and finally set $u^N(x) = u^1$.

We have made several simplifying assumptions and glossed over many technical obstacles to reach this conclusion. The purpose of the foregoing sketch was more to increase the reader's familiarity with the concepts of multiperson utility theory than to present any particular rigorous theorem.

The emergence of transitivity for the group is nonetheless quite remarkable. At first sight, our only substantive assumptions seem to have been (1) complete orderings for individuals, (2) the Pareto principle, in its weak and strong forms, and (3) a "decisive" binary preference relation for the group. The Voters Paradox is not excluded by these assumptions, and yet it exhibits a sharp intransitivity. This apparent contradiction is best explained by our assumption of continuity, which we needed in order to get indifference hypersurfaces that could not cross. The moral seems to be that the Condorcet intransitivity, while a real enough phenomenon in finite models, cannot be sustained if there are continuous gradations between outcomes, along which the individual and group preferences behave in a suitably continuous manner.[8]

EXERCISES

5.1. *Let $n = 3$ and $\mathcal{D} = \mathcal{U}_N$, and let the group decision be by majority vote as in the Voters Paradox. Describe the sets $P_N(\alpha)$, $Q_N(\alpha)$, and $I_N(\alpha)$ for a general α in $\mathcal{U}_N$, and explain why $>_N$ is not transitive.*

5.2. *Each person has a real-valued utility scale, but to other people he becomes "noticeably happier" only at a discrete set of points in his scale, which we may take to be the integers. Group preference occurs only when there is unanimous individual preference, as in (5.1), or when at least one person is noticeably happier and none of the others is noticeably less happy. Sketch the sets* $P_N(\alpha)$ *and* $Q_N(\alpha)$ *for the case* $|N| = 2$ *and three choices of* α*—(1/2, 1/2), (1/2, 1), and (1, 1)—taking special care with the boundaries of the preference sets.*

5.4 The Extended Pareto Principle

A more comprehensive form of the principle of unanimous preference is especially congenial to the game-theoretic approach. We may imagine that each subset S of the "society" N is at least potentially a social unit to which a preference relation $>_S$ can be imputed. It is then plausible to extend the postulates (5.1) and (5.2) to include not only groups composed of individuals but groups composed of smaller groups. To make this idea precise, let S be any subset of N, and let $\{S_1, S_2, \ldots, S_p\}$ be any *partition* of S (i.e., each member of S belongs to one and only one S_j). Then the extended or "coalitional" forms of the Pareto principle are

$$x >_{S_j} y \; (j = 1, 2, \ldots, p) \Rightarrow x >_S y \tag{5.6}$$

and

$$x \gtrsim_{S_j} y \; (j = 1, 2, \ldots, p) \Rightarrow x \gtrsim_S y \tag{5.7}$$

for each x, y in $\mathcal{D}$. In words, if there is any way to break up S into fractions such that each fraction prefers x to y, then S as a whole will also prefer x to y.[9] The previous statements (5.1) and (5.2) are of course included in (5.6) and (5.7): simply take $S = N$, $p = n$, and $S_j = \{j\}$.

With the Pareto principle thus strengthened, we can often weaken some of the other hypotheses and still obtain a social utility function. For example, we can make the assumptions of section 5.3 only for two-member groups, and thereby obtain pairwise utility functions in each of the two-dimensional utility spaces $\mathcal{U}_{\{i, j\}}$, $i, j \in N$, $i \neq j$. With the aid of (5.6) and (5.7) we can then derive utility functions for all other subsets of N, including N itself. The argument is constructive, the essential step being illustrated in figure 5.3. Starting with families of indifference curves in $\mathcal{U}_{\{1, 2\}}$ and $\mathcal{U}_{\{2, 3\}}$ and a typical point α in $\mathcal{U}_{\{1, 2, 3\}}$,

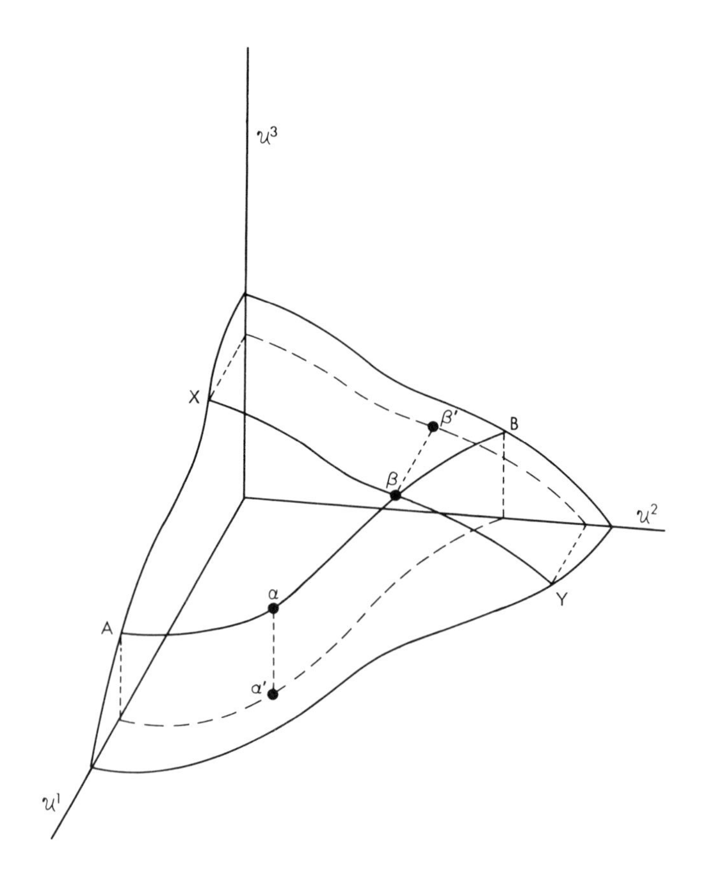

Figure 5.3
Three-person indifferences determined from two-person indifferences.

we build up the {1, 2, 3}-indifference surface through α as follows: First, we project α down to α' in $\mathcal{U}_{\{1,2\}}$; next we draw the curve AB through α by copying the {1, 2}-indifference curve through α', holding the u_3 coordinate constant; finally we fill out the surface with analogously determined copies of {2, 3}-indifference curves such as XY, one curve through every point β in AB.[10]

Even stronger conclusions can sometimes be drawn when we are working with conditions that lead to cardinal utility (Fleming, 1952; Harsanyi, 1955). For example, let us suppose that $\mathcal{D}$ is a mixture set or mixture space (see appendix A.4) and that the individuals have cardinal utility scales $\mathcal{U}^i$ and risk-linear utility functions $u^i(x)$. Let us also make the fairly innocent assumption that whenever the group N is indifferent between x and y, it is indifferent between x and any mix of x and y. When preferences have been translated to the utility space $\mathcal{U}_N$, as in section 5.3, it will follow that the indifference sets $I_N(\alpha)$ are convex and so, if they are surfaces, that they are hyperplanes. Moreover, they will be parallel hyperplanes, because any one of them can be used to determine the $\{i, j\}$-indifference sets for all pairs $\{i, j\}$, which will perforce be families of parallel straight lines, and these will in turn make all the $I_S(\alpha)$ linear and parallel for every set $S \subseteq N$. In fact, barring certain degenerate situations, positive coefficients c_i can be found that enable us to represent each subgroup's preferences as a weighted sum of the individual utilities:

$$u^S(x) = \sum_{i \in S} c_i u^i(x). \tag{5.8}$$

If we like, we can adjust the individual scales to make all $c_i = 1$, giving us the simplest possible utility aggregation rule.

In a similar vein, let us suppose that individual utilities are based on monetary valuations, and let us add the assumption that the society N regards as equally desirable any two outcomes having the same *total* money value. (This belief might be rationalized by the possibility of future transfers of money within the society.) Then the form (5.8) again results, with the aid of the extended Pareto principle, this time with all coefficients already equal to 1 because of the common unit of measurement.

EXERCISE

5.3. *Show that conditions (5.6) and (5.7) follow from the same conditions restricted to partitions with $p = 2$. Confirm that there are $(3^n - 2^{n+1} + 1)/2$ cases with $p = 2$ and that they are all logically independent.*

5.5 Social Welfare Functions: The Arrow Theorem

Kenneth Arrow's "general possibility theorem" is a famous landmark in the application of rigorous mathematical methods to the social sciences. Arrow's goal was to analyze the possibilities for preference aggregation rules or *social welfare functions* when the domain of alternatives is finite. His definition: "By a social welfare function will be meant a process or rule which, for each set of individual orderings $R_1, R_2, \ldots, R_n$ for alternative social states (one ordering for each individual), states a corresponding social ordering of alternative social states, R" (Arrow, 1951, p. 23).

The thrust of the theorem is that no such function exists that satisfies all of a certain set of arguably reasonable or desirable properties. There is now a very considerable literature dealing with extensions, clarifications, and elaborations of the theorem, and we do not propose to enter very deeply into the subject here.[11] We do wish, however, to make one rather specific methodological observation about the use of models in this area—an observation that may help some readers to put the more destructive or negative side of Arrow's "possibility theorem" (which is in fact frequently called an "impossibility theorem") into perspective.

We shall do this by applying the "sensitivity analysis" procedure for model validation that we sketched at the end of chapter 1. Two models will be set forth, different in their structure but having roughly the same verbal interpretation in ordinary language. By showing that diametrically opposite results flow from the two models, we make plain the need to ask more searching questions about the real-world situation being modeled before coming to any conclusions.

The first model provides a sufficient setting for a simple version of the Arrow theorem. The second describes a very similar situation, differing most conspicuously in its use of (ordinal) utility scales instead of rankings. In both models we postulate a finite set N consisting of n individuals—say, voters—and a finite domain $\mathscr{D}$ consisting of m alternatives—say, candidates for an office. Denote by Ω the set of all possible complete orderings of $\mathscr{D}$; there are $m!$ of these. If ω is an ordering and d a candidate, then $\omega(d)$ will mean the "ω-rank" of d.

In Model I a social welfare function is defined as a function ϕ_{I} from $\Omega \times \Omega \times \ldots \times \Omega$ (n times) to ω, that is, a function that takes n-tuples of rankings into a single ranking:

$$\omega = \phi_{\mathrm{I}}(\omega^1, \omega^2, \ldots, \omega^n). \tag{5.9}$$

Here ω may be regarded as a sort of consensus, derived from the individual opinions ω^i, while ϕ_{I} may be thought of as a "constitution" or a general method for reaching consensus.

There are, of course, many possible such functions ϕ_{I}, but certain plausible restrictions may be imposed. *Monotonicity* states that if the rank of some particular candidate changes in one of the individual orderings, then his rank in the social ordering, if it changes at all, changes in the same direction. *Independence* states that if some candidates are eliminated from $\mathscr{D}$, then the ranking induced on the remainder by (5.9) does not depend on how the voters ranked the missing candidates. *Unanimity* is just our friend, the Pareto principle: if one candidate is ranked higher than another by all voters, then he is also ranked higher in the consensus. Finally, *nondictatorship* states that there is no single voter whose opinion prevails over all others; that is, there is no i_* in N for which $\phi_{\mathrm{I}}(\omega^1, \omega^2, \ldots, \omega^{i_*}, \ldots, \omega^n) \equiv \omega^{i_*}$.

These simple, almost transparent conditions have the following remarkable property (assuming that $n \geq 3$): *no function ϕ_{I} exists that satisfies them all.* As might be expected, Condorcet configurations plays a central role in the proof.[12]

In Model II we postulate utility scales $\mathscr{U}^i$ for the voters i in N and utility functions u^i from $\mathscr{D}$ to $\mathscr{U}^i$ that express their judgments or "estimations" of the candidates, and that incidentally define orderings (or weak orderings) on $\mathscr{D}$. A social welfare function may now be defined to be a function ϕ_{II} from $\mathscr{U}_N = \mathscr{U}^1 \times \mathscr{U}^2 \times \ldots \times \mathscr{U}^n$ (the group utility space) to an ordered set $\mathscr{U}^N$ (the group utility scale). This function immediately gives us a *social utility function* u^N, namely,

$$u^N(d) = \phi_{\mathrm{II}}(u^1(d), u^2(d), \ldots, u^n(d)), \tag{5.10}$$

defined for all d in $\mathscr{D}$.

Let us try to restate, in this new setting, the restrictions that we imposed on ϕ_{I}. For *monotonicity,* let ϕ_{II} be weakly monotonic in each variable, so $x^i \geq y^i$ implies $\phi_{\mathrm{II}}(x^1, x^2, \ldots, x^i, \ldots, x^n) \geq \phi_{\mathrm{II}}(x^1, x^2, \ldots, y^i, \ldots, x^n)$. *Independence* is automatically satisfied, since dropping some candidates from $\mathscr{D}$ does not affect either the individual or social utilities of the remainder. For *unanimity,* we can again use the strong Pareto principle: if $x^i > y^i$ for all i in N, then $\phi_{\mathrm{II}}(x^1, x^2, \ldots, x^n) > \phi_{\mathrm{II}}(y^1, y^2, \ldots, y^n)$. Finally, *nondictatorship* may be rendered by the statement that for no voter i_* in N does the inequality $x^{i_*} > y^{i_*}$ imply $\phi_{\mathrm{II}}(x^1, x^2, \ldots, x^{i_*}, \ldots, x^n) > \phi_{\mathrm{II}}(y^1, y^2, \ldots, y^{i_*}, \ldots, y^n)$ regardless of the values of x^i and y^i for $i \neq i_*$.

These simple, almost transparent conditions have the following unremarkable property (assuming that there are at least two nontrivial $\mathcal{U}^i$): *there are many functions ϕ_{II} that satisfy them all.* For example, if the scales $\mathcal{U}^i$ are realizable as subsets of the real line, then simple summation will do:

$$\phi_{\text{II}}(x^1, x^2, \ldots, x^n) = \sum_{i=1}^{n} x^i.$$

There is, of course, no reason to prefer this consensus to any other one, since, as we must emphasize, the utility scales of Model II are purely ordinal. So we are not brought any closer to solving the problem of utility aggregation by this model; instead of no social welfare function, we have too many. It is interesting, nevertheless, that the famous "impossibility" result is so sensitive (and writers on the subject so insensitive) to the particular formal setting into which the axioms that have been so thoroughly scrutinized are placed.

In summary, we have seen here how two translations of roughly the same verbal description into mathematical language can yield radically different formal conclusions. The verbal description must therefore be refined to bring out the implicit assumptions and hidden distinctions that could allow us to separate Model I from Model II. The very existence of the latter shows that the "impossibility of a social welfare function" is not a robust social principle. It is a theorem pertaining to a particular formal system. In order to carry this theorem over to real societies, one must establish that the distinctive conditions of the formal system are met—in particular, that the individuals in the real situation are not only ordinalists but relativists, able to discern better from worse but not good from bad. Few of the writers who purpose to "apply" the Arrow theorem to real problems in economics, politics, or sociology have troubled to do this, or indeed to inquire at all into the "structural" sensitivity of the Arrow model.[13]

5.6 Approaches to Social Welfare: The Anthropomorphic Trap

The now-classical approach of Bergson and Samuelson to the construction of the social welfare function differs in several respects from the foregoing. The most noticeable contrast is in the emphasis on the role of quantitative economic variables. Thus, after some discussion of more general forms, Samuelson writes the following social welfare function:

$$W = W[u^1(x_1^1, x_2^1, \ldots, x_n^1; v_1^1, v_2^1, \ldots, v_m^1), u^2(x_1^2, x_2^2, \ldots, x_n^2; v_1^2, v_2^2, \ldots, v_m^2), \ldots, u^s(x_1^s, x_2^s, \ldots, x_n^s; v_1^s, v_2^s, \ldots, v_m^s)],$$

where the *x*s and *v*s refer to commodities and productive services, respectively (Samuelson, 1948, p. 229). Arrow, with a quite different object in view, avoids particularizing the nature of the "social states" that make up $\mathscr{D}$, although he makes the not-quite-innocent assumption that they are finite in number (Kirman and Sonderman, 1972; D. Brown, 1974). Of course he does not exclude economic variables, nor do Bergson and Samuelson exclude noneconomic factors from their general scheme. But while each approach may in principle permit representation of virtually anything that the other permits, the difference is not merely a matter of notation. This is because so much depends on the subjective plausibility both of the axioms and of the many secondary modeling assumptions, technical or substantive, that are made in tailoring the theory to an applied situation. An assumption that is difficult to swallow in one formal framework may appear intuitive and natural—or no worse than a harmless technicality—in another setting.[14] Indeed part of the power of an elegant formal system resides in its ability to suggest or propose "natural" conditions and to point to previously unsuspected regularities in the real-life counterpart. Unfortunately, formal elegance has also the power to dazzle and blind its worshipers.

Arrow (1963, p. 106) brings out another contrast: "All the writers from Bergson on agree on avoiding the notion of a social good not defined in terms of the values of individuals. But where Bergson seeks to locate social values in welfare judgments by individuals, I prefer to locate them in the actions taken by society through its rules for making social decisions." I cannot resist a comment on Arrow's surprising espousal, here, of what might be called the revealed-preference principle in social welfare. I have already stressed several times the need to differentiate clearly "between the motion and the act," that is, between the inward value judgment and the outward, revealed decision (see also May, 1954). In the context of individual choice we have already seen that strategic considerations may trap the unwary observer into misjudging the true preferences of the respondent; three examples are "sophisticated" voting in a legislature, buying stock in order to push up the market price, and demanding more than you want at the outset of a negotiation. With group preferences there is also a more subtle "anthropomorphic" trap. The shaky analogy between individual and group psychology is easily abused. It may be meaningful, in a given setting, to say that a group "chooses" or "decides" something. It is rather less likely to be meaningful to say that the group "wants" or "prefers" something. Moreover, the correlation between preference and choice will be weaker for groups than for individuals. Admittedly a group might not so readily indulge in sophisticated voting or deceptive bargaining stances, at least in the absence of strong leadership. But even a well-organized group, having a clearly articulated "purpose," must still contend with the pressures of individual and factional self-interest. For these reasons "revealed preference" seems an inappropriate philosophical basis for social values.

In conclusion, the approaches to utility or welfare aggregation set forth in this chapter would be seen by most game theorists as too narrow in scope. This is not due only to the slighted strategic factor, though that is important. Game theory makes a special point of *not* requiring "society" to be a generalized person, capable of making choices and judgments among actions or outcomes on the basis of some sort of welfare function. The *set of players* engaging in a game is a different and more elusive kind of "animal" than the individual player. In our pluralistic theory, different solution concepts will view that animal differently. Granted, it will often be legitimate (and interesting) to ask whether a given solution concept permits a set of players to be aggregated, conceptually, into a single superplayer. But in general it is fair to say that multiperson game theory represents an attempt to attack and solve the utility aggregation problem by means other than those touched upon in this chapter.

Notes to chapter 5

1. Foundations for a theory of games played by committees are laid in the work of Owen (1964), Shapley (1964a, 1967c), Megiddo (1971, 1974a,b), Billera (1970b, 1971a), and Bixby (1971). Also see D. Brown (1975).

2. See Granger (1956), Jamison and Luce (1972), or, for a survey with a full bibliography, Plott (1971).

3. See Eisenberg (1961). In other words, the observed behavior of an aggregate is an unreliable indicator of the desires of the aggregate. For example, an observer of a laissez-faire capitalistic society might fancy that it behaves like a rational organism that "desires" monopolistic concentrations of power. This anthropomorphic metaphor might help him predict the

course of events in the unregulated system, but it would afford him no insight into questions of real social welfare.

4. In experimental games, as in some real-life situations, it is all too easy to mislead people into attaching special importance to a zero point that has no intrinsic meaning (see Shubik, 1975b, chaps. 4 and 11).

5. The strong form does not actually imply the weak form in the absence of further assumptions. In fact, we shall often find (5.2) more powerful than (5.1). For the connection between preference ($\succ$) and indifference ($\sim$), see appendix A.1. Preference is not necessarily transitive at this stage of the discussion.

6. This is the interchangeability property of ($\succ$, $\sim$) discussed in appendix A.1.

7. Without continuity, $I_N(\alpha)$ might, for example, be just the point α, with $\succ_N$ ordering the boundary of $P_N(\alpha)$ in some lexicographic fashion.

8. I do not know of a sharp formulation of this last assertion, that is, a minimal necessary condition on continuity, etc., such that the three stated assumptions would imply group transitivity.

The general question of how the transitivity of a relation can be inferred from its topological properties (e.g., continuity) has been treated, in a different context, by Sonnenschein (1965); see also Eilenberg (1941), Debreu (1960), Rader (1963), Lorimer (1967), Sonnenschein (1967), and Schmeidler (1971).

9. There is considerable redundancy in stating (5.6) and (5.7) for partitions (see exercise 5.3). Even after removing the redundancy, however, there are still $(3^n - 2^{n+1} + 1)/2$ independent conditions being imposed on the $2^n - 1$ relations $\succ_S$ or $\succsim_S$, as a simple counting argument shows.

10. Note that the {1, 3}-indifference curves are not used in this construction. The {1, 2, 3}-indifference surface, once determined, tells us what they must have been. This shows that the intergroup consistency required by (5.6) and (5.7) is so restrictive that we cannot independently choose utility-aggregation rules for all $(n^2 - n)/2$ pairs of individuals. Indeed the whole package of preference relations can in general be determined from those of $n - 1$ pairs of individuals if that set of pairs "connects" all members of N.

11. See especially the added pages in the second edition (1963) of Arrow's book; also, more recently, Sen (1970).

12. Vickrey (1960) has given perhaps the most efficient and elegant proof. The hypotheses given here can be substantially weakened without affecting the result. Wilson (1972c) has even found a version that avoids using the Pareto principle; see also his game-theoretic approach to Arrow's theorem (Wilson, 1972b).

13. This confrontation between the two models was first pointed out to Lloyd Shapley many years ago by Norman Dakley; I have not seen anything quite like it in the literature. The usual ways of circumventing the paradox have been either to restrict the diversity of individual preferences or to cardinalize the utilities by introducing additional structure such as intensities of preference or probabilitistic outcomes (see section 4.4). In fact, the Arrow theorem is sometimes adduced as an argument for adopting cardinal utility in social welfare theory.

For some other viewpoints see the illuminating colloquies between Coleman, Park, and Mueller in the 1966–1967 *American Economic Review,* and between Hansson and Fishburn in the 1969–1970 *Econometrica.* In the latter, Fishburn writes: "In the 1968 . . . election I preferred Abraham Lincoln to every one of the feasible candidates: should that preference have been taken into account by the social choice process?" Although the rhetorical question might demand a "no," the answer in light of our Model II might well be "yes," if the phrase "Abraham Lincoln" is understood to denote an abstract standard of comparison—a collection of personal and political attributes—rather than a hypothetical flesh-and-blood third-party candidate.

14. For example, having accepted the structure and discipline of an underlying commodity space, Bergson and Samuelson plunge easily and naturally into the differential calculus. In the Arrow type of model, however, one might be hard put to say what it means for the social welfare function to be differentiable.

6 Characteristic Function, Core, and Stable Sets

"The good old rule
Sufficeth them, the simple plan,
That they should take, who have the power,
And they should keep who can."
William Wordsworth, *Rob Roy's Grave*

6.1 Descriptions and Solutions

A theory of games can be regarded as composed of two parts, a descriptive theory and a solution theory. The descriptive part concerns the representation of the players and their preferences, the rules and strategic possibilities, and the outcomes and payoffs. The solution part concerns the end results of rational, motivated activity by the players. Since the objectives of a solution theory are less definite than those of a descriptive theory, it is not surprising to find several distinct solution concepts built on the same descriptive foundation. Indeed it is an advantage to have consistency at the descriptive level when we are comparing different types of solutions to the same game.

The next six chapters will review a great number of solution concepts at varying levels of detail. Stress will be placed on those that have shown most applicability to models of economic activity, but the discussion will be limited for the most part to the basic definitions and relationships. In the course of the survey, however, reasonably complete bibliographic coverage will be given to more extended mathematical developments and economic applications.

The solution concepts to be treated in this chapter are the *core* and the von Neumann–Morgenstern "solutions," which we call *stable sets.* We start, however, by presenting some concepts that are basic to much of cooperative game analysis—specifically the characteristic function, the Pareto set, and the space of imputations. We shall sometimes refer to these constructs as *presolutions,* in recognition of the contribution that they can make to the elucidation of the cooperative game, even were the analysis to go no further.

6.2 The Characteristic Function

The cornerstone of the theory of cooperative n-person games is the *characteristic function,* a concept first formulated by John von Neumann in 1928. The idea is to capture in a single numerical index the potential *worth* of each coalition of players. The characteristic function is, in a sense, the final distillation of the descriptive phase of the theory. Simultaneously it is a "presolution"—a first, rudimentary attempt at a solution.

For games that are "inessential"—intuitively, those where no profitable grounds exist for cooperation among players—the characteristic function *is* the final solution. The values of the function tell what each player can expect to get if he plays optimally, while the calculation of those values will (usually) reveal how to play optimally. For essential games the characteristic function can be regarded as a generalization of the "solution" as this term is generally understood in connection with two-person zero-sum games, in that it provides values and (implicitly) optimal strategies for coalitions, as well as for individuals, on the assumption of no cooperation with outside players.

Mathematically the characteristic function—traditionally denoted by the letter v—is a function from subsets of players to the real numbers. Hence only a finite list of numbers is involved if we are considering just finite-person games. But the list may be long, since there are $2^n - 1$ coalitions to consider in an n-person game, and we can often find an easier way to convey the information by exploiting special properties of the game.

There need be nothing trivial about the determination of the characteristic function. In Chess, for example, the characteristic function, were it known, would state whether the game is a win for White, a win for Black, or a draw. But our interests in this volume will lead us most often to games whose characteristic functions are arrived at easily; the analytical difficulties, if any, come later.

With the characteristic function in hand, all questions of tactics, information, and physical transactions are left behind. The characteristic function is primarily a device for dividing difficulties—for eliminating as many distractions as possible in preparation for the confrontation with the indeterminacy of what we have called the n-person problem. Engrossed with this problem, many authors writing after von Neumann and Morgenstern have begun by basing their solution concepts on the characteristic function above, with no initial concern for the concrete rules of the games in strategic or extensive form.[1]

Unfortunately, not all games admit a clear separation between strategic and coalitional questions, and for those that do not, the characteristic-function approach must be modified or abandoned. We shall return to this matter in section 6.2.2, after an example.

6.2.1 A first example

Let us return to the three-person Shooting Match of chapter 2 (see also section 3.3.5) and compute its characteristic function, using the accuracy parameters $a = 0.8$, $b = 0.6$, $c = 0.4$. For numerical convenience we shall assume that the prize to the winner is \$27. The method of calculation is similar to that in chapter 2, and the result is shown in table 6.1.

Several features of this example are noteworthy. First, the function v is *superadditive*: any set of players can do at least as well in coalition as in any subcoalition. This is a general property of characteristic functions. Formally it is expressed by the inequality

$$v(S \cup T) \geq v(S) + v(T), \tag{6.1}$$

which holds whenever S and T have no members in common.

Second, this particular game is *essential*: some coalitions, at least, do strictly better by sticking together. For example, A and B in coalition have an expectation of \$24.60, while separately their expectations come to just \$13.40. The difference, \$11.20, is a measure of their incentive to cooperate.

Table 6.1
Characteristic function for the three-person Shooting Match

S	$v(S)$
A	\$ 8.00
B	5.40
C	2.40
AB	24.60
AC	21.60
BC	19.00
ABC	27.00

Third, this particular game is *constant-sum*: the characteristic value of every coalition, when added to that of its complement, totals \$27. This is typical for parlor games, although the total amount of wealth created by a real economic process generally depends on the pattern of collaboration.

A fourth feature of this example is its *symmetry*. This may not be apparent to the eye, but it becomes very plain if we normalize the game by redefining the zero level for each player to be his expectation when playing alone. The normalized characteristic function is shown in table 6.2.[2] The worth of any coalition now depends only on its size. In particular, each pair of players has the same incentive to cooperate. Despite their unequal skills in the shooting gallery, the players are on an equal footing in negotiating for coalition partners. This kind of symmetry is always present in essential three-person constant-sum games—there is, in effect, just one such game—but it is rare in larger games.

6.2.2 The concept of c-game

The characteristic function as defined here is clearly not universally applicable. For one thing, to represent a coalition's worth by a single number implies freely transferable utility. For the nontransferable case a different notion of characteristic function will be required. More importantly, many games do not allow the clean separation that we need between questions of strategic optimization and questions of negotiation. For example, a player may have a number of possible threat strategies against a bargaining opponent. Although we can still define the characteristic function for such variable-threat situations,

Table 6.2
Normalized characteristic function for the three-person Shooting Match

S	$v(S)$
A	\$ 0.00
B	0.00
C	0.00
AB	11.20
AC	11.20
BC	11.20
ABC	11.20

we cannot analyze them properly without additional information about the actual rules of play—information that is lost when one passes to the characteristic-function form.

To simplify discussion, we have coined the term "c-game" for a game that is adequately represented by its characteristic function. We do not attempt a categorical definition of this term. What is adequate in a given instance may well depend on the solution concept that we intend to employ. To say that a game is a c-game is merely a way of asserting that nothing essential to the ultimate purpose of the model is lost in the process of condensing the extensive or strategic description into a characteristic function.

Although our definition of c-game is not categorical, two important and easily recognized classes of games qualify quite generally as c-games, no matter what solution concept is adopted. The first are the *constant-sum games,* in which the total payoff is a fixed quantity, regardless of the strategies chosen.[3] In such a game the characteristic value of a coalition accurately represents its true worth, because the worst threats against the coalition are precisely the plays that maximize the payoff to the players outside the coalition. The characteristic function pessimistically assumes the worst, and in this case "the worst" is a reasonable assumption.

Of more immediate interest to economics is another class of c-games, called *games of consent* or *games of orthogonal coalitions.* The idea here is that nothing can happen to change a player's fortune (payoff) unless he himself is a party to the action. Either you cooperate with someone or you ignore him; you cannot actively hurt him. In this case the only threat by outsiders against a coalition is not to belong to it—that is, the boycott. To sum up, in a constant-sum game "you're either with us or against us"; in a game of consent "you're either with us or we don't care what you do." In both cases a coalition might as well assume the worst and equate the outsiders' intentions with their capabilities.

While constant-sum games are relatively uncommon in economic theory, games satisfying the consent condition arise in many places, most notably in models of pure competition without externalities.

6.2.3 More examples: Symmetry considerations

THE DISGRUNTLED LABORER. *A workman says to his employer, "Since you make a profit from my labor, I demand that you share this profit with me; if you refuse to do so, I shall maim myself in such a way that I shall henceforth be unable to work, and you will get no profit at all from me."*[4]

This is a famous early challenge to the adequacy of the characteristic function; it is obviously not a c-game. The following payoff matrix will serve to quantify the example for later discussion:[5]

		Employer (E) resist	Employer (E) accede
Workman (W)	insist	$-1000,\ 0$	$5,\ 5$
	retract	$0,\ 10$	$0,\ 10$

In other words, \$10 is the employer's profit, and \$1000 is the monetary estimate of the contemplated injury. The characteristic function is

$$v(W) = 0, \quad v(E) = 5, \quad v(WE) = 10.$$

Observe that there is no clue here to the fatal weakness in the laborer's bargaining position, namely, the disastrous consequences *to himself* if he carries out his threat. Indeed, after normalization the function is perfectly symmetric. We shall return to this game in the next chapter.

GARBAGE GAME (Shapley and Shubik, 1969b). *Each player has a bag of garbage, which he must dump in someone's yard. The disutility of having b bags dumped in one's yard is kb, where k is a positive constant.*

For this game the characteristic function is best written with the aid of an auxiliary function f (figure 6.1), thus:[6]

$$v(S) = f(s) = \begin{cases} 0 & \text{if } s = 0, \\ -k(n-s) & \text{if } 0 < s < n, \\ -kn & \text{if } s = n. \end{cases}$$

Here n is the total number of players, s is an integer, and S is any coalition having s members. The cases $s = 0$ and $s = n$ are exceptions to the linear function $-k(n - s)$ for obvious reasons: the empty set cannot be hurt, and the all-player set cannot avoid "fouling its own nest."

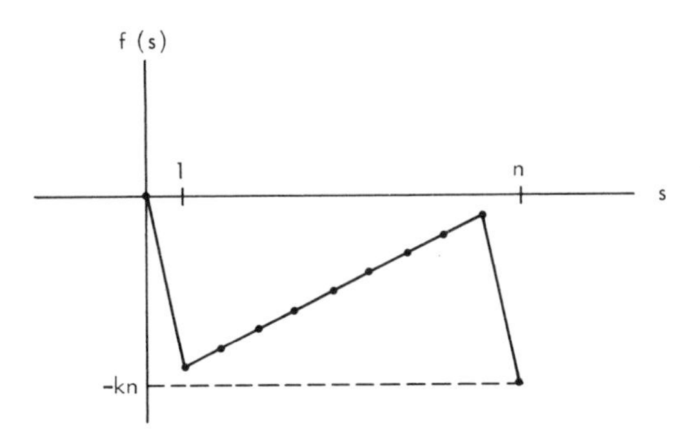

Figure 6.1
Characteristic function for the Garbage Game.

This is certainly not a game of consent; indeed it is a simple model of "directed pollution" in which externalities of production or consumption are imposed upon unwilling outside parties. We can still claim that it is a c-game, however, because of the happenstance of a constant-sum characteristic function.

Symmetry makes games like this especially easy to handle (Gelbaum, 1959; Ahrens, 1962). The reduction of a function of sets, $v(S)$, to a function of numbers, $f(s)$, allows many graphical and analytical aids. Few games encountered in applications are so completely symmetrical, but the idea of basing the characteristic function on a numerical function transcends symmetry. For example, the players in the Garbage Game might start with different numbers m_i of bags. Then we could write the characteristic function as follows:

$$v(S) = f\left(\sum_{i \in S} m_i\right), \tag{6.2}$$

where now

$$f(x) = \begin{cases} 0 & \text{if } x = 0, \\ -k(M - x) & \text{if } 0 < x < M, \\ -kM & \text{if } x = M. \end{cases}$$

Here M denotes the total number of bags. Games of this general type are called *measure games,* since the worth of a coalition is determined by the measure of some resource that the coalition members can pool. More generally, there may be several poolable resources (a *vector measure game*). Measure games and vector measure games play a prominent role in economic game theory.[7]

6.2.4 The generalized characteristic function

In games without u-money, a single index number $v(S)$ cannot completely express the worth of the coalition S. For one thing, it makes no sense to lump together the payoffs to the members of S, since they are no longer free to redistribute the total as they please. In fact the total may be meaningless if the payoffs are in incomparable units. But even a vector of payoffs will not usually be enough to represent the "best" efforts of the coalition; a set of vectors will generally be required—defining a sort of "Pareto" or maximal surface—to capture the true worth or effectiveness of S.

There are several ways to define this *characteristic set,* which is traditionally denoted by $V(S)$. Some authors place it in the space of S-vectors, since only the payoffs to members of S are significant (Aumann, 1967; Billera, 1970a,c, 1972; Kalai, 1972; Peleg, 1963c, 1969; Scarf, 1967). This has an inconvenient aspect, however, in that the characteristic sets for different coalitions will be in different spaces. Other authors treat each space of S-vectors as a linear subspace of the space of N-vectors (where N denotes the all-player set); this has the effect of making each element of $V(S)$ a complete payoff vector of the game by arbitrarily assigning zero to the players outside S. Of course zero has no intrinsic meaning in this formulation (Billera and Bixby, 1973). The most popular device, however, which we adopt here, is to make $V(S)$ a full-fledged subset of N-space, defining it as though the players outside S could get all possible payoffs. This makes $V(S)$ a "cylinder," that is, the Cartesian product of a subset of the S-vectors with a linear space of dimension $|N - S|$.[8] This expedient simplifies several basic formulas and definitions, such as those of superadditivity and balancedness, and also provides a more coherent geometrical picture for the mind's eye (see figure 6.2). We stress, however, that the differences between these ways of formulating the characteristic sets are purely notational.[9]

Finally, it is most convenient not to work with the maximal surface alone, but to include all points that lie "below" it; again that is a matter of notation, not substance. The foregoing conventions can all be summed up by the following condition: for any two payoff vectors α and β, if $\alpha \in V(S)$ and if $\alpha_i \geq \beta_i$ for each $i \in S$, then $\beta \in V(S)$.

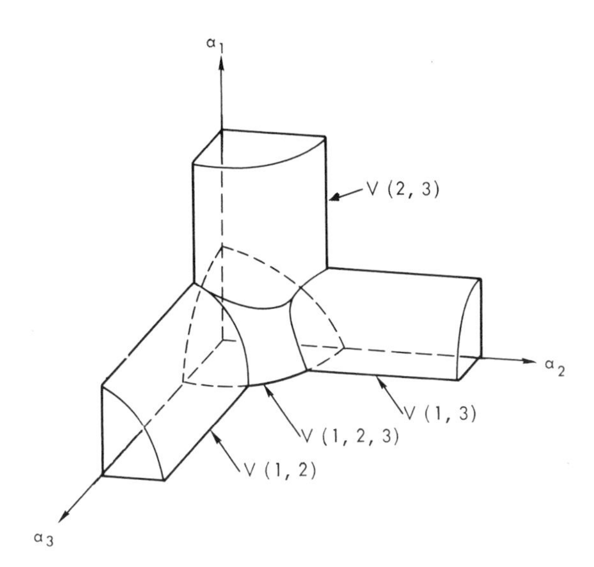

Figure 6.2
A geometrical representation of characteristic sets.

We shall always define $V(S)$ so that it is a closed set. Its *interior* will be denoted by $D(S)$; intuitively $D(S)$ is the set of all payoff vectors that the coalition S can improve upon. In many applications $V(S)$ and $D(S)$ will be convex sets, but this will not be a requirement.[10] Similarly, in virtually all applications *superadditivity* will hold; this condition takes the following simple form:

$$V(S \cup T) \supseteq V(S) \cap V(T) \quad \text{if } S \cap T = \varnothing, \tag{6.3}$$

which should be compared with (6.1).

Not surprisingly, if we try to express a transferable-utility game in this notation, we find that the resulting sets $V(S)$ have an especially simple form. Transferability means that with any payoff vector α in $V(S)$ we must include all other vectors $\alpha + \pi$, where $\Sigma_S \alpha_i = 0$. Thus $V(S)$ is a half-space whose upper boundary (which is all that really matters) is a hyperplane oriented in a direction that depends only on S. The equation of this hyperplane has the form $\Sigma_{i \in S} \alpha_i$ = constant, and the constant is, of course, the number $v(S)$ of the transferable theory.

6.2.5 Alpha and beta theories

When we come to write down a mathematical definition of the sets $V(S)$, starting from the game in strategic form, we find that there are two ways to proceed in defining the effectiveness of a coalition. The distinction lies between what a coalition can achieve and what it cannot be prevented from achieving. Let us try to make this precise.

On the one hand, we could say that a payoff vector lies in $V(S)$ if and only if the coalition S has a strategy that assures all its members that amount or more (in expected value), regardless of the behavior of the players outside S. We shall call this *alpha theory*.

On the other hand, we could say that a payoff vector lies in $V(S)$ if and only if the players outside S have no strategy that makes it impossible for all members of S to get that amount or more. We shall call this *beta theory*.

Lest the reader protest that this is a distinction without a difference, we append an elementary example:[11]

A COALITION PROBLEM. *A and B must jointly decide which of them will send \$200 to C. If they cannot agree, each forfeits \$300. Simultaneously C must send \$200 to A or B. Side payments are forbidden.*

The payoff matrix facing the coalition AB is as follows:

		Outsider's strategies	
		$C \to A$	$C \to B$
Coalition's strategies	$A \to C$	0, 0	−200, 200
	$B \to C$	200, −200	0, 0
	Disagree	−100, −300	−300, −100

The coalition's first strategy guarantees only $(-200, 0)$, the second only $(0, -200)$, and the third only $(-300, -300)$. More generally the mixed strategy (p_1, p_2, p_3), where p_i is the probability that the coalition will pursue strategy i, ensures $(-200p_1 - 300p_3, -200p_2 - 300p_3)$. Setting $p_3 = 0$, the coalition can enforce any payoff along the diagonal line in figure 6.3. Hence under the alpha definition $V(AB)$ is the crosshatched region at the lower left.

On the other hand, the outsider's first strategy prevents B from getting more than zero, while his second prevents A from getting

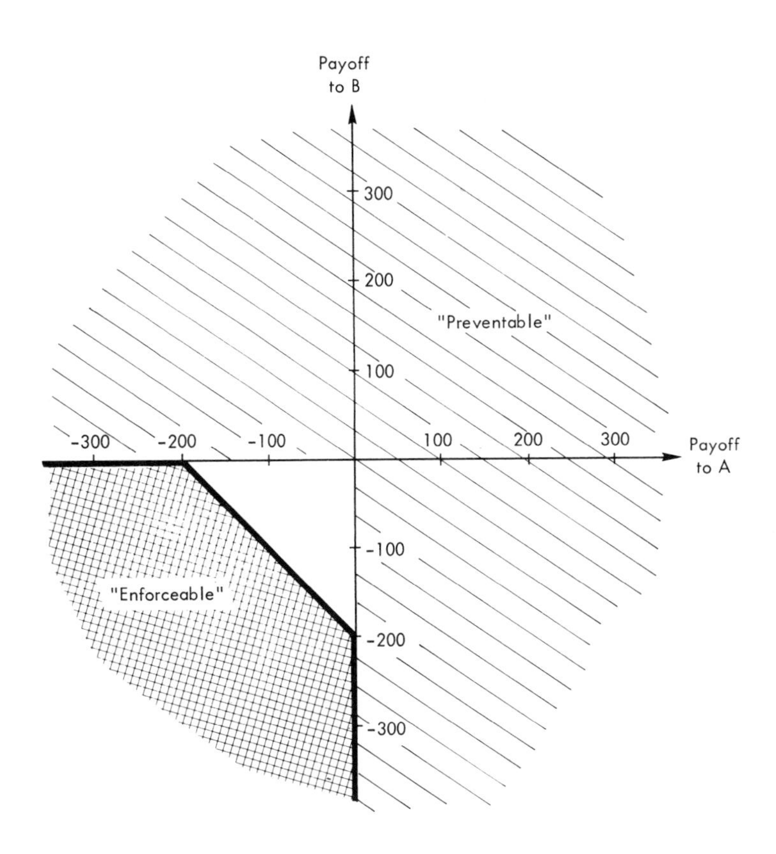

Figure 6.3
Alpha and beta characteristic sets for the Coalition Problem.

more than zero. Hence any point in the shaded region can be prevented. His mixed strategies prevent nothing more, however. If he plays his two strategies with probabilities p_1 and p_2, the coalition can counter with $(p_1, p_2, 0)$, and the expected payoff (0, 0) will result since (200, -200) and (-200, 200) will both have probability p_1p_2. Thus under the beta definition $V(AB)$ is the complement of the shaded region, that is, the (closed) negative quadrant. Points in the unshaded triangle can neither be guaranteed by AB nor blocked by C.[12]

This example illustrates the general principle that beta effectiveness is "easier" than alpha effectiveness, in the sense that the beta characteristic sets always include the alpha characteristic sets. It is easily shown that in a game of orthogonal coalitions the two characteristic functions coincide.

The key to the example just discussed was the fact that the outsider could *selectively* influence the fortunes of the coalition members. The coalition was not allowed to compensate by redistributing its income. We thus have in some sense a variable-threat situation, and it might be thought that games in which the alpha and beta theories diverge can never properly be regarded as c-games. This is not quite the case; it may remain useful to consider them c-games provided that the cost of the threat (to the outsider) does not vary.[13] For such games the choice between the two functions—alpha or beta—will depend on the rationale of the solution concept subsequently applied, and a solution concept that somehow depends on both characteristic functions is not inconceivable (see, e.g., Scarf, 1971).

6.2.6 Simple games

A special class of games, which may be thought of as "games of control," are important tools in the modeling of organizational and group decision processes. They are called *simple games* and are distinguished by the property of having just two kinds of coalition, namely, *winning* and *losing*. In the presence of transferable utility they are c-games, and after suitable normalization they give rise to an especially simple type of characteristic function, to wit:

$$v(S) = \begin{cases} 0 & \text{if } S \text{ is losing,} \\ 1 & \text{if } S \text{ is winning.} \end{cases}$$

Simple games were first defined and studied by von Neumann and Morgenstern; there is now an extensive literature dealing with their descriptive classification, their structural and other mathematical properties, their cooperative solutions of all kinds, and their applications to political and other models.[14] Let us consider an example.

WEIGHTED MAJORITY VOTING. *There are n players, having $w_1, w_2, \ldots, w_n$ votes to cast, respectively. A total of at least Q votes is required to effect a decision.*

The characteristic function of this measure game is given by

$$v(S) = \begin{cases} 0 & \text{if } \Sigma_{i \in S} w_i < Q, \\ 1 & \text{if } \Sigma_{i \in S} w_i \geq Q. \end{cases}$$

In the most familiar case n is odd, the w_i are all 1, and $Q = (n + 1)/2$, yielding simple majority rule. The more general "weighted" game provides a model for stockholders' control of a corporation or bloc voting by political parties in an assembly (see, e.g., Shapley and Shubik, 1954; Riker, 1959; Shapley, 1961a; Leiserson, 1968; Owen, 1971).

We stress that this is just one kind of simple game. Most simple games with many players cannot be brought into the weighted-majority format (a bicameral legislature is an example). Even when there are just a few players, it is remarkable how many different control schemes are possible, as the following tabulation reveals:[15]

$n =$	1	2	3	4	5
Number of simple games	1	2	5	20	180
Number of weighted-majority games	1	2	5	17	92

The theory of simple games is often presented without reference to a characteristic function. The usual axioms are the following:

A. Every coalition is either winning or losing.

B. The empty set is losing.

C. The all-player set is winning.

D. No losing set contains a winning set.

An additional axiom is required for superadditivity:

E. The complement of any winning set is losing.

Yet another one is needed to make the game constant-sum:

F. The complement of any losing set is winning.

A game satisfying axiom E is said to be *proper*; a game satisfying axiom F is said to be *strong*. Axiom E prevents the confusion that might result from allowing separate winning coalitions to make simultaneous decisions. Axiom F prevents the paralysis that might result from allowing a losing coalition to obstruct the decision.[16] Thus the strong proper games, also called *decisive games*, represent efficient group decision rules, whereas the other simple games represent procedures that may be plagued by inconsistencies, deadlocks, or both.

It will be observed that simple games, at least in their descriptive theory, enable one to sidestep many of the controversies of utility theory. Indeed simple games may be regarded as dispensers of political "power" rather than economic "utility." But if an economic basis must be provided, one need only assume a single lump-sum prize and have the players strive to maximize their chances of winning it. Payoffs can be then expressed in units of probability, which is a "substance" that can readily play the role of u-money, within the obvious limits.

The theory of simple games without transferable utility is somewhat more complex. It is still based on a system of winning and losing coalitions, satisfying axioms such as A–E above (Bloomfield and Wilson, 1972), but we must also specify a set $\mathscr{A}$ of payoff vectors representing the *available outcomes.* The idea is that a winning coalition can select any outcome it pleases from $\mathscr{A}$, whereas a losing coalition has no control over its fate. A characteristic function $V(S)$ can now be determined, but we should stress that the underlying model is *not* in general a c-game.

An example should make this clear. Suppose that there are three players using majority rule, so that the winning coalitions are just the two- and three-person sets. Suppose that the available outcomes yield the payoff vectors (0, 0, 0), (3, 3, 0), (3, 0, 3), and (2, 3, 3) together with all their probability combinations, making $\mathscr{A}$ a convex polyhedron in the payoff space. In this game the first player can reasonably expect to get at least 2, even if he is left out of the winning coalition. He cannot guarantee this, of course, but his position is certainly not adequately represented by the pessimistic value 0 given by the characteristic function. The number 2 in the vector (2, 3, 3) is obviously significant in this game, but the characteristic function is not sensitive to it.

Nevertheless, although the characteristic function by itself may be insufficient for analyzing simple games without transferable utility, an equally concise mathematical structure is presented by the set $\mathscr{A}$ of available outcomes, together with the specification of the winning coalitions. This structure is for many solution concepts just as tractable as the characteristic-function form.[17]

EXERCISES

6.1. *What is the formal distinction between a dictator and an individual having absolute veto power?*

6.2. *Show that a bicameral legislature is not decisive, regardless of the internal voting rule in each chamber.*

6.3 Imputations

Cooperative solutions take various forms. Solution theories have been developed in terms of single payoff vectors, sets of payoff vectors, single coalitions, sets of coalitions, hierarchies of coalitions, and even more elaborate configurations. Under some definitions there are many solutions to the same game, or many outcomes in the same solution, or both. Uniqueness is always welcome in a solution, but it is just one of many desirable properties. Others that we might mention, in general terms, are enforceability of the outcome, stability of the solution against collusive counteraction, derivability from an intuitively plausible social process, and compatibility with experimental evidence. Unfortunately, the many desiderata are often in conflict, and the enlightened theory builder must be selective and willing to compromise.

Three conditions that are almost always in demand, however, are *feasibility, Pareto optimality,* and *individual rationality.* The special term *imputation* will be used for the payoff vectors that enjoy these three properties.[18] Formal definitions are easily stated in terms of the characteristic function. Let N denote the all-player set. In the transferable case a payoff vector α is *feasible* if it satisfies

$$\sum_N \alpha_i \leq v(N); \tag{6.4}$$

it is *Pareto-optimal* if it satisfies both (6.4) and

$$\sum_N \alpha_i \geq v(N); \tag{6.5}$$

it is *individually rational* if it satisfies

$$\alpha_i \geq v(\{i\}) \quad \text{for all } i \in N; \tag{6.6}$$

and it is an *imputation* if it satisfies all three conditions. In the nontransferable case we merely replace these conditions by

$$x \in V(N), \tag{6.7}$$

$$x \notin D(N), \tag{6.8}$$

and

$$x \notin D(\{i\}) \quad \text{for all } i \in N, \tag{6.9}$$

respectively (recall that $D(S)$ is the interior of $V(S)$). Note that none of these definitions depend on our having a c-game, since they do not involve $v(S)$ or $V(S)$ for coalitions S other than the all-player set N and the singletons $\{i\}$. Also note that for $S = N$ or $S = \{i\}$ there is no distinction between the alpha and beta interpretations of $V(S)$.

Conditions (6.7) and (6.8) define what is sometimes known as "weak" Pareto optimality: it is not possible for everyone to improve. The "strong" statement—that no one person can be better off without hurting another—would lead in general to a smaller Pareto set.[19] If the weak Pareto set is regarded as depicting "social efficiency," then the strong Pareto set may be regarded as adding a pinch of altruism to the picture: grant other people as much as you can when it doesn't cost you anything.

The *imputation space* (to which similar weak/strong considerations apply in the nontransferable case) is the individually rational portion of the Pareto set. Like the characteristic function, the imputation space can be regarded as a crude presolution concept. It is the territory over which the coalitions do battle. In the inessential game this battle is trivial, since there is but one point in the space. In the essential game with transferable utility, the imputation space takes the shape of an *n-simplex,* that is, an $(n - 1)$-dimensional polyhedron with n vertices—one for each player. Without transferable utility the imputation space is generally a hypersurface in n-space, which may be more or less simplicial in appearance. When $V(N)$ is convex, the surface will be concave when viewed from "below," that is, from the reference point v^0 whose coordinates are $v_i^0 = v(\{i\})$, $i \in N$.

Let us consider an example, which is a simple prototype of the general exchange economy that will be studied extensively in coming chapters:

THE EDGEWORTH BOX. *Trader A starts with a units of one good, trader B with b units of a second good, and they trade at will. Their utility functions are* u_A, u_B; *thus, if A gives t units of the first good to B in return for r of the second, then their respective payoffs are* $u_A(a - t, r)$ *and* $u_B(t, b - r)$.

There being only two traders and no production, one person's final holding determines everything, and the well-known Edgeworth Box results (Edgeworth, 1881). In figure 6.4 A's position is measured on the solid axes and B's on the broken, inverted axes. Thus the initial point R has "solid" coordinates $(a, 0)$ and "broken" coordinates $(0, b)$. The shapes of the utility functions u_A and u_B are indicated by the solid and broken contours, respectively. For the present we shall assume these functions to be strictly increasing and strictly concave. It follows that their contours are strictly convex when viewed from the origin, as shown in the figure.

The Edgeworth Box depicts outcomes, not payoffs. It is possible, however, to locate in the Edgeworth Box the sets corresponding to the various presolutions that we have been discussing. First, the shaded area in figure 6.4 represents the outcomes that are preferred to the initial point R by both players. This area, including its boundary, is therefore the individually rational set. Next, it is evident that for an outcome to have no joint improvement, the two contours, solid and broken, that touch at the point must not cross. The locus of tangency points, $DCC'D'$, therefore belongs to the Pareto set (heavy line). But the boundary segments OD and $D'O'$ are also in the Pareto set, since they too cannot be jointly improved upon. Finally, the outcomes that determine the imputation space are just those that lie in the intersection of the heavy line with the shaded region—in other words, the arc CC'. This is traditionally called the *contract curve* or *contract set*.

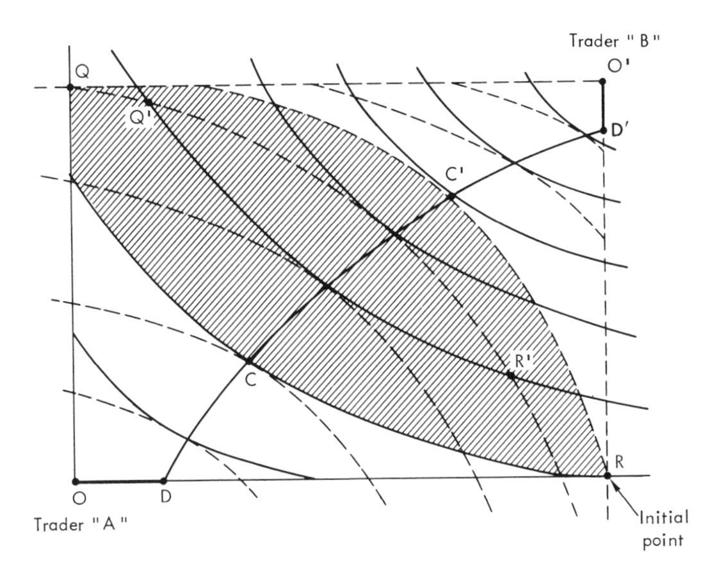

Figure 6.4
The Edgeworth Box.

Note that if the initial point were placed elsewhere, for example at R', then the individually rational set and the contract curve would be changed but the Pareto set would remain the same.

Let us now shift our attention from the Edgeworth Box, which is the allocation-of-commodities space, to the payoff or utility space. Figures 6.4 and 6.5 stand in the same relation to each other as an outcome matrix to a payoff matrix (see section 3.4). The mapping

$$(x, y) \rightarrow (u_A(x, y), u_B(a - x, b - y)) \tag{6.10}$$

takes each point of figure 6.4 (using "solid" coordinates x, y) into a point in figure 6.5. Some of the latter points are covered twice; in effect the Edgeworth Box gets folded along the curve $DCC'D'$ in the course of applying (6.10). We should note that figure 6.5 depends on the actual utility functions and not just on their contours.

In figure 6.5 the true feasible set is just the distorted rectangle $ORO'Q$, but in accordance with our convention the characteristic set $V(AB)$ includes everything below and to the left of that image as well,

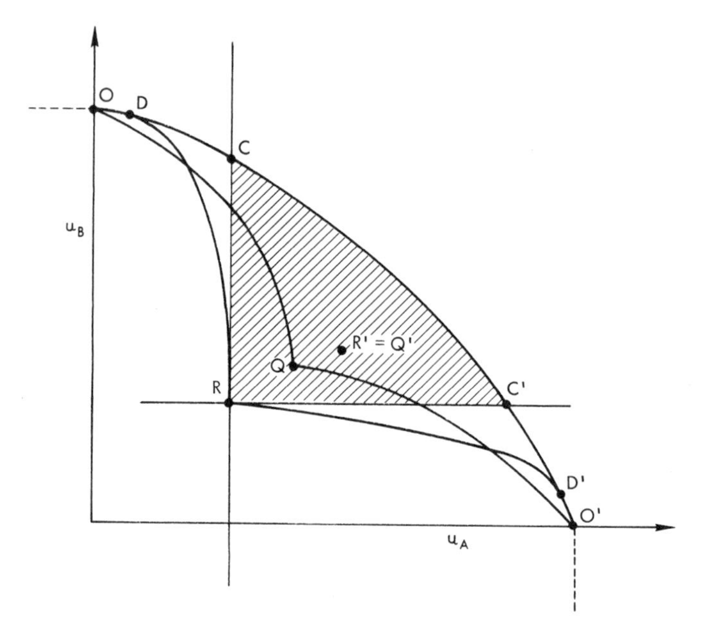

Figure 6.5
The image of the Edgeworth Box in the payoff space.

as indicated by the dashed boundaries. The individually rational set is obtained by subtracting from $V(AB)$ the open half-planes $D(A)$ and $D(B)$ lying to the left of and below the point R, leaving just the shaded area. The Pareto set is the entire northeast (upper right) boundary of $V(AB)$, including the broken lines. Finally, the imputation space is the arc CC'.

It will be noted that the true feasible set is not convex on its bottom. In fact the four sides of the box, RO, OQ, QO', and $O'R$, are mapped into curved lines that are concave when viewed from below.[20] This illustrates a general rule: If we take a straight line in the Edgeworth Box and if A's utility steadily increases and B's steadily decreases along that line, then the image of that line under (6.10) will be concave from below. It can also be shown that the Pareto set itself, though not in general the image of a straight line, is concave from below, ensuring the convexity of the set $V(AB)$.

These remarks depend, of course, on the assumed concavity of the utility functions, which is a cardinal-utility property. Quasiconcavity—that is, convexity of the level sets in figure 6.4—is not enough. If we replaced the given functions by, say, their cubes, we might well destroy convexity in the utility space, though nothing would change in the box diagram, which is purely an ordinal device.

EXERCISES

6.3. *Show that the image under (6.10) of the straight line joining R' and Q' in figure 6.4 has a cusp or a loop. Is more than one loop possible?*

6.4. *Prove that $V(AB)$ is a convex set.*

6.5. *Let the indifference curves for A and B be as indicated in figure 6.6a, as might occur if the two goods could only be consumed in certain ratios (the Gin and Tonic game: see Shapley and Shubik, 1966; or Debreu and Scarf, 1963), and let $a = b$. Show that the contract curve consists of four solid triangles arranged in a ring (figure 6.6b). What is the image of this set in the payoff space? What happens if $a \neq b$?*

6.4 The Core

Perhaps the most immediate way to grasp the n-person problem—the inherent indeterminacy of multiperson games—is to observe that it may be impossible to satisfy all coalitions at the same time. For example, in the three-person game of table 6.1 we can show very

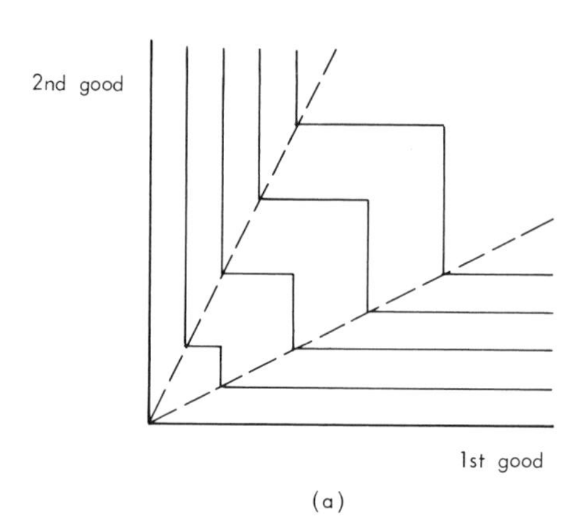

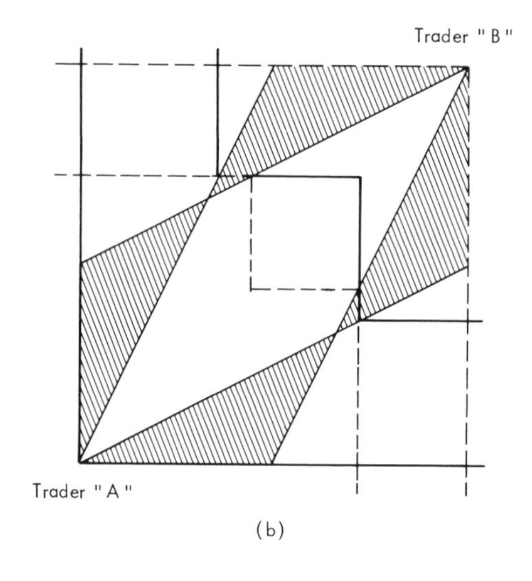

Figure 6.6
Gin and Tonic.

easily that no feasible payoff vector fulfills the potentials of all three pairs AB, AC, and BC. Indeed, if $\alpha = (\alpha_A, \alpha_B, \alpha_C)$ were such a vector, then we would have

$$\alpha_A + \alpha_B \geq \$24.60,$$
$$\alpha_B + \alpha_C \geq \$21.60,$$
$$\alpha_B + \alpha_C \geq \$19.00.$$

Adding these three inequalities gives

$$2\alpha_A + 2\alpha_B + 2\alpha_C \geq \$65.20 \quad \text{or} \quad \alpha_A + \alpha_B + \alpha_C \geq \$32.60.$$

In other words, to satisfy all coalitions we would have to distribute more than the \$27 prize offered. A similar impasse can be demonstrated for any other essential constant-sum game.

Nevertheless there are important classes of games, both with and without transferable utility, in which it is possible to satisfy all coalitions at once. A solution concept called the *core* has accordingly been introduced with this motif.[21] The core can be described intuitively as the set of imputations that leave no coalition in a position to improve the payoffs of all its members.

A formal definition can be given in terms of the characteristic function.[22] The core is the set of feasible payoff vectors α such that

$$\sum_S \alpha_i \geq v(S) \quad \text{for all } S \subseteq N. \tag{6.11}$$

For games without transferable utility we merely replace (6.11) by

$$\alpha \notin D(S) \quad \text{for all } S \subseteq N. \tag{6.12}$$

The conditions for the core are a direct generalization of—and unification of—the conditions for Pareto optimality and individual rationality. [Compare (6.11, 12) with (6.5, 8) and (6.6, 9).] While those conditions eliminate all outcomes that can be improved upon by society as a whole, or by individuals acting alone, the core condition eliminates all outcomes that *any* set of players can improve upon.

Geometrically the core in the transferable case is a closed, convex, polyhedral set of imputations. Figure 6.7a shows a typical three-person core. In the nontransferable case the core is still a closed set, but it need not be convex, polyhedral, or even connected. Figure 6.7b shows the core for the game illustrated in figure 6.2, and figure 6.7c shows how a disconnected core can result when the Pareto

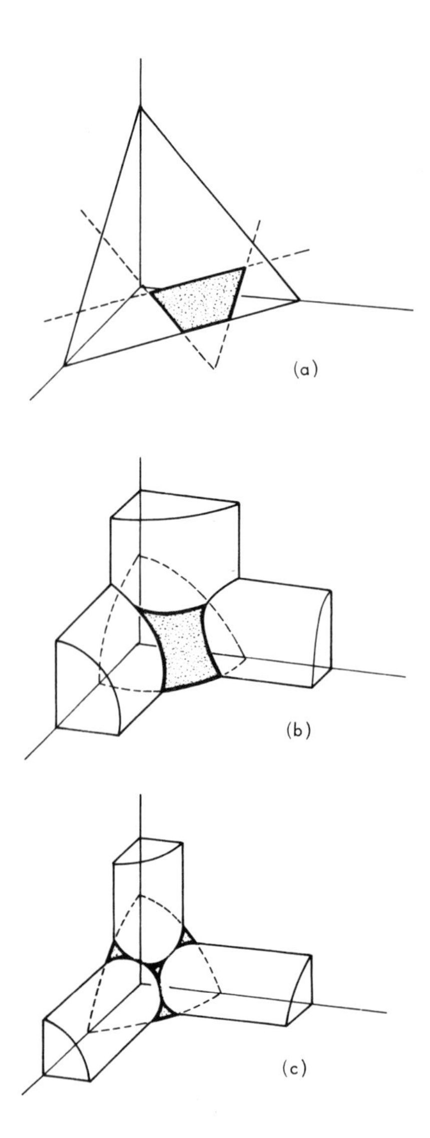

Figure 6.7
Examples of cores in three-person games.

surface is "flatter" for the all-player set than for the smaller coalitions.[23]

In a symmetric game, where $v(S)$ depends only on the size of S (see section 6.2.3), it is an easy matter to discover whether a core exists, since the core in such a game always includes the symmetric or "equal-shares" imputation that gives each player $v(N)/n$.[24] This imputation must therefore satisfy all coalitions if the core exists, and so we must have

$$\frac{f(s)}{s} \leq \frac{f(n)}{n} \quad (s = 1, 2, \ldots, n - 1),$$

where $f(s)$ denotes the characteristic value of coalitions having s members. Intuitively this means that no coalition is stronger per capita than the all-player set. Geometrically it means that the point $(n, f(n))$ is "visible" from the origin. Thus the game graphed in figure 6.8a

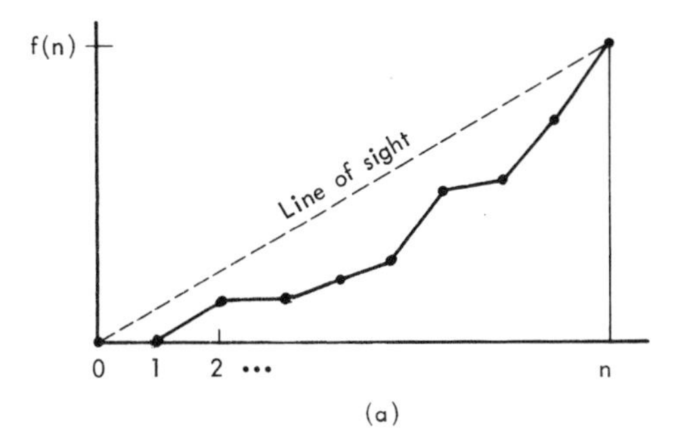

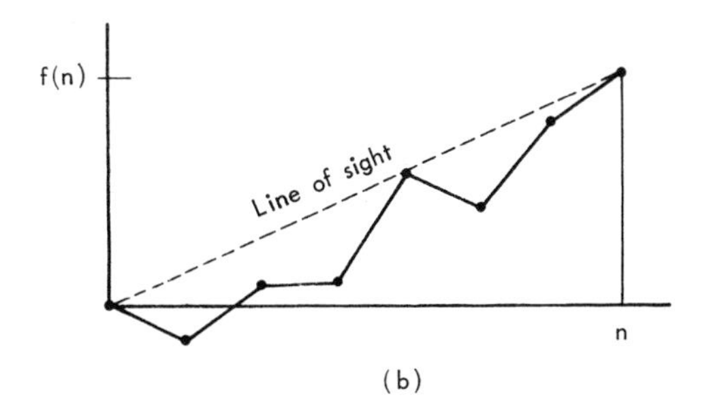

Figure 6.8
Existence of a core in two symmetric games.

has a rather large core, and the Garbage Game (figure 6.1) has none. In a borderline case like figure 6.8b the core exists but contains only the symmetric imputation. These diagrams illustrate vividly how the existence of a core depends on the relative weakness of coalitions of intermediate size.

EXERCISE

6.6. *Show that in a simple game (see section 6.2.6) the core consists of the imputations that award everything to the veto players. If there are no veto players, there is no core.*

6.4.1 Interpretation

A game that has a core has less potential for social conflict than one without a core, since every cooperative or collusive effective demand can be granted. A coreless game, on the other hand, must leave at least one coalition unsatisfied, that is, short of its potential. At least one group of players can always do better by dropping out and going it alone. If they try it, though, another group will be able to make a better offer to some of the dropouts in a new alignment, and the bargaining may continue at great length. In a coreless game, then, some constraints on coalitional activity must be operative in the society or else they will be engendered during the play of the game.

The condition (6.11, 12) of "coalitional rationality" that defines the core has great intuitive appeal. It does not follow, though, that when there is a core all other outcomes are bad and should be rejected out of hand. Social forces or taboos powerful enough to deny classes of players their birthright do exist in the world, and such forces must figure in one way or another in any solution theory capable of solving coreless games. Why should these forces necessarily become impotent in the presence of a core?[25]

Nevertheless the core, or lack of core, is an undeniably important feature of any cooperative game. Its existence, size, shape, location within the space of imputations, and other characteristics are crucial to the analysis under almost any solution theory. The core is usually the first thing we look for after we have completed the descriptive work.

The core has been extensively studied in reference to economic models. In particular, the core properties of markets with large numbers of traders are strikingly related to the properties of price systems in these models.

6.4.2 A simple market game

In any two-person game the core is exactly the imputation space. An example is the curve CC' in figure 6.5, the image of Edgeworth's contract curve. Since there are no intermediate coalitions between the singletons and the all-player set, (6.11, 12) adds nothing to (6.5, 8) and (6.6, 9). Only with three or more players does the core concept develop significance in its own right.

A THREE-CORNERED MARKET (adapted from von Neumann and Morgenstern, 1944). *A farmer's land is worth \$100,000 to him for agricultural use; to a manufacturer it is worth \$200,000 as a plant site; a subdivider would pay up to \$300,000.*

Let us first determine the characteristic function. Denote the players by F, M, S; denote coalitions by FM, FS, MS, FMS. Assuming u-money and letting zero represent the landless state for each player, we obtain table 6.3. To achieve $v(FM)$, for example, F must transfer the property to M, receiving in return some payment p. His final utility level will then be p, and M's will be \$200,000 $-$ p, so that the coalition is worth \$200,000, regardless of what p is.

We observe in passing that v is *not* constant-sum, in contrast with the examples in section 6.2. Depending on which two-against-one game is played, the sum may be \$100,000, \$200,000, or \$300,000.

The imputation space for this game is the set of all vectors $(\alpha_F, \alpha_M, \alpha_S)$ that majorize the initial utility vector (100,000, 0, 0) and add up to 300,000. These form a triangle in 3-space, as shown in figure 6.9a. Restricting ourselves to the plane of this triangle, we can drop to two dimensions to display more clearly the regions in which the various

Table 6.3
Characteristic function for the Three-Cornered Market

Coalition	Worth
F	\$100,000
M	0
S	0
FM	200,000
FS	300,000
MS	0
FMS	300,000

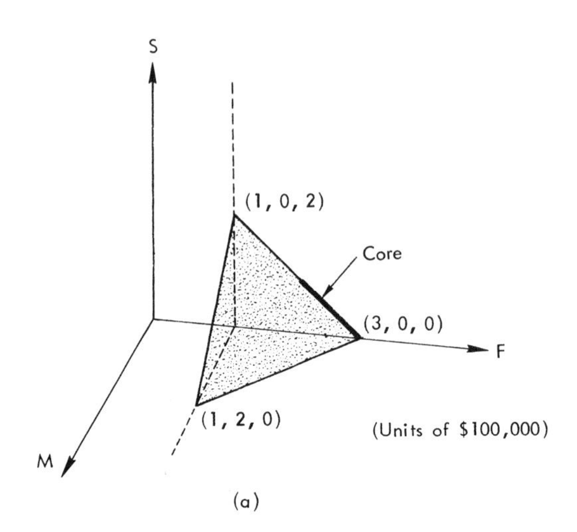

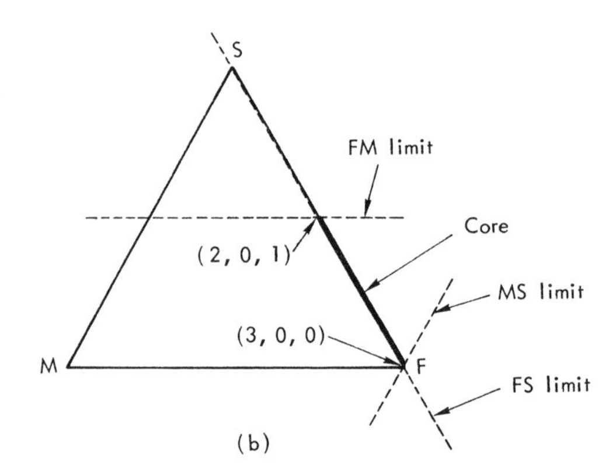

Figure 6.9
The core of the Three-Cornered Market.

two-person coalitions are effective. Thus in figure 6.9b the coalition *FM* can improve on any imputation above the dashed line labeled *FM*, which is the line on which they receive exactly $200,000. But between that line and the bottom edge (on which they would share $300,000) *FM* is ineffective. In contrast, *FS* is a stronger coalition that is satisfied (unable to improve its fortunes) only along the right edge of the triangle. Finally, *MS* is an inessential coalition that can do nothing that *M* and *S* could not do individually. By inspection we see that the core is the heavy line segment at the lower right, having end points (2, 0, 1) and (3, 0, 0).

The core in this simple example may be interpreted as making the prediction that the farmer will sell the land to the subdivider at a price somewhere between $200,000 and $300,000.[26] We shall return to this example several times as new solution concepts are considered.

EXERCISE

6.7. *Suppose that money can only be passed from the final buyer to the seller in the Three-Cornered Market. Determine the (generalized) characteristic function and core.*

6.4.3 Extensions of the core concept

One may visualize a geometric search for the core that begins with the set of all feasible payoff vectors and then trims off the unwanted subsets. Each cut eliminates the points for which a particular coalition is effective, in the sense of (6.11, 12). What remains at the end, if anything, is the core.[27] It was this imagery that inspired the term "core" in the first place.[28]

From this geometrical viewpoint it is easy to see why the core is not likely to be a very sharp solution. If the "bites" are too small, then the core will be large and unwieldy. The principle of coalitional improvement, implicit in (6.11, 12), will spend all its force before the possibilities are narrowed down to a unique, final outcome or a small class of outcomes. Other solution concepts will have to be invoked to shed light on the remaining distributional questions and their possible resolution.

On the other hand, if the "bites" are too large, there may well be no core at all, as we have already seen. But here, at least, there is some hope of extracting useful insights from the core concept. We do this by retaining the idea of coalitional improvement but making it more difficult to implement.

Two related definitions, among the many that could be devised, are of basic interest here. Both require transferable utility. For any positive number ϵ, the *strong* ϵ*-core* is the set of Pareto-optimal α, if any, such that

$$\sum_{S} \alpha_i \geq v(S) - \epsilon \quad \text{for all } S \subseteq N. \tag{6.13}$$

The *weak* ϵ*-core* is the set of Pareto-optimal α, if any, such that

$$\sum_{S} \alpha_i \geq v(S) - s\epsilon \quad \text{for all } S \subseteq N, \tag{6.14}$$

where s denotes the number of players in S.[29] In the first case we require a coalition to have a total excess of at least ϵ before we rule out the outcome in question. In the second case we require a per capita excess of ϵ.

These quasicores, as they are sometimes called, provide a crude way to reflect the costs or frictions associated with forming real coalitions. Under the weak definition (6.14) the costs depend on the size of the coalition; in the strong case (6.13) there is a fixed charge. Alternatively we might regard organizational costs as negligible or as already included in $v(S)$, but view the number ϵ or $s\epsilon$ as a threshold, below which a dissatisfied coalition will not consider it worth the trouble to upset the established order of things. Using ϵ-cores is, however, no substitute for serious modeling of the formation and operation of a real coalition; at best they only provide a suggestion of the direction in which the core solution might be modified if such considerations were included in the model.

The main applications of ϵ-cores to date have been as approximations to the core itself, as ϵ approaches 0. In fact, there are classes of games of economic interest which exhibit ϵ-cores, for arbitrarily small positive values of ϵ, even though true cores are absent.[30] In other words, we can say that some games "nearly" have cores, while in others the core is "far away." We can even define a numerical measure of corelessness. Let $\epsilon_{\min}$ be the smallest ϵ for which the strong ϵ-core exists. This is well-defined because the set of ϵ for which the strong ϵ-core exists is closed, nonempty, and bounded below. Then $\epsilon_{\min}$ represents the amount that would have to be deducted from the characteristic values $v(S)$ (excepting $v(\emptyset)$ and $v(N)$) for the core to appear.

Pursuing this idea further, we may define the *least core* (or *near core*) of a coreless game to be the smallest strong ϵ-core—in other words, the $\epsilon_{\min}$-core. It is of lower dimension than the other strong ϵ-cores, and it is often a single point as in figure 6.10a.[31] The least core may be regarded as revealing the latent position of the nonexistent core; it has also been proposed as a solution concept in its own right.[32]

It is of some interest, when the game has a core, to extend these definitions to negative values of ϵ.[33] Letting ϵ go negative has the effect of moving the walls of the core closer together, reducing the indeterminacy of the predicted outcome. Figure 6.10b illustrates this for $\epsilon = 0, -0.5, -1, -1.5, -2$, and -2.5, the last value yielding the least core. This core-shrinking process can be given some heuristic justification, and technically it is very useful in the development of another solution concept, the "nucleolus" (see chapter 11).

EXERCISES

6.8. *Determine the weak and strong ϵ-cores of the Three-Cornered Market for $\epsilon = \$1000$.*

6.9. *Determine the least core for the characteristic function in table 6.1.*

6.10. *Verify the following chain of set inclusions:*

weak ϵ-core $\supset$ strong ϵ-core $\supset$ weak (ϵ/n)-core $\supset \ldots \supset$ core,

assuming that $n > 1$, $\epsilon > 0$, and the core exists.

6.4.4 The inner core

We can construct a variant of the core for no-side-payment games in the following way. Consider the no-side-payment game illustrated in figures 6.7b and 6.7c. We can construct associated side-payment games for each point in the core of the game by using the direction cosines of the supporting hyperplane at the point to provide the exchange rates for the transfer of utility among the players.[34] We can then solve for the cores of the newly constructed side-payment games. We define the set of points in the core of the no-side-payment game which are also in the cores of their related side-payment games as the *inner core* of the no-side-payment game.

The inner core may be empty even though the no-side-payment game has a core. It has been shown, however, that for market games the inner core will always be nonempty. We may view a side-payment game as a special case of a no-side-payment game. Applying the construction noted above to the game shown in figure 6.7a, we see immediately that the core and inner core coincide.

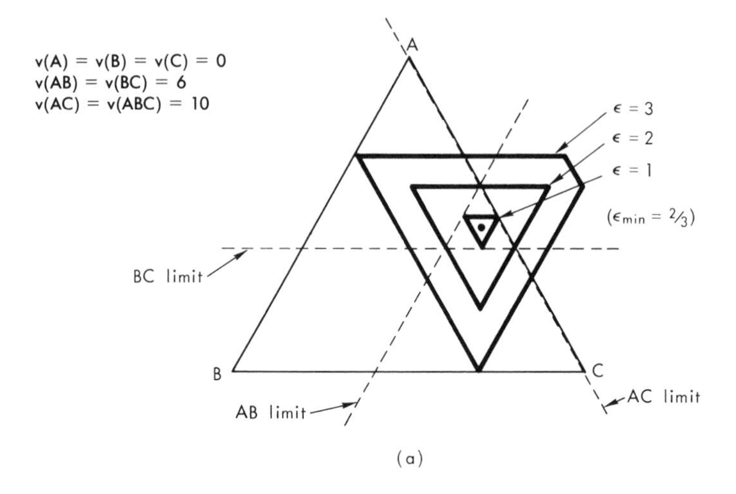

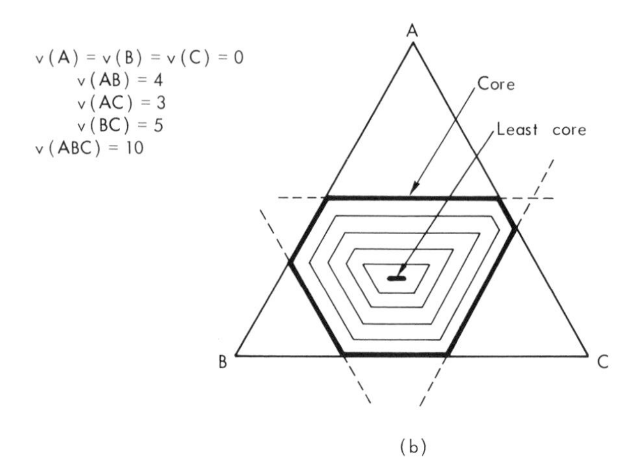

Figure 6.10
Strong ϵ-cores.

EXERCISE

6.11. *Show that the inner core for the following three-person no-side-payment game is empty. The three-person coalition can achieve any outcome* $x_1 + x_2 + x_3 \leq 10$, $x_i \geq 0$ $(i = 1, 2, 3)$. *Any two-person coalition i, j can achieve any amount on the line* $x_i + 10x_j = 10$, *or less, where* x_i, $x_j \geq 0$ *and individuals can obtain no more than zero individually.*

6.5 Stable Sets: The Theory of von Neumann and Morgenstern

One way to describe the core is in terms of a relation between payoff vectors called *domination*. Informally, one payoff vector dominates another if it is preferred by some coalition that can enforce its preference. Formally, α is said to dominate β if there is a coalition S such that

$$\alpha_i > \beta_i \quad \text{for all } i \in S \tag{6.15}$$

(i.e., S prefers α to β) and

$$\sum_{i \in S} \alpha_i \leq v(S) \tag{6.16}$$

(i.e., S is effective for α). In the nontransferable case (6.16) becomes

$$\alpha \in V(S). \tag{6.17}$$

From these definitions it can be shown that the core is just the set of imputations that are not dominated by other imputations, with respect to any coalition.[35]

This does not mean that the core imputations necessarily dominate everything outside the core. Usually, in fact, they do not. In disputing an unacceptable outcome, a coalition may have nothing better to propose than something that would be unacceptable to another coalition. Such a core lacks the property called *external stability*, which is defined as the ability of the members of a set of imputations to dominate all imputations outside the set.

External stability does not make for a useful solution concept by itself, since it is far too easy to achieve. For example, the set of all imputations is externally stable. A counterbalancing principle, called *internal stability*, can be defined as the absence of any domination between members of the set. Internal stability is similar to, but weaker than, the property of nondomination that characterizes the core. The core is internally stable, obviously, but other sets having no particular relationship to the core are also internally stable. For example, any single imputation is internally stable.

The solution of a game, as defined by von Neumann and Morgenstern, is any set of imputations that is both externally and internally stable. Put another way, it is any subset of the imputation space that dominates its complement. In our pluralistic approach we have reserved the term "solution" for more general use, as we shall refer to the von Neumann–Morgenstern solutions as *stable sets.*

The external and internal conditions are so nicely balanced in this definition that either increasing or decreasing a stable set by a single point destroys its stability. Hence, although there may be (and usually are) many stable sets, sometimes intersecting each other, no stable set lies completely within another one. A stable set is, simultaneously, a minimal externally stable set and a maximal internally stable set.

If there is a core, every stable set must include it, since points in the core are not dominated. If the core happens to be externally stable, then it is the unique stable set.[36] But when there is no core, the stable sets may be quite unrelated to each other and quite dissimilar in form.[37] Some simple examples will be discussed in the next section.

The definition of stability is in a sense circular,[38] and there is no straightforward method for calculating stable sets, or even for determining when they exist. The general existence question, in fact, was a major outstanding problem of game theory for many years, not settled until 1964 for nontransferable utility games and 1968 for transferable utility games (Stearns, 1964b, 1965; Lucas, 1968b, 1969b). Stable sets have fascinated mathematicians, and there is a large body of literature dealing with the search for stable sets that solve particular classes of games such as four-person games, simple games, quota games,[39] or symmetric games, or that enjoy special properties such as finiteness, symmetry, discrimination,[40] or unusual pathological features. A guide to this widely scattered literature will be found in appendix B.

While stable-set theory is admittedly difficult in its many ramifications, experience has shown that most games are abundantly endowed with stable sets, and several proposals have been advanced for modifications in the definition, or for further stability conditions, with the aim of reducing the multiplicity (Shapley, 1952a; Luce and Raiffa, 1957; Vickrey, 1959; Harsanyi, 1974). Since none of these have gained widespread acceptance, we shall not discuss them here.

EXERCISES

6.12. *Show that every stable set is closed, that is, includes its limit points.*

6.13. *Show that an imputation that is dominated by a core point cannot belong to any stable set.*[41]

6.14. *Show that a stable set is Pareto-optimal.*

We close this section with the Lucas (1968b) game without a stable set. It has ten players, and $v(S) = 0$ for all $S \subset N$ other than the following:

$v(12) = v(34) = v(56) = v(78) = v(9, 10) = 1,$

$v(357) = v(157) = v(137) = 2,$

$v(359) = v(159) = v(139) = 2,$

$v(1479) = v(3679) = v(5279) = 2,$

$v(3579) = v(1579) = v(1379) = 3,$

$v(13579) = 4,$

$v(N) = 5.$

The core is the convex hull of the six imputations:

$(1, 0, 1, 0, 1, 0, 1, 0, 1, 0)$, $(0, 1, 1, 0, 1, 0, 1, 0, 1, 0)$,

$(1, 0, 0, 1, 1, 0, 1, 0, 1, 0)$, $(1, 0, 1, 0, 0, 1, 1, 0, 1, 0)$,

$(1, 0, 1, 0, 1, 0, 0, 1, 1, 0)$, $(1, 0, 1, 0, 1, 0, 1, 0, 0, 1)$.

This example with no stable-set solution can be interpreted as having arisen from an exchange economy with no obvious pathologies and may be viewed as a market game (see section 6.6).

It has recently been proved by means of a fourteen-person counterexample that there exists an n-person game which has neither a stable set nor a core (Lucas and Rabie, 1980).

6.5.1 Examples

Every stable set includes the core, if there is one, but the core by itself need not be stable. This is illustrated by a game we considered earlier: In figure 6.9b the core dominates all points directly to its left—that is, below the middle line—but none of the points on or above that line. To obtain a stable set we must tack on a "tail"—a curve traversing the triangular region (shaded in figure 6.11a) that the core fails to dominate. Since this appendage can be more or less arbitrary, this game has an infinity of of different stable sets. Moreover, each stable set contains an infinity of imputations.

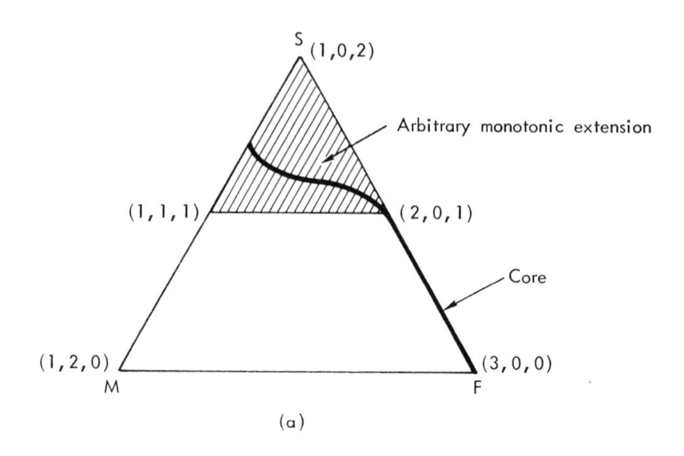

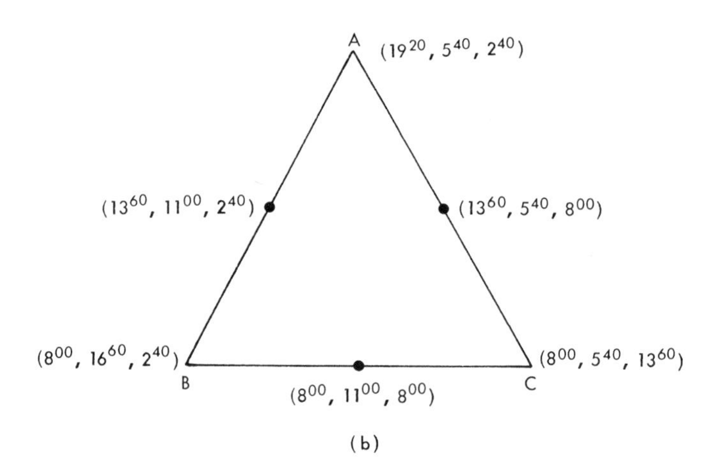

Figure 6.11
Examples of stable sets in three-person games.

For contrast, let us consider the three-person constant-sum case. (Recall that there is essentially just one such game.) One of its stable sets consists of exactly three points: the midpoints of the three sides of the imputation triangle. Thus, if we use the Shooting Match data (table 6.1), we obtain the three imputations

(13.60, 11.00, 2.40), (13.60, 5.40, 8.00), (8.00, 11.00, 8.00),

as shown in figure 6.11b.[42] It is easy to verify the stability of this set. Call the set Y. The points of Y obviously do not dominate each other, since only one player would ever want to switch from one to another and it takes two to make an effective coalition. Hence Y is internally stable. On the other hand, any imputation outside Y will give less than the upper payment (13.60, 11.00, or 8.00, respectively) to at least two players,[43] so these players will prefer the imputation in Y that gives them their upper payments; moreover they will be effective for this preference. Hence Y is externally stable.

We should emphasize that finite stable sets are rather unusual occurrences and should not be regarded as typical of the theory. Even the three-person game just considered has many other stable sets, all of them infinite.

EXERCISES

6.15. *Determine the set of imputations dominated by a typical point on the curve in figure 6.11a, and so find the conditions that the extended curve must satisfy.*

6.16. *In the three-person constant-sum game, show that the set of imputations in which a given player receives his minimum is stable.*

6.5.2 Interpretation

The defining conditions for stability have a plausible ring, but there remains the task of putting together a coherent interpretation. Von Neumann and Morgenstern proposed that a stable set be viewed as a *standard of behavior*—or a *tradition, social convention, canon of orthodoxy,* or *ethical norm*—against which any contemplated outcome can be tested.[44] Because this standard of behavior is assumed to be well known to the community of players, the elements of the stable set are constantly in the public mind, and whenever a "heretical" imputation is suggested, violating the canon, it can be instantly refuted by an "orthodox" imputation that dominates it.

In practical applications this elaborate interpretation is most successful when it can be related to some familiar cultural constraint or institutional form. The three-point solution just described, for example, could arise from an egalitarian tradition which says that in situations of this kind a two-player coalition will form and split the proceeds, but which says nothing about the identity of the successful pair. Another stable set of the same game consists of one complete edge of the imputation triangle (see exercise 6.16 above). This stable set represents a nonsymmetric, "discriminatory" standard of behavior. It identifies a particular pair of players as the presumptive winners, but it says nothing about how they should divide the prize.

In the Three-Cornered Market (which is a junior version of several of our later economic models), one possible societal constraint might declare it unethical to compensate a buyer for staying out of the market, even though by doing so he may help another buyer get a better price. The stable set corresponding to this standard of behavior is a straight line having for its "tail" (figure 6.11a) just the linear continuation of the core. In this discriminatory solution player *M* gets nothing, regardless of the arrangement reached between *F* and *S*. In another society, however, a successful bidder might be expected to share his good fortune with any potential buyer who could have bid against him but refrained. A stable set more like the one illustrated in figure 6.11a would then be indicated.

Stable-set theory, then, does not predict a standard of behavior. Nor does it predict an outcome when the standard of behavior is known. Rather, it tells us whether a given set of social or economic procedures is stable, by investigating the domination properties of the set of imputations they give rise to. We can only get something approaching a prediction from this branch of game theory when we are able to discern some general properties enjoyed by all stable sets of the game.

A case in point is provided by the "Böhm-Bawerk games"—two-sided markets in a single, indivisible commodity. The Three-Cornered Market represents the smallest nontrivial example of this type of game (Böhm-Bawerk, 1891, p. 203ff.; von Neumann and Morgenstern, 1944, p. 560ff.; Shapley, 1961b; Shapley and Shubik, 1972a). In these games the core is a straight line segment, and every stable set extends the core (in both directions if necessary) along a continuous, monotonic "bargaining curve," until limits are reached where all the profit goes to the players on one side of the market.[45]

A stable set can here be explained as a pattern of special pricing arrangements, possibly including side payments to third parties. The one free parameter, identifying positions along the bargaining curve, can be taken to be the marketwide average price, from which the individual transactions can deviate only within narrow limits.

6.5.3 Pathologies

Unfortunately a plausible interpretation is not always so ready to hand. In fact a rogues' gallery of horrible examples of stable-set "misbehavior" has accumulated over the years in the literature and the unpublished folklore of game theory. There are some very simple n-person games, for instance, with n any number greater than 3, in which one can start with an arbitrary closed subset of a certain $(n - 3)$-dimensional region in the imputation space and build it up into a stable set, adding only points remote from the starting set (Shapley, 1959a; Shapley and Shubik, 1969a). Such a degree of arbitrariness is somewhat harder to explain away than the arbitrariness in the bargaining curves of Böhm-Bawerk games. More recently a procession of counterexamples has emerged out of the work of William Lucas (1968c). Thus there is a partially symmetric eight-person game, none of whose stable sets have the symmetry of the game. There are games with unique stable sets which are not cores, or even corelike, since they are nonconvex (Lucas 1968b, 1969b).[46] One innocent-appearing game, with twenty players, was discovered *all* of whose stable sets are infinite aggregations of two-dimensional "flakes," attached densely yet sparsely to the line-segment core (Shapley, unpublished). After that monstrosity, Lucas's climactic discovery of a ten-person game having no stable sets at all came almost as a relief to those working in this field.[47]

The detection and diagnosis of stable-set pathologies has a fascination of its own, but the question naturally arises whether they are mere mathematical curiosities or whether they could actually arise in application. Since the most notable applications of this branch of solution theory have been to models of trade, or trade and production, a more specific question can be asked: Are there markets, or other basic economic systems, whose representations as n-person games exhibit the various stable-set pathologies? Unfortunately the answer is yes in almost all known cases. There appear, however, to be no outward signs of trouble in the economic models so afflicted (Shapley and Shubik, 1969a).

The explanation for these assertions lies in some very fundamental relationships between markets, cores, and stable sets, which will be discussed in the next section.

6.5.4 Extensions

Since it uses only the notions of imputation and domination, which both generalize readily, the definition of a stable set has been applied to several extensions and variants of the "classical" model—the finite-person c-game with transferable utility—for which it was originally intended. The extension to games without transferable utility has already been discussed. Stable sets have also been studied for infinite-person games, for games in partition-function form (see chapter 11), and for abstract mathematical structures. While some of these extensions contain at least the seeds of potential application to economics and other fields, they have also made their contributions to the rogues' gallery. In fact, solutionless games were found in each of the categories noted above before they were found in the classical case, which proved the hardest nut to crack.

6.6 Balanced Sets and Market Games

The modeler of markets and other economic institutions must usually, as we have stressed, attend closely to the special features of the concrete situation being analyzed, as well as to the special purposes of the investigation. There is, however, such a fundamental connection between cooperative n-person games and certain marketlike processes that a brief consideration of some very abstract market models can give valuable insights into the concepts we have been discussing.

The traders in these abstract markets have, in addition to u-money, a number of idealized commodities that they can trade among themselves without restriction. The payoff to a player is just his utility for his final holding of goods and money, without regard for the particular sequence of trades he may have engaged in and without regard for the holdings of the other players. His utility function is assumed to be continuous, concave, and expressed in monetary units.

Such a model reduces very quickly to characteristic-function form. The *worth* $v(S)$ of any subset of traders is simply the maximum total utility that they can achieve by trading only among themselves. Clearly this model is a "game of orthogonal coalitions" and hence a c-game. Any characteristic function arrived at in this way is called a *market game.* The key to the analysis of market games lies in an important technical concept, known as balancedness, which we shall now describe.

6.6.1 Balanced games

The characteristic function of our abstract market model is, of course, superadditive; this can be expressed by saying that the inequality

$$v(S_1) + v(S_2) + \cdots + v(S_m) \leq v(S) \tag{6.18}$$

holds whenever the family $\{S_j\}$ of coalitions is a *partition* of S (i.e., each member of S belongs to exactly one of the S_j). This is nothing more than the observation that if S breaks up into smaller trading groups, they may be able to do as well as before, but certainly will not do any better.

In a market game, however, something less drastic than breaking up into completely independent groups is possible, pointing to a generalization of (6.18). A player can do some business with one group, some with another. Let $\{S_j\}$ be a family of "actively trading" subsets of S, including all players in S but not necessarily forming a partition. Let each set S_j demand the fraction f_j of the resources of each of its members. If these numbers f_j can be chosen so that they total 1 for each player, thereby accounting for all his resources, then $\{S_j\}$ is said to be a *balanced family* of subsets of S.[48] In that case the inequality

$$f_1 v(S_1) + f_2 v(S_2) + \cdots + f_m v(S_m) \leq v(S) \tag{6.19}$$

will be satisfied when v comes from a market game. The reason for this is that if each coalition S_j uses just the fraction f_j of its members' resources, it can generate at least $f_j v(S_j)$ (because the utility functions are concave), whereas the coalition S acting as a whole can put together these partial efforts (because the family is balanced) and thereby achieve at least the sum of the $f_j v(S_j)$.

A game in characteristic-function form is said to be *totally balanced* if (6.19) holds for every S. It is said to be *balanced* if its characteristic function satisfies (6.19) in the special case where S is the all-player set N. In other words, a game is totally balanced if it is balanced and if all of its subgames are balanced as well.

It is not hard to see that some superadditive characteristic functions are not balanced (see exercise 6.18) and that some balanced characteristic functions are not totally balanced. For the reasons just sketched, however, every market game is not only balanced but totally balanced (Shapley and Shubik, 1969a).

A similar theorem holds for market games without side payments (i.e., markets without u-money); we merely replace the numerical inequality (6.19) by the set-theoretical condition

$$V(S_1) \cap V(S_2) \cap \cdots \cap V(S_m) \subseteq V(S), \tag{6.20}$$

again asserted for all balanced families $\{S_j\}$ of subsets of S.[49]

A sort of converse is available. It can be shown, at least in the side-payment theory, that every totally balanced characteristic function is the characteristic function of some market game. Indeed we can construct a market from any set function v in the following way:

THE DIRECT MARKET.[50] *Associated with each subset S there is an activity called Project S, and each individual must distribute one unit of labor (say) among the projects that include him. The amount of (transferable) utility generated by Project S is $f(S)v(S)$, where $f(S)$ is the minimum of the individual contributions to the project by the members of S and v is a given set function.*

These rules suffice to define an abstract market with as many commodities as traders; and if v happens to be totally balanced, then the characteristic function of the market is precisely v. To see this, note that the members of S can attain $v(S)$ by operating Project S full time; but because of (6.19) they can do no better than $v(S)$ by diverting effort to other projects under their control—that is, to Project R for $R \subset S$.

If the given function v is not totally balanced, however, the resulting direct market has for its characteristic function not v but a new function $\bar{v}$, called the *cover* of v. This function turns out to be the smallest totally balanced set function that is everywhere greater than or equal to v (Shapley and Shubik, 1969a).

EXERCISES

6.17. *Show that if a family of coalitions is balanced by virtue of more than one set of fractions $\{f_j\}$, then some proper subset of that family is also balanced. In other words, show that the balancing fractions for a minimal balanced family are unique.*

6.18. *By considering the balanced family consisting of all sets of the form $N - \{i\}$, show that no essential constant-sum game can be balanced.*

6.19. *Show that a market without u-money is totally balanced, in the sense of (6.20).*

6.20. *Determine the cover of the normalized three-person constant-sum game displayed in table 6.2.*

We provide below an example to contrast the two kinds of balanced games, ordinal and cardinal. Let

$F(\overline{123}) = \{x \geq 0 : \sum_i x_i \leq 10\}$,

$F(\overline{12}) = \{(4, 4, 0)\}, \quad F(\overline{13}) = \{(4, 0, 5)\}, \quad F(\overline{23}) = \{(0, 4, 1)\}$,

$F(\overline{1}) = F(\overline{2}) = F(\overline{3}) = \{(0, 0, 0)\}$;

then

$V(S) = F(S) - E_+^S$.

This game is ordinally balanced but not balanced, since we have

$\frac{1}{2}(4, 4, 0) + \frac{1}{2}(4, 0, 5) + \frac{1}{2}(0, 4, 1) = (4, 4, 3) \notin V(\overline{123})$.

It has a good-sized core, as shown in figure 6.12, but it has no inner core.

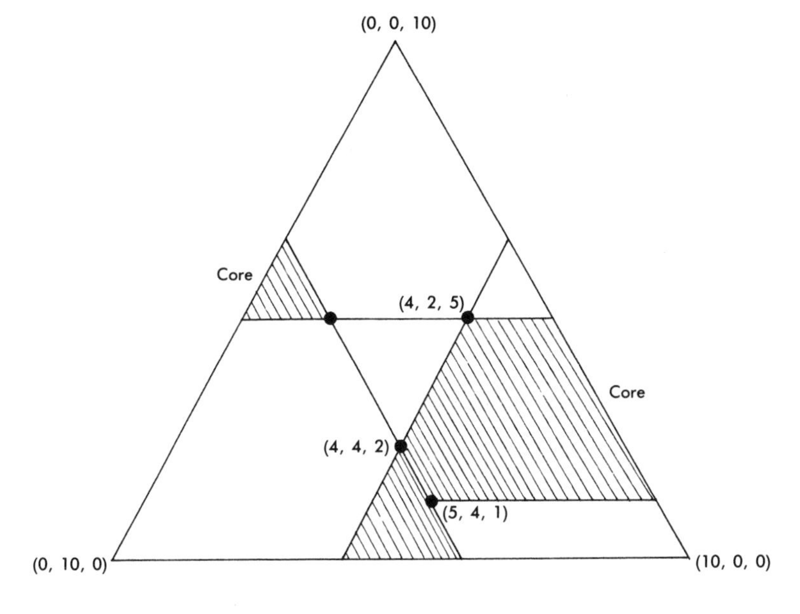

Figure 6.12
An ordinally balanced game.

Now apply the following concave order-preserving transformation to u_3:

$$\hat{u}_3 = T(u_3) = \min\left(u_3, 2 + \frac{1}{10}(u_3 - 2)\right).$$

The new game $\hat{F}$ is unchanged except for $\hat{F}(\overline{13}) = (4, 0, 2.3)$ and $\hat{F}(\overline{123})$ being "bent" at $x_3 = 2$, as shown in figure 6.13.

The game is now balanced in both senses, since the above calculation now gives us the point (4, 4, 1.65), which does belong to $\hat{V}(\overline{123})$. The inner core turns out to be the small triangle whose vertices are (5, 3, 2), (5, 3.7, 1.3), and (4.3, 3.7, 2).

A handy illustrative example of a three-person game with nontransferable utilities has been provided by E. W. Paxson (personal communication) to illustrate the core, stable set, and bargaining set.

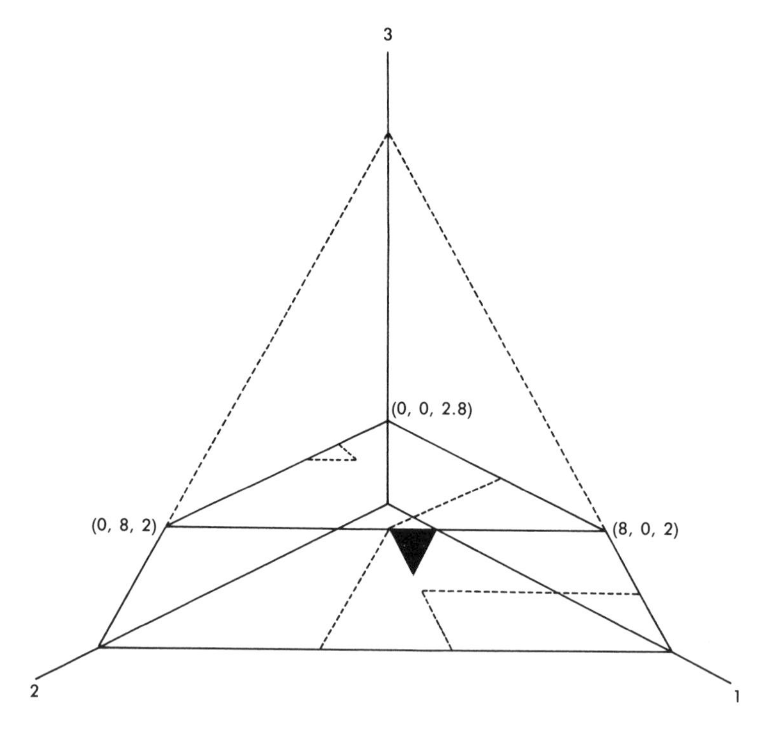

Figure 6.13
Ordinal and cardinal balance.

A fixed budget A is to be allocated to three line items, $A = A_1 + A_2 + A_3$. This may be thought of as incremental changes in last year's budget, or $a_1 + a_2 + a_3 = 0$, $|a_1| \leq 1$, $|a_2| \leq 1$, $|a_3| \leq 1$. The players' preferences (utilities) are expressed by

$$u_1 = a_1 - a_2, \quad u_2 = 2a_2 - a_3, \quad u_3 = -a_1 - a_2 + 2a_3.$$

With $a_3 = -a_1 - a_2$, this becomes

$$u_1 = a_1 - a_2, \quad u_2 = a_1 + 3a_2, \quad u_3 = -3(a_1 + a_2)$$

over the region R: $|a_1| \leq 1$, $|a_2| \leq 1$, $|a_1 + a_2| \leq 1$. For convenience in plotting, add 3 to all utilities. Then for the boundary of R we have

a_1	a_2	u_1	u_2	u_3	
1	0	4	4	0	P^1
0	1	2	6	0	P^2
−1	1	1	5	3	P^3
−1	0	2	2	6	P^4
0	−1	4	0	6	P^5
1	−1	5	1	3	P^6

The game is played over linear combinations of the P^j. Any two players can enforce their desired allocation. The strong Pareto sets (see figure 6.14) are

$\{1, 2\}$: $\overline{P^6P^1P^2}$,

$\{1, 3\}$: $\overline{P^5P^6}$,

$\{2, 3\}$: $\overline{P^2P^3P^4}$.

This is most quickly seen by projecting on the coordinate planes.

There is no core since the Pareto sets have a vacuous intersection. The stable set is the union of the Pareto sets. The game is not balanced. The bargaining set consists of the three vectors α^1, α^2, α^3. This set has strong stability in that each player gets the same amount in the two winning coalitions to which he belongs. In fact $\alpha^1 = (13/3, 3, 1)$, $\alpha^2 = (13/3, 1/3, 5)$, $\alpha^3 = (5/3, 3, 5)$, and the set is unique.

6.6.2 Solution properties

Using the concept of balancedness, some fundamental relationships between markets, cores, and stable sets can be stated quite simply. Indeed balanced families seem to have been first introduced for the

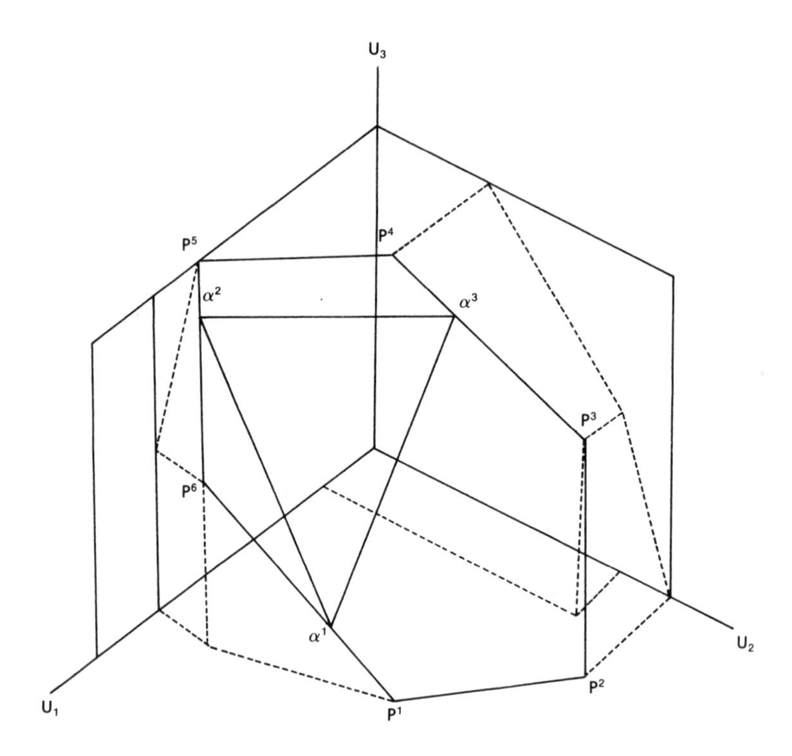

Figure 6.14
A three-person game with its bargaining set.

purpose of expressing the existence conditions for the core (Shapley, 1967a). A fundamental theorem states that *a game with transferable utility has a core if and only if it is balanced.*[51] Since totally balanced games are a fortiori balanced, we are assured of the existence of a core for any economic model that can be represented in the form of an abstract market. This is the first basic fact that we wish to stress.

The second basic fact involves the relationship between a game v and its cover $\bar{v}$, as defined above. If v is not balanced, then there is little connection between the solutions of v and $\bar{v}$. For example, $\bar{v}$ has a core and v does not. But if v is balanced, then the two games v and $\bar{v}$ turn out to have identical imputation spaces, identical cores, and identical stable sets.[52] Thus the operation of "taking the cover" gives us a way of going from balanced games to totally balanced games without disturbing either the core or any of the stable sets.

This has a direct bearing on the question of pathologies in the stable-set theory, since it means that every game with a core is "solution-equivalent" to some abstract market. Remarkably almost all of the rogues in the gallery of games with ill-behaved stable sets possess cores. (This does not mean that games with cores are more prone to pathologies, but merely that they are mathematically more tractable.) As a result, all of the anomalies that we described in section 6.5.3, including the nonexistence of stable sets, can and do occur in market models, and in relatively simple ones at that (see Shapley and Shubik, 1969a, for some explicit examples).

EXERCISE

6.21. *Show that any abstract market is equivalent, in terms of its characteristic function, to a direct market.*

6.7 Applications of the Core

Possibly the major application of core theory has come in attempts to exploit the relationship between economies with price systems and market games with cores. Because of this two-way relationship it is possible to consider an abstract game and ask, Could this game have arisen from an underlying economic structure?

Making use of the relationship between trading economies and totally balanced games, we can, for example, investigate some aspects of the political problem of logrolling. "Logrolling" refers to the exchange of votes on different issues among legislators indulging in the quid pro quo of cooperative political behavior. A natural question to ask is whether the exchange of votes can be regarded as similar to the exchange of goods in a market with a price system. This question can be answered by core theory if we rephrase it to ask if the model of logrolling can be translated into a market game. If it can, then there will exist a price system, perhaps not for votes themselves, but for some compound commodities dependent upon votes. If the game is not a market game, then no matter how "commodities" are defined there will be no market-clearing price system, and hence the idea of a market for votes as a description of the political process will not suffice. Following this approach, Shubik and van der Heyden (1978) have shown that the phenomenon of logrolling does not lend itself to a direct interpretation as a market for votes similar to a market for commodities.

Many voting problems are represented by simple games, whereas many economic problems can be modeled as market games. A key distinction appears to be that in simple games, unless there are veto players or a dictator present, there is no core, whereas for markets (without externalities) the representative game always has a core.

The operation of an economy with private corporations is characterized by a blend of voting and economic procedures. In particular, control is determined by vote, but earnings are a result of economic activity. Suppose that each stockholder holds one share of some joint activity whose net revenue is one; and for simplicity assume that the number n of stockholders is odd. The simple majority voting game can be described by the characteristic function

$$f(s) = \begin{cases} 0 & \text{if } s < n/2, \\ 1 & \text{if } s \geq n/2, \end{cases}$$

where because of symmetry we need describe coalitions only by their size. Figure 6.15a shows a representation of the characteristic function. This well-known simple majority voting game has no core. In politicoeconomic terms this is because the winners take all. There are no minority rights in this game. In contrast, corporate law is designed

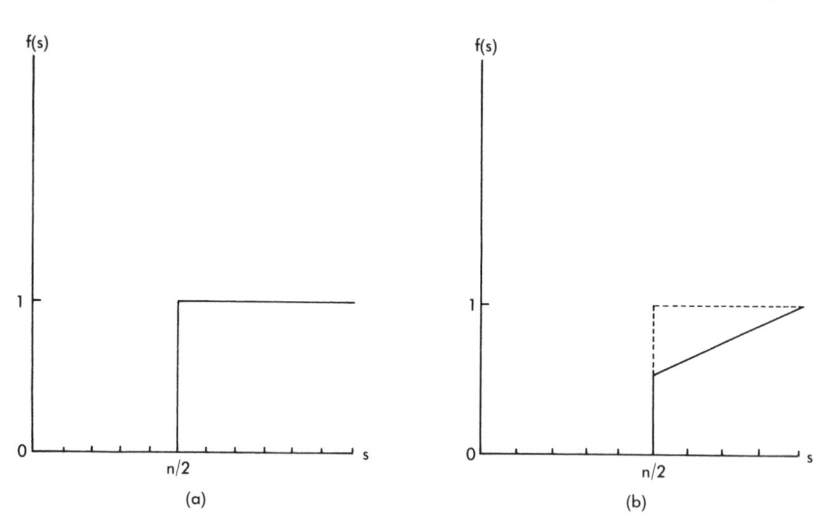

Figure 6.15
Voter protection.

to protect the rights of the minority; in particular, although control of the corporation may go to the majority, the payout of earnings must be made in proportion to shares held. The new characteristic function is

$$f(s) = \begin{cases} 0 & \text{if } s < n/2, \\ s/n & \text{if } s \geq n/2. \end{cases}$$

This is shown in figure 6.15b. We can see from figure 6.8 that this protection of minority rights is precisely what is required to introduce a core into a voting game (Shubik, 1973b). The resultant game is no longer simple.

Possibly the most striking application of core theory to economics is the demonstration of the convergence of the core as the number of competitors grows (see, e.g., Edgeworth, 1881; Shubik, 1959a; Debreu and Scarf, 1963; Scarf, 1967; Telser, 1972; Hildenbrand, 1974; Hildenbrand and Kirman, 1976).

Notes to chapter 6

1. Examples include the Shapley value, to be discussed in chapter 7, as well as a number of others to be discussed in chapter 11. Subsequent developments have freed many of these solutions from an exclusive dependence on the characteristic function.

2. To normalize $v(AB)$, for example, we subtract $v(A)$ and $v(B)$ from it, thus: \$24.60 − \$5.40 − \$8.00 = \$11.20.

3. It usually suffices that the characteristic function be constant-sum, that is, that it satisfy $v(S) + v(N - S) = v(N)$ for all S. (Note that games with non-constant-sum payoff functions may nevertheless have constant-sum characteristic functions.)

4. Quoted from McKinsey (1952a, p. 352); see also McKinsey (1952b, p. 606), Luce and Raiffa (1957, p. 190), and Isbell (1959, p. 363). McKinsey comments that "the workman could hardly expect anything but rude and stubborn resistance to such a demand."

5. McKinsey's formulation differs in detail, but it makes the same case for the inadequacy of the characteristic function.

6. The shape of the curve in figure 6.1 should dispel any idea that superadditivity, which the function $f(s)$ does enjoy, is somehow synonymous with convexity, or monotonicity, or "increasing returns to scale," which $f(s)$ certainly does not enjoy.

7. See Aumann and Shapley (1974). There is a technical trick whereby any game with a finite number of players can be represented in the form of a measure game. One merely selects the numbers m_i so that the sums $\Sigma_{i \in S} m_i$ are all different; one then tailors the function f so that (6.2) is satisfied. The concept of measure game becomes fruitful when regularity conditions of some sort can be imposed on f.

8. For a reasonably complete overview of work on "games without side payments" in characteristic-function form see Aumann and Peleg (1960), Aumann (1961b), Peleg (1963a,c, 1969), Stearns (1964a,b, 1965), Scarf (1967), Billera (1970a,c, 1971b, 1972a, 1974), Kalai (1972), Vilkov (1972), Billera and Bixby (1973, 1974), Scarf and Shapley (1973), and Shapley (1973a).

9. Billera (1972a) suggests yet another formulation in terms of support functions; see also Rosenthal (1970), Bloomfield and Wilson (1972), and Baudier (1973).

10. Convexity is not a meaningful condition if we are only using ordinal utilities, since it is not invariant under general order-preserving transformations of the utility scales (see section 4.3.1).

11. The importance of this distinction was first noted by Aumann and Peleg (1960) and, independently, by Jentzsch [see Aumann's preface to Jentzsch's posthumous paper (1964)]; the idea was foreshadowed by Aumann (1959) and, in a different (but still game-theoretic) context, by the approachability-excludability theorems of Blackwell (1956).

12. The existence of such a region may be explained intuitively by saying that the "minimax" theorem does not hold for coalitions, in the absence of transferable utility (see Jentzsch, 1964).

13. Note that the payoff to the outsider is not an issue in the alpha/beta dichotomy, whereas it is the central issue in the c-game/non-c-game dichotomy.

14. References concerning solutions and applications will be given later. For the descriptive theory see von Neumann and Morgenstern (1944), Shapley (1954, 1962b, 1964a, 1967c), M. Richardson (1956), Isbell (1956, 1957b, 1958, 1960a, 1964), Gurk (1959), Owen (1964, 1968a), Peleg (1968), and Lapidot (1972).

A simple game is just one possible interpretation of the underlying set-theoretical structure, which has been studied by mathematicians for many years under a variety of other names. These include *coherent system, Sperner collection, monotonic Boolean function, switching function,* and, most recently, *clutter*. This redundant terminology is indicative of the many widely separated applications of the basic structure; only recently have there been attempts to unite the literature and avoid further duplication of effort. For references see Dedekind (1897), Sperner (1928), Church (1940), Gilbert (1954), Birnbaum, Esary, and Saunders (1961), Birnbaum and Proschan (1963), Meshalkin (1963), Birnbaum and Esary (1965), Korobkov (1965), Hansel (1966), Lubell (1966), Fulkerson (1968), Hanisch, Hilton, and Hirsch (1969), Kleitman (1969), Billera (1970b, 1971a), Edmonds and Fulkerson (1970), Hochberg and Hirsch (1970), Levine and Lubell (1970), Bixby (1971), Butterworth (1972), and Dubey and Shapley (1979).

15. We have excluded games with dummies and games that are permutations of games already counted. If only proper games are counted, then the top row would read 1, 1, 3, 9, 72, the bottom 1, 1, 3, 9, 48.

The so-called Dedekind problem is the problem of determining for each n the number of simple games *without* eliminating permutations; for $n = 1, 2, 3, 4, 5, 6$ these numbers are 1, 4, 18, 166, 7579, 7,828,352, respectively. For a history of this problem see Hanisch, Hilton, and Hirsch (1969).

16. The complement of a losing coalition is called a *blocking coalition.* Axiom E states that every winning coalition is blocking; axiom F states that every blocking coalition is winning. An example of an improper simple game is the rule whereby only four votes out of nine are required in the Supreme Court to grant a writ of certiorari (to accept a case for hearing).

17. See, e.g., Wilson (1971a) and Bloomfield and Wilson (1972). A generalization of this structure, called the "effectiveness form" of a cooperative game, has been developed by Rosenthal (1970). See also Vorobyev (1970c).

18. See von Neumann and Morgenstern (1944, p. 263). Not all authors consistently use the term in this narrow sense; sometimes it means any payoff vector, or any feasible payoff vector.

19. For example, suppose that a finite set A of payoff vectors is given, and suppose that $V(N)$ consists of all payoff vectors β such that $\beta \leq \alpha$ for some α in A. Then the strong or "altruistic" Pareto set contains only points in A, whereas the weak Pareto set is the entire boundary of $V(N)$. Characteristic sets of this kind are called *finitely generated* (see, e.g., Scarf, 1967; Scarf and Shapley, 1973).

20. This means that by coordinated randomization the traders could achieve payoff vectors lying outside the feasible set, though not outside $V(AB)$.

21. The core concept in economics goes back to Edgeworth (1881) and Böhm-Bawerk (1891), after which it lay dormant for three-quarters of a century. In game theory the core was introduced and named by Gillies (1953b,c) and Shapley (1953b) as an adjunct to studies of the von Neumann–Morgenstern stable-set solutions. It was then developed as a general game-theoretic solution concept in its own right in lectures by Shapley at Princeton University in 1953–1954 (unpublished). Early applications to economic game models (paralleling those of Böhm-Bawerk and Edgeworth, respectively) will be found in Shapley (1955a,b) and Shubik (1955c, 1959a).

22. This definition is phrased to exclude the possibility of the core being the empty set; either there are feasible payoffs satisfying (6.11, 12) or the game has no core. We have found this way of speaking more natural in informal discourse, and no less rigorous in formal discourse, than the common alternative wherein the core always exists but may be empty.

23. Disconnected cores arise in the study of economies with multiple competitive equilibria. For general information concerning cores of games with nontransferable utility see Aumann (1961b), Stearns (1964a), Scarf (1967), Billera (1970a), Yasuda (1970), Vilkov (1972), Scarf and Shapley (1973).

24. *Proof:* Let α be any point in the core, and let $\alpha(\pi)$ be the same point with coordinates subjected to the permutation π. By symmetry all points $\alpha(\pi)$ are in the core; by convexity so is their center of gravity $\bar{\alpha}$. But $\bar{\alpha}$ is just the equal-shares imputation.

25. Economic examples in which it seems quite plausible to look outside the core for a solution have been discussed by von Neumann and Morgenstern (1944, p. 564ff), Shapley (1959b, 1961b), and Shapley and Shubik (1969a,b, 1972a), among others.

26. This will be recognized as the classical solution propounded by Böhm-Bawerk (1891).

27. Thus to arrive at the core in figure 6.9 we might first make the all-player cut, reducing the feasible set to two dimensions; then make the three singleton cuts, reducing it to the triangle of imputations; then make the three cuts for the coalitions *FM*, *FS*, and *MS* as shown by the dashed lines in the figure. Of course, the end result does not depend on the order of cutting.

28. "The central part of anything, esp. as being the residue or unused portion . . . ," *Webster's New International Dictionary,* 2nd edition.

29. Some authors use "feasible" in place of "Pareto-optimal" in these definitions.

30. See Shapley and Shubik (1966). Kannai (1968) introduces some additional varieties of ϵ-core. Analogous modifications of other solution concepts have been suggested by R. M. Starr (1969), Shishko (1970), Henry (1972), and Kalai (1972).

31. The coordinates are $(4\frac{2}{3}, 4\frac{2}{3}, \frac{2}{3})$. The reader is invited to verify that each coalition is "$\epsilon_{\min}$-satisfied" here.

32. See, e.g., Baudier (1969), who derived the least core from a bargaining argument.

33. It is necessary now to except the cases $S = N$ and $S = \emptyset$ in (6.13) and (6.14); these are vacuous when $\epsilon \geq 0$.

34. For simplicity in exposition we limit our description here to no-side-payment games in which there is a unique supporting hyperplane for each point in the imputation set.

35. The proof of this remark depends on the superadditivity of v or V or, more precisely, on a weaker condition that is implied by superadditivity (see Shapley and Shubik, 1969a).

36. This is the case, for example, for games whose characteristic functions are *convex*, that is, satisfy $v(S \cup T) \geq v(S) + v(T) - v(S \cap T)$ for all S and T [compare (6.1)]; see Shapley (1971a). Other examples are given by Gillies (1959) and Bondareva (1963).

37. It was once conjectured that the intersection of all stable sets is the core, that is, that only a core point could belong to *every* stable set (Shapley, 1953b). However, Lucas (1967b) found a five-person game with a unique stable set strictly larger than the core. Subsequently, other counterexamples have been found (see Lucas, 1968c, 1969a, 1972).

38. If we write A for the set of all imputations, and dom X for the set of points that are dominated by points in X, then the stable sets are the solutions to the equation $X = A - \text{dom}\, X$. The core, in this notation, is given by $C = A - \text{dom}\, A$.

39. A *quota game* is one in which there exists a payoff vector that exactly satisfies all two-person coalitions (or more generally, in an m-quota game, all m-person coalitions); see Shapley (1953a), Kalisch (1959), Bondareva (1963), and Peleg (1963b).

40. A stable set is *discriminatory* if there is a player whose payoff is constant throughout the set.

41. More generally, using the notation of an earlier note, we have for any stable set X:

$$C_1 \subseteq C_2 \subseteq C_3 \subseteq \cdots \subseteq X \subseteq \cdots \subseteq A_3 \subseteq A_2 \subseteq A_1,$$

where $A_1 = A$, $C_k = A - \text{dom}\, A_k$, and $A_{k+1} = A - \text{dom}\, C_k$.

42. After normalization (see table 6.2), the corresponding imputations would be (5.60, 5.60, 0), (5.60, 0, 5.60), and (0, 5.60, 5.60).

43. *Proof:* Suppose that $\alpha_A \geq 13.60$ and $\alpha_B \geq 11.00$. Since $\Sigma\alpha_i = 27$, we have $\alpha_C \leq 2.40$. But if $\alpha_C < 2.40$, then α is not an imputation, whereas if $\alpha_C = 2.40$, then α is in Y.

44. Copeland (1945) describes a stable set as a set of *trusted* imputations.

45. Shapley (1955a, 1959b) treated the symmetric Böhm-Bawerk game in detail. B. Peleg (private communication) then observed that the qualitative results carry over to the general case.

46. Nonconvexity can sometimes be an obstacle to interpretation, since chance moves or mixed strategies can be presumed to occur in the underlying strategic model.

47. See Lucas (1968b, 1969b), and also his 1972 survey. The existence ques-

tion which he thereby settled in the negative had for 25 years been regarded as the outstanding unsolved problem of cooperative game theory.

48. For example, $\{\overline{12}, \overline{13}, \overline{23}\}$ is a balanced family of subsets of $\overline{123}$, by virtue of the fractions 1/2, 1/2, 1/2. Similarly $\{\overline{123}, \overline{14}, \overline{24}, \overline{34}\}$ is a balanced family of subsets of $\overline{1234}$, by virtue of the fractions 2/3, 1/3, 1/3, 1/3. But $\{\overline{12}, \overline{13}, \overline{234}\}$ is not balanced. (See Shapley, 1967a, 1973a; Peleg, 1965a.)

49. See Scarf (1967) and Scarf and Shapley (1973). Note that the fractions f_j do not appear explicitly in (6.20), which depends only on ordinal utilities. A more restricted definition of balanced games, based on cardinal utilities, is used by Billera and Bixby (1973); both concepts specialize to (6.19) in the presence of u-money.

50. Based on an idea of D. Cantor and M. Maschler (private communication); see also Shapley and Shubik (1969a).

51. See Bondareva (1962, 1963); also Shapley (1967a) and Charnes and Kortanek (1967). Without transferable utility, balancedness (in the sense of (6.20) with $S = N$) is sufficient but not necessary for the existence of the core (see Scarf, 1967; Billera, 1970a, 1971b; Billera and Bixby, 1973; Shapley, 1973a).

52. This is because the two games exhibit the same pattern of domination, which is all that matters. The equivalence of patterns of domination is discussed by Gillies (1959) and Shapley and Shubik (1969a).

7 The Value Solution

"The value or WORTH of a man, is as of all other things, his price; that is to say so much as would be given for the use of his power."
Thomas Hobbes, *Leviathan*

7.1 The Need for a Value Solution

The solutions and presolutions discussed in chapter 6 (cores, stable sets, the characteristic function, etc.) make only modest and cautious predictions about outcomes. On many occasions, however, it is important to have a specific "one-point" solution—a single payoff vector that expresses the value of the game to each of the players. This is desirable not only for the intellectual satisfaction of coming up with a clear, sharp answer to a complex problem. It is desirable also because many applications of game theory—such as writing a political constitution or regulatory legislation, settling a case out of court by arbitration, or dividing the assets of a bankrupt—by their very nature seem to demand a straight answer, free from sociological hedging, to the question of what a particular competitive position is actually worth.

It is basic to our methodology that every prospective outcome that is considered by a player have a definite rating in his scale of preferences. But we cannot afford to maintain that there are no further games in prospect; the end of a game is not usually the end of a player's competitive career. Of course, if the future games interact significantly with the primary game we are interested in, they should be included in the model. But if the future games are essentially separate events, if the outcomes of the primary game are "game-theoretically inert," then we must be able to assign specific utility values to those future games. This is necessary as a matter of both principle and practical modeling.

The following sections describe a series of value theories; each is developed for a particular domain of application, but all of them are manifestations of a single, unified valuation concept. The fact that this concept can be developed in such a stepwise fashion will be a great convenience for exposition; each of the major difficulties will be confronted and conquered separately as the valuation concept is extended to wider classes of games.

A reader's initial impression may well be one of multiplicity and confusion.[1] To counteract this, in section 7.5 we shall recapitulate as concisely as we can the different values discussed and their domains of validity. The reader with some acquaintance with the subject may wish to turn at once to that section for orientation.

Value theory developed from two separate source ideas. One was the "maximize the product of the gains" solution to the two-person bargaining problem obtained first by F. Zeuthen (1930, chap. IV) and later (in a different way) by John Nash (1950a) and also John Harsanyi (1956). The other was the Shapley value formula as a solution for n-person games representable by von Neumann–Morgenstern characteristic functions.[2] Soon Nash (1953) was able to extend the bargaining solution to the general two-person case, but even then the two solution concepts were only barely in contact with one another, having in common only the trivial "split-the-difference" solution to the two-person, transferable-utility game. It was Harsanyi (1959) who made the decisive extension that united the two branches of the theory within a coherent single model. Subsequent developments in the field can largely be viewed as modifications, extensions, or alternative justifications of Harsanyi's synthesis.[3]

In the following pages we shall review, somewhat cursorily, the conceptual foundations of the various interrelated value theories. As in the other chapters of this volume we do not attempt to recount the theoretical developments beyond the foundational or definitional level; and although some examples are provided, we do not attempt to discuss in detail the applications of value theory that have been made in economics and political science.[4]

7.2 c-Games with Transferable Utility

When an n-person game is presumed to be adequately represented by an ordinary numerical characteristic function, then the value payoffs to the several players can depend on only the $2^n - 1$ numbers $v(S)$, as S ranges over all nonempty subsets of the all-player set. Is there any way we can deduce the functional form of this dependence?

On technical grounds we can expect the value formula to be invariant with respect to the choice of zero points on the individual utility scales and to changes in the common unit of utility measurement.[5] We can also expect the value to be externally symmetric (see section 2.1), since any differences between the players ought to make themselves felt through the numbers $v(S)$, not in the functional form into which these numbers are inserted.

Other properties, perhaps not so absolutely necessary, are nonetheless highly desirable for either technical or heuristic reasons. Thus linearity of the value formula might be assumed outright, as a technical convenience; alternatively it might be deduced from a heuristic consideration—for example, from a postulate saying that the value of a probability mix of two games should lie between the values of the separate games. Also important are feasibility and Pareto optimality, if the value is to sustain the interpretation that we wish to place on it. Similarly it is difficult to object to a postulate that a zero dummy—defined as a player d such that $v(S) = v(S - \{d\})$ for all $S \cap d$—gets nothing.

Remarkably, no further conditions beyond those noted in the last two paragraphs are required in order to determine the value exactly (Shapley, 1951b, 1953e). The sought-after formula can be written in several equivalent ways, for example,

$$\phi_i = \frac{1}{n} \sum_{s=1}^{n} \frac{1}{c(s)} \sum_{\substack{S \ni i \\ |S|=s}} [v(S) - v(S - \{i\})], \tag{7.1}$$

where $c(s)$ is the number of coalitions of size s containing the designated player i:

$$c(s) = \binom{n-1}{s-1} \equiv \frac{(n-1)!}{(n-s)!(s-1)!}.$$

An inspection of (7.1) reveals that the value of the game to a player is his *average marginal worth* to all the coalitions in which he might participate. Note, however, that all coalitions do not receive equal weight in this average.[6]

A simple procedure that realizes the value vector ϕ (in expectation) is the following: take the players in random order and pay to each his marginal worth to the coalition consisting of those who come before him (Shapley, 1951b, 1953e). Another process that realizes the value is due to Harsanyi (1959, 1963; see also Davis, 1963a): every set S of players is imagined to form a "syndicate" that declares a dividend (positive or negative) and distributes evenly among its members the difference between the syndicate's characteristic value,

$v(S)$, and the total dividend declarations of all its subsyndicates. The algebraic sum of all dividends received by a player can be shown to be exactly the value given by (7.1).

In establishing a solution concept such as the value, two complementary methods are useful. A deductive approach, such as was outlined at the beginning of this section, works with "necessary" conditions and tends to show that there is *at most* one definition that is intuitively acceptable. A constructive approach, such as Harsanyi's, works with a plausible model and endeavors to show that there is *at least* one intuitively satisfactory definition. Each method helps to justify and clarify the other (Nash, 1953, pp. 128–129).

7.2.1 Examples

The Shooting Match of section 2.2.2—or any other three-person constant-sum c-game for that matter—is easy to evaluate because of the underlying symmetry of the players. The cooperative gain must be split equally, and the value solution is simply the midpoint of the triangle of imputations. Referring to table 6.1, we see that we must add one-third of \$11.20 to each of the individual characteristic values; the result is

$$\phi_A = \$11.73, \quad \phi_B = \$9.13, \quad \phi_C = \$6.13.$$

In other words, if played as a cooperative game with side payments, the Shooting Match (with accuracies $a = 0.8$, $b = 0.6$, $c = 0.4$) has an a priori valuation that divides the prize roughly in proportion to the numerical hit probabilities.[7]

The Three-Cornered Market has no such underlying symmetry. Applying (7.1) for $i = F$, we calculate the value as follows:

Increment	Weight
$v(F) - v(\varnothing) = \$100{,}000$	1/3
$v(FM) - v(M) = \$200{,}000$	1/6
$v(FS) - v(S) = \$300{,}000$	1/6
$v(FMS) - v(MS) = \$300{,}000$	1/3

Multiplying and summing, we find that

$$\phi_F = \$216{,}667.$$

Similarly,

$$\phi_M = \$16{,}667, \quad \phi_S = \$66{,}667.$$

Thus the cooperative gain in this example ($200,000) goes 7/12 to the farmer, 4/12 to the subdivider, and 1/12 to the manufacturer—the last a measure, perhaps, of his "nuisance value." If we suppose that the purchaser makes this payment to M, then the actual selling price is $216,667.

Figure 7.1 shows the location of this solution in relation to the core and the stable sets determined in chapter 6. In this case it happens that the value is not in the core, nor in any stable set.

7.2.2 Simple games and the power index

The special class of *simple games* is an important tool in the modeling of organizational and group decision structures. These are in effect "games of control," distinguished by the property that each coalition is either "winning" or "losing." The winning coalitions are able to control the outcome of the game; and since the winning coalitions are assumed to overlap each other, only one can form. In the transferable-utility case the outcome can be regarded as a money prize that the winning coalition can distribute as it sees fit. If the model is normalized so that the amount of the prize is equal to 1, then the characteristic function is given by

$$v(S) = \begin{cases} 0 & \text{if } S \text{ is losing,} \\ 1 & \text{if } S \text{ is winning.} \end{cases} \tag{7.2}$$

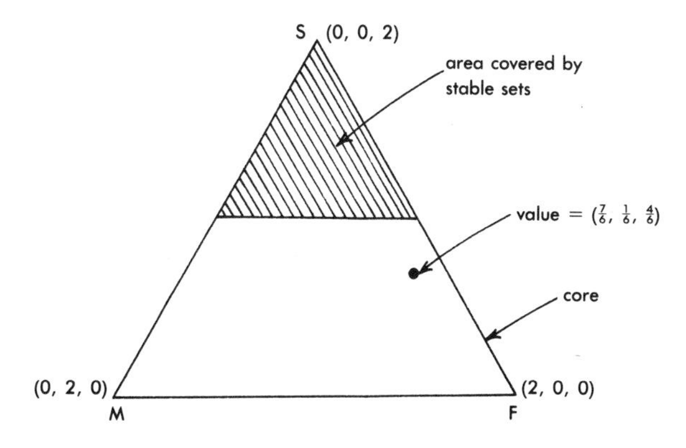

Figure 7.1
Location of the value imputation for the Three-Cornered Market.

The value solution (7.1) has an interesting interpretation when applied to simple games. Most of the marginal worths $[v(S) - v(S - \{i\})]$ are zero; only when a coalition is very close to the borderline between winning and losing will the adherence or defection of a single member make any difference. The value, then, is in effect a way of counting the opportunities a player has of being pivotal in the group decision process.

The random-order approach to the value solution can be used to make this idea precise. In any ordering of the players the coalition that is being built changes at some point from losing to winning.[8] The player who effects this change is called the *pivot* of that ordering. Now shuffle the players, like a deck of cards, so that all orderings have an equal chance. The value (7.1) to any player is precisely the probability that he will be the pivot.

It will be observed that simple games, at least in their descriptive theory, do not depend heavily on utilitarian assumptions. Simple games may be regarded as dispensers of political power more than of economic value. If the solution (7.1) is interpreted, as above, as a probability rather than as a share of a divisible prize (or even as the mathematical expectation of a lottery for an indivisible prize), then it too is virtually freed from utility considerations and becomes not a value of the game but a dimensionless measure of the distribution of power among the players that is inherent in the constitution or other group decision rule represented by the simple game. In this interpretation the value solution is called the *power index* (Shapley and Shubik, 1954).

CORPORATE CONTROL. *One man holds 40 percent of the stock in a corporation, while the rest is scattered among 600 small holders, with 0.1 percent apiece. Issues are settled by a simple majority (i.e., 50.1 percent) of the shares.*

The power indices are readily determined by the random-order method. There are 601 positions in the order. If the big shareholder finds himself in position 1–101, he is not pivotal because he and his predecessors do not have a majority of the shares. If he finds himself in position 502–601, he is again not pivotal because his predecessors already have a majority without him. In all other cases he is pivotal. Therefore his power index is 400/601, or about 66.6 percent. The little players divide the rest, so each one has a power index of (200/601)/600, or about 0.0555 percent. The big player therefore has proportionately more power than his equity position, the small players less.[9]

EXERCISE

7.1. *Generalize the preceding calculation to the case where the major player has the fraction f of the total shares, $0 < f < 1$, and the minor players are infinitesimally small.*

7.3 Non-c-Games with Transferable Utility

The foregoing value concept must be extended in two directions if it is to be generally useful, namely, (1) to strategically rich games, in which the possibilities for complicated threats or other maneuvers invalidate the c-game hypothesis, and (2) to games without transferable utility.

The first of these extensions is accomplished by a value solution first discovered and described by Harsanyi (1959, 1963), as a special case of some valuation schemes aimed more generally at the nontransferable case, and later axiomatized by Selten (1964). Selten considered games in extensive form and imposed a number of plausible requirements on the behavior of the value under various manipulations of the information and payoff structure. For example, one axiom states that a player's value cannot increase if the rules are changed to give more information about his moves to another player. We shall not attempt to do justice here to either Harsanyi's or Selten's derivation but shall merely describe a few of the properties of the result.

7.3.1 The two-person case

In the two-person case Selten's value coincides with Nash's cooperative solution, specialized to the case of transferable utility (Nash, 1953; Selten, 1960). The mathematical formulas for the Nash–Selten value can be expressed in a very simple and intuitive way that is worth sketching here.

Let Γ denote a two-person non-constant-sum game in strategic form, with payoffs (x_1, x_2) given as functions of the strategies chosen. We split the game into two parts, as follows:

$$\Gamma = \frac{1}{2}\Gamma_{\text{sum}} + \frac{1}{2}\Gamma_{\text{diff}}.$$

The strictly cooperative game Γ_{sum} has the payoff $(x_1 + x_2, x_1 + x_2)$, and the strictly competitive game Γ_{diff} has the payoff $(x_1 - x_2, x_2 - x_1)$. The Nash–Selten value $\phi = (\phi_1, \phi_2)$ of Γ is then given by the easy-to-remember pair of equations

$$\begin{aligned} \phi_1 + \phi_2 &= \text{val}\,\Gamma_{\text{sum}}, \\ \phi_1 - \phi_2 &= \text{val}\,\Gamma_{\text{diff}}. \end{aligned} \tag{7.3}$$

Here the first "val" stands for the joint maximum solution of Γ_{sum}, that is, the maximum of $x_1 + x_2$ over all strategies of both players. This ensures the Pareto optimality of Γ. The second "val" stands for the standard minimax solution of the zero-sum game Γ_{diff}. The optimal strategies of Γ_{diff}—which is a sort of "beat-the-average" or "status" game (see Shubik, 1971c)—were interpreted by Nash as *optimal threats* for use in the original game Γ. Indeed it can be argued that in a transferable-utility situation the only reasonable basis for evaluating threats is according to the difference between the damage they do to the victim and the costs they entail for the one who threatens.

The word "threat" is fairly accurate in this context but is sometimes misleadingly hostile in connotation. A threat can be thought of as merely a contingency strategy—what the player publicly intends to do if negotiations fail.

7.3.2 Examples

In the case of the Disgruntled Laborer (see section 6.2.3), we have

Γ_{sum}

	E	
W	−1000	10
	10	10

Γ_{diff}

	E	
W	−1000	0
	−10	−10

By inspection, the maximum value of the first matrix is 10, and the minimax (saddlepoint) value of the other matrix is −10. Equations (7.3) yield

$$\phi_W + \phi_E = 10, \qquad \phi_W - \phi_E = -10,$$

or

$$\phi_W = 0, \quad \phi_E = 10.$$

This is an intuitively reasonable solution. The employer's optimal threat is to resist the demand, the workman's is to forget the whole thing.

The solution changes if we reduce the self-cost of the injury, replacing -1000 by some number between -10 and 0, say -8. The workman now has a tenable bargaining position and should insist on his demand. The resultant value solution, which splits the joint gain over the "threat point" $(-8, 0)$, comes out $\phi_W = 1$, $\phi_E = 9$.

In contrast to these threat-sensitive solutions, the value obtained directly from the characteristic function, as though we were dealing with a c-game, is given by

$$\phi_W = 2.5, \quad \phi_E = 7.5,$$

regardless of the cost assigned to the laborer's threat.

For another example consider the following bimatrix game:

Γ

8, 4	0, 0
0, 0	8, 4

Γ_{sum}

12	0
0	12

Γ_{diff}

4	0
0	4

Here

$$\text{val}_s = 12, \qquad \text{val}_d = 2,$$

$$\phi_1 = \frac{12 + 2}{2}, \quad \phi_2 = \frac{12 - 2}{2}, \quad \phi = (7, 5),$$

and the optimal threat for each player is to play randomly (1/2, 1/2). The optimal bargain is for players to coordinate strategies, receive (8, 4), then transfer one unit from P_1 to P_2.

A parametric study demonstrates the sensitivity of the solution to threats. Consider the following game:

Γ

8, 4	0, 0
0, 0	8, x

Γ_{sum}

12	0
0	$8 + x$

Γ_{diff}

4	0
0	$8 - x$

Here

$$\text{val}_s = \max(12, 8 + x), \quad \text{val}_d = \begin{cases} \dfrac{4(8 - x)}{12 - x} & \text{if } x \leq 8, \\ 0 & \text{if } x \geq 8. \end{cases}$$

This yields the results in table 7.1.

7.3.3 The modified characteristic function

Returning to the case of n players, we again find a simple, almost facile way of expressing the mathematical definition of the value. We should emphasize once more that the mathematical definition given here does not stand as a basis for the theory, but follows as a consequence of more substantial analyses that are beyond the scope of this chapter.

Table 7.1

x	ϕ_1	ϕ_2	$\phi_1 + \phi_2$	$\phi_1 - \phi_2$	Optimal threat for P_2
12	10.00	10.00	20	0	0, 1
11					
10	9.00	9.00	18	0	0, 1
9					
8	8.00	8.00	16	0	0, 1
7					
6	7.67	6.33	14	1.33	1/3, 2/3
5					
4	7.00	5.00	12	2.00	1/2, 1/2
3					
2	7.20	4.80	12	2.40	3/5, 2/5
1					
0	7.33	4.67	12	2.67	2/3, 1/3
−1					
−2	7.43	4.57	12	2.86	5/7, 2/7
−3					
−4	7.50	4.50	12	3.00	3/4, 1/4
−5					
−6	7.56	4.44	12	3.11	7/9, 2/9
⋮					
$-\infty$	8.00	4.00	12	4.00	1, 0

To set up the definition, we consider each coalition S in turn and pit it against its complement, $N - S$, in a two-person game. But instead of accepting just the characteristic values $v(S)$, $v(N - S)$, which pessimistically decline to allocate the cooperative gain to either party, we use the cooperative values ϕ_S, ϕ_{N-S} from (7.3). In this way we obtain a new function for evaluating the worth of coalitions, which we shall denote by h. Unlike $v(S)$, $h(S)$ is sensitive to asymmetries in threat potential; it is, however, not a superadditive function—forming coalitions may be costly.[10] The ultimate importance of h lies in the fact that we can calculate the Harsanyi–Selten value for games with transferable utility merely by replacing v by h in the basic value formula (7.1).

In the restricted domain of zero-sum and constant-sum games, v and h are identical functions. In fact, h may be regarded as the correct extension of this restricted function *for the purposes of value theory,* just as v itself was shown by von Neumann and Morgenstern (1944, pp. 504–527) to be the correct extension *for the purposes of stable-set theory.* For non-constant-sum games, v and h are different, but in the important class of "games of consent" (to which many of the economic models of exchange belong), the difference is of no account; the same numerical result is obtained whether we use v or h in (7.1).[11]

7.4 Interpersonal Utility Comparisons

As stressed in chapter 5, bargaining or other negotiatory processes characteristically make use of interpersonal comparisons of preference intensity in one form or another. Two distinct kinds of comparisons may arise; the key to the distinction is in the relative directions of the contemplated utility changes.

The first mode of comparison, in which the utilities move together, is exemplified in remarks such as the classic slogan of parental discipline, "This hurts me more than it hurts you," or "The alliance means more to him than to anyone else," or even "Let's settle this fairly!" Gains are compared with gains, or losses with losses.

There is a contrasting mode of comparison in which a gain for one person entails a loss for someone else. Thus: "Do me a favor! It would trouble you only a little and would help me a lot." The point of this plea is that society as a whole would be better off if the favor were granted. We might say the first mode concerns *relative* welfare, the second *total* welfare; and the two principles involved might be termed *equitability* and *efficiency*.

Whether or not interpersonal comparisons are openly used during the play of a cooperative game, the outcome that is reached carries implications concerning the rates of comparison that would have had to prevail in order to explain the result as, say, equitable or efficient. In general the two principles may imply different comparison rates. A simple example will illustrate the essential point.

7.4.1 The two-person pure bargaining game

In the bargaining game depicted in figure 7.2 the two players can have any payoff vector on the curve CC' if they agree, but must take the point D if they disagree. Utilities are assumed to be cardinal but not transferable and not extrinsically comparable.

Suppose the point V were proposed as a solution. Its utilities are (0.3, 0.6), and those of D are (0, 0); therefore V implies "equitability" comparison weights in the ratio 2:1. That is, before saying that the players are sharing equally at V, we must change the ratio between their utility units by defining, say,

$$u_1' = 2u_1, \qquad u_2' = u_2.$$

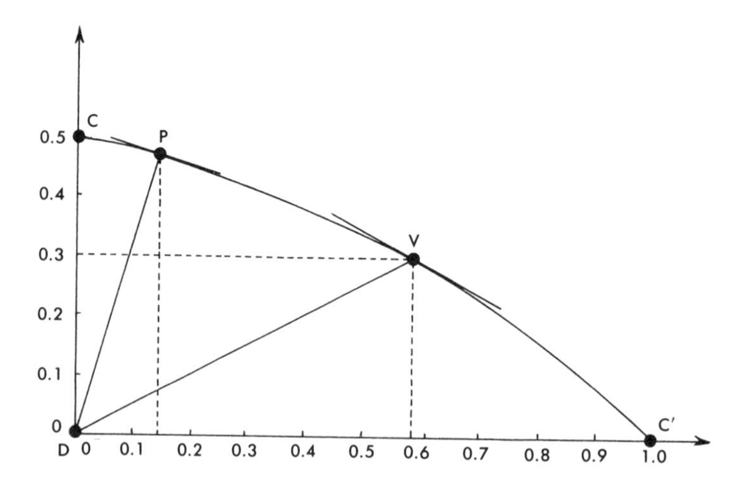

Figure 7.2
The principle of equivalence in the bargaining problem.

This rescaling gives V the new coordinates (0.6, 0.6) and leaves D at (0, 0). The new slope of the line DV is $+1$.

In contrast, the outcome P implies "efficiency" comparison weights in the ratio 1:3, since the slope of the tangent to the bargaining curve at P is $-1/3$. That is, before claiming that the players are maximizing total welfare at P, we must rescale, say, by

$$u_1'' = u_1, \qquad u_2'' = 3u_2.$$

In the new coordinates the slope of the bargaining curve at P is -1, so that P indeed maximizes $u_1'' + u_2''$.

7.4.2 The principle of equivalence

If utility is transferable, the two modes of comparison inevitably get locked into the u-money scale and hence into each other. Without transferable utility, however, the comparison rates may well disagree for a given Pareto-optimal outcome, as we have just seen. By demanding agreement, in fact, we impose a strong restriction on the class of acceptable outcomes. This is the controlling idea in the extension of the value-solution concept from transferable games to nontransferable games. It has been called the Principle of Equivalence (Shapley, 1969), and it may be stated as follows: *For an imputation to be acceptable as a value solution, there must exist positive linear transformations of the players' utility scales under which the imputation is simultaneously "equitable" and "efficient."*

In the two-person problem typified by figure 7.2, it can be shown that only one outcome is "acceptable" according to this stricture. Indeed V is the unique point on CC' at which the tangent to the curve and the line from D have slopes that are equal in magnitude and opposite in sign. This is the well-known Zeuthen–Nash bargaining solution, which can also be characterized by the property that it maximizes the product of the utility gains (Zeuthen, 1930; Nash, 1950a; Harsanyi, 1956). This solution can be given a much more solid derivation than the one sketched here; the virtue of our present argument is that it generalizes to more complex games, where strategies and coalitions play essential roles.

7.4.3 The general case

For the restricted class of pure bargaining games (i.e., games with any number of players in which unanimous consent is required for any agreement), neither variable threats nor coalitions play a role, and the Principle of Equivalence gives the answer, single-handedly as it were. For the general case, however, some more discussion is required, mainly to elucidate the idea of equitability. Without the symmetry of the pure bargaining game, the simple idea of sharing equally no longer works; it may not even be well defined. We shall invoke instead the more general notion of equity embodied in the Harsanyi–Selten value for games with transferable utility.

In order to pass from the transferable to the nontransferable theory, we shall have to contemplate the effect of introducing u-money into a game whose rules do not otherwise change. We can then speak of the "transfer value" of a game that was first formulated in the nontransferable theory, but with specific cardinal utility scales given for the players. Of course, there is no assurance that the transfer value is a feasible outcome in the original formulation. Assuming for the moment that it happens to be feasible, though, we would argue that it is also equitable, since it is equitable (by assumption) in the presence of side payments, and disallowing side payments only removes some irrelevant alternatives from the set of outcomes. Moreover, we would argue that no efficient outcome, other than the transfer value, can be equitable, since it would give more than the transfer value to at least one player and less to at least one other player, and this is true whether or not the transfer value is feasible.[12] (Figure 7.3 suggests the reason for this.) Finally, we note that the transfer value, if it is feasible, is automatically efficient.

Thus, under the principle of equivalence, the problem of evaluating a game without transferable utility is equivalent to the problem of rescaling the individual utilities so that the transfer value can be achieved without using side payments. The existence of such rescaling can be proved by topological arguments (Shapley, 1969). It is not always unique, but a naive counting of equations and variables indicates that, barring "degeneracy," there should not be more than a zero-dimensional set of solutions (cf. Debreu, 1970).

There is a strong analogy, though no formal equivalence that we know of, between the comparison weights that we must introduce in order to obtain a feasible transfer value and the prices in a competitive market. It is as though the players in the nontransferable game introduced a money-of-account as an aid to rational bargaining. Treating their personal utilities as commodities, consumable only by themselves, they set the prices in such a way that, when the dust has settled and all the bargaining is over, all accounts are miraculously in

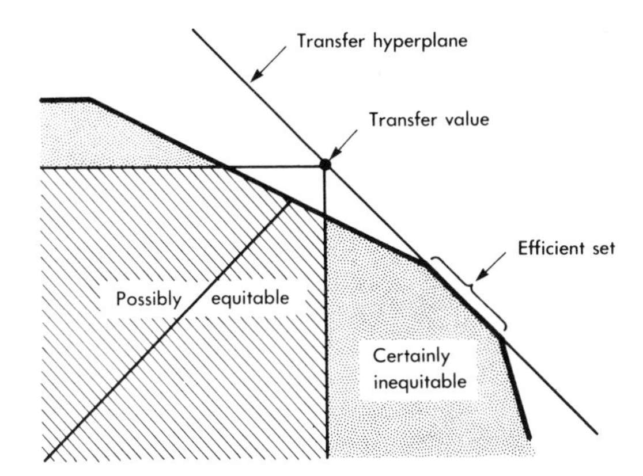

Figure 7.3
Illustrating the proposition that if the transfer value is not feasible, then the feasible efficient points are not equitable.

balance. As in so many other parts of value theory, there are several alternative approaches to, and justifications for, this final extension, having much the same net mathematical result.[13]

7.5 Recapitulation

Not surprisingly, several other solution concepts have been proposed that do not fit into the general scheme outlined in this chapter; these will be taken up in chapter 11. But even those that do fit into the present scheme go under a variety of different names and may give the impression of great multiplicity. To help reduce confusion, and also to underline the essential unity of the theory, we shall summarize the chapter to this point with an organized list of capsule descriptions.

In the simplest possible case, in which two individuals with transferable utilities and a fixed disagreement point bargain over the division of a specified gain, "split the difference" is the only reasonable solution. Moving out from this basic case, there are three distinct complications that we may encounter, singly or in combination:

I. More players can be added, raising the question of coalitions.

II. More strategies can be added, raising the question of threats.

III. Utility can be made nontransferable, raising the question of interpersonal comparability.

In figure 7.4 this is shown three-dimensionally: I points into the page, II points up, and III points to the right. The shaded blocks represent new territory covered by each of the successive extensions.

The *Shapley value* extends "split the difference" in direction I only. To obtain it we calculate the average marginal worths of the players (7.1).

The *Nash–Selten value* (not shown in figure 7.4) extends in direction II only. It can be determined by solving the "sum" and "difference" games as in (7.2).

The *Zeuthen–Nash value* extends in direction III only. It can be obtained by maximizing the product of the utility gains.

The *Harsanyi–Selten value* covers the combination I and II. It may be obtained by applying formula (7.1) to the modified characteristic function h.

The *Nash cooperative value* (Nash, 1953) covers the combination II and III. To obtain it we split the game into a noncooperative threat game and a cooperative pure bargaining game, generalizing the procedure described in section 7.3.1.

The *general value* covers I, II, and III together; to find it we introduce weights and search for a feasible Harsanyi–Selten value.

The brief indications on how to obtain the various values are intended mainly as mnemonic aids. In particular cases some other method of calculation may be more efficient or more explanatory as regards the meaning of the solution.

The value has been extended to many classes of infinite-person games by a variety of different procedures. Most of the work to date has dealt with c-games with transferable utility (Milnor and Shapley, 1961; Shapley, 1961a, 1962c; Kannai, 1966; Aumann and Shapley, 1968, 1969, 1970a, 1974; Rosenmüller, 1969; Kohlberg, 1970; Aumann, 1975; Dubey, Neyman, and Weber, 1981).

7.6 Other Approaches to Two-Person Bargaining

The two contributions by Nash (1950a, 1953) giving axioms for the two-person bargaining game without side payments and with constant or variable threats provided a considerable step forward in our understanding of the subtleties of what is meant by "fair" division. In the modeling of behavior, however (as we have already seen in chapter 5 when considering the Arrow impossibility theorem), certain results hinge with considerable delicacy on an axiom which may

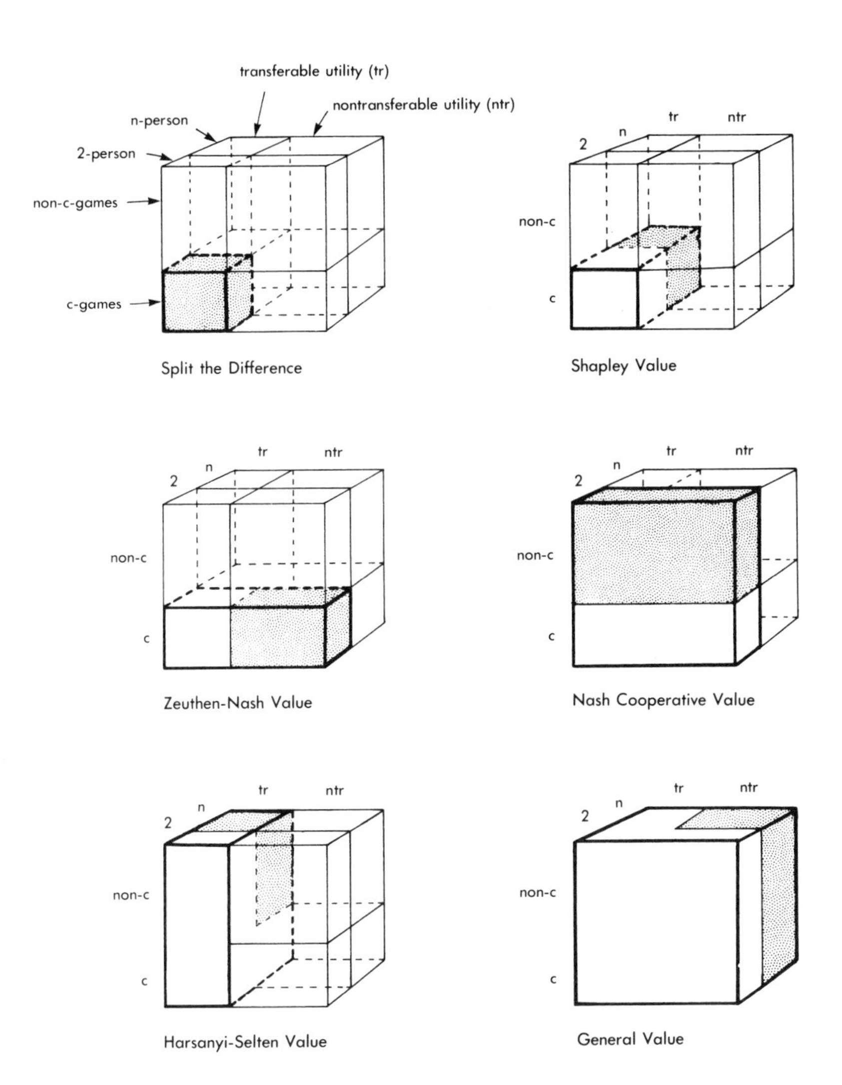

Figure 7.4
Coverage of the various value definitions.

appear to be reasonable but which could be replaced by another axiom that appears to be equally reasonable. The replacement of one reasonable axiom by another may give radically different results.

7.6.1 The Kalai–Smorodinsky Bargaining Game

Two further two-person fair-division schemes with somewhat different axioms have been suggested. Kalai and Smorodinsky (1975) have proposed a fair division based upon the following five assumptions:

1. invariance of the solution to linear transformations of the utility function;
2. Pareto optimality or efficiency of the solution point;
3. the existence of a natural no-settlement point;
4. symmetry;
5. monotonicity with respect to the solutions of related games.

The first four assumptions are the same as those used by Nash in his 1950 paper. The fifth assumption differs from Nash and is used instead of his assumption of the independence of irrelevant alternatives. It is illustrated in figure 7.5. Start with a game with Pareto-optimal surface A_1CA_2. Now consider a new game with Pareto-optimal surface A_1DA_2, differing from the previous game in that, for any fixed payoff to one player, the other player will obtain as much as in the first game in all instances or more in at least some instances. The assumption of monotonicity is such that in the second case the solution will assign at least as much as or more than it does in the first.

It can be shown that a unique point satisfying these assumptions exists. It is the point C (or D) given by the intersection of the diagonal joining the origin to the northeast vertex of the rectangle.

This scheme is essentially the same as that suggested by Raiffa (1953) who, without giving axioms, described a finite and a continuous process for selecting a final division point. Figure 7.6 illustrates the finite process for the following game:

6, 8	2, 9	0, 9
10, 0	9, 4	2, 2

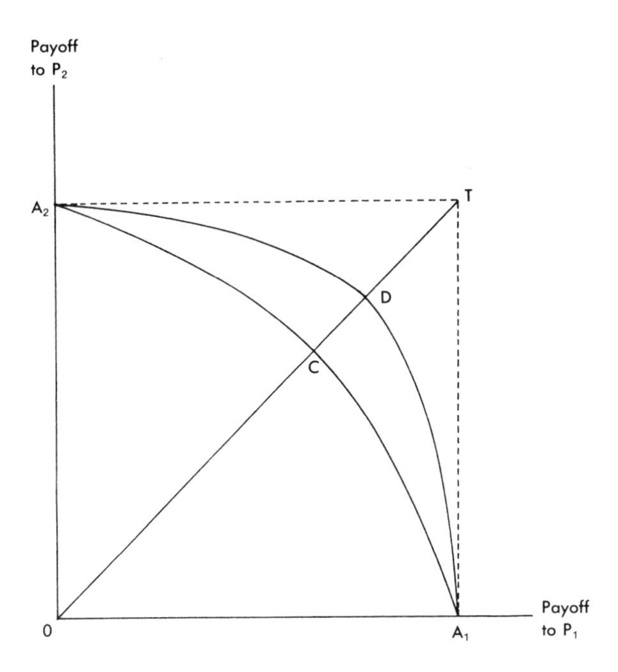

Figure 7.5
Illustrating the monotonicity assumption.

The natural zero is at (2, 2), which is the minmax of the difference game based on the above matrix, that is,

				min
	−2	−7	−9	−9
	10	5	0	0
max	10	5	0	

The players each see which point on the optimal surface they would prefer; here P_1 would like ($9\frac{1}{2}$, 2) and P_2 would prefer (2, 9). They average to a new point S_1 ($5\frac{3}{4}$, $5\frac{1}{2}$), after which they average to the midpoint of ($5\frac{3}{4}$, $8\frac{1}{16}$) and ($7\frac{7}{8}$, $5\frac{1}{2}$), which is S_2 ($6\frac{13}{16}$, $6\frac{24}{32}$), and finally after one more step this leads to S_3 ($6\frac{221}{256}$, $6\frac{163}{192}$).

We observe that this process depends upon the extreme points of

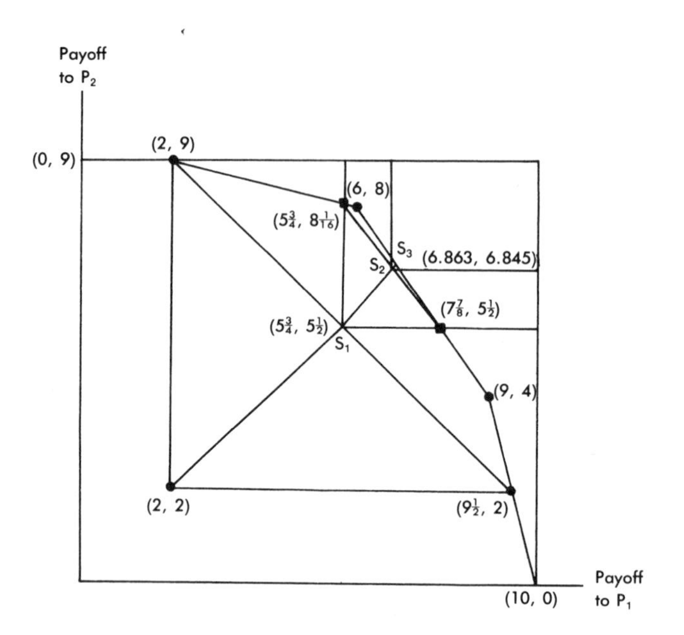

Figure 7.6
The Kalai–Smorodinsky Bargaining Game.

the Pareto-optimal surface and certainly violates the axiom of the independence of irrelevant alternatives used by Nash.

7.6.2 The Maschler–Perles Bargaining Game

Maschler and Perles (1978) use the same first four axioms but replace the fifth with a "superadditivity axiom" of the form:

5′. If the value of one game is given by u^S and that of another by u^T, and if the two games are considered as one, then $u^{S+T} \geq u^S + u^T$.

It can be proved that this scheme also leads to a single-point solution in which equal areas are swept out in the same time as the settlement closes. An example is shown in figure 7.7, which uses the same game as figure 7.6.

We may imagine two donkeys starting out at the ends of the Pareto-optimal surface, A_1 and A_2, respectively. They walk along the Pareto-optimal surface at variable speeds but in such a manner that the area encompassed by the product of each donkey's speeds in the horizontal and vertical directions is constant.[14] This description must be

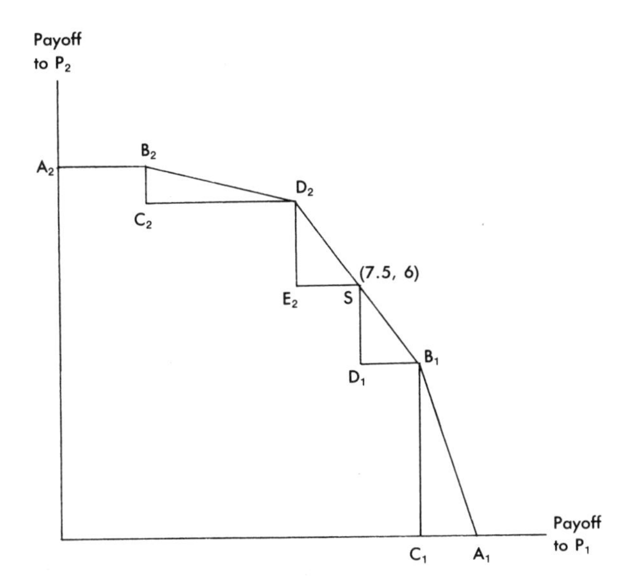

Figure 7.7
The Maschler–Perles Bargaining Game.

supplemented by the convention that a completely flat or vertical portion of the optimal surface is traversed in zero time. Thus the second donkey essentially starts at B_2.

We observe that the area $B_2C_2D_2$ equals $A_1B_1C_1$, and D_2E_2S equals B_1D_1S; hence the solution is at the point S with components (7.5, 6). The immediate equality of the areas of the first two triangles, each spanning the distance between two vertices (B_2 and D_2, A_1 and B_1), is an accidental property of this simple example.

The Maschler axiom 5′ is in the spirit of or even stronger than the Shapley axiom of additivity noted in section 7.2. Maschler's justification for the axiom is that if an individual faces a possibility of being confronted with two completely separate games, then as an inducement for him to settle by arbitration before Nature has moved, the expected value from arbitration in the compound game should be as high as or greater than the sum of the expected settlements from the two separate games.

It may be argued that this justification for axiom 5′ owes its appeal to a fuzziness in the specification of the strategic form of the game to be arbitrated. In particular, we must decide whether the choice to arbitrate before or after Nature has selected the final game is part of the strategic possibilities available to the players.

7.6.3 Summary

Nash originally suggested five axioms for the two-person bargaining game with fixed threats. They led to a fair-division scheme calling for a maximization of the product of the individuals' utilities. The one axiom that has been challenged has been the axiom of the independence of irrelevant alternatives. In figure 7.8 the Nash solutions for these two games are the same, yet some might feel that intuitively P_2 is in a stronger position in the second game than in the first. The challenge to this axiom has taken the form of its replacement by one of two others noted in sections 7.6.1 and 7.6.2. These three systems will give the same prediction for a totally symmetric bargain, but even for a relatively simple matrix game such as the one discussed above they will all differ.

It is simple to calculate the Nash outcome by extending the line B_1D_2 in figure 7.7 and taking the midpoints of the base and altitude of the resulting triangle with the axes. This is the point (6, 8). Thus we have for the three models:

Nash:	(6, 8),
Kalai–Smorodinsky:	(6.86, 6.85),
Maschler–Perles:	(7.5, 6).

7.7 Variations of the Value

A fruitful domain of application of the value has been to problems in voting. For the most part it has been possible to use the class of simple games with side payments to provide a basis for the models considered in such applications.[15]

7.7.1 The Banzhaf power index

The Shapley value was originally constructed for application to general games in characteristic-function form. Shapley and Shubik (1954) considered its application to the class of simple games and proposed it as an a priori measure of power in a voting system. This measure has come to be known as the *Shapley–Shubik power index.*

Prior to the Shapley–Shubik index, von Neumann and Morgenstern (1944) had considered a specific type of stable-set solution

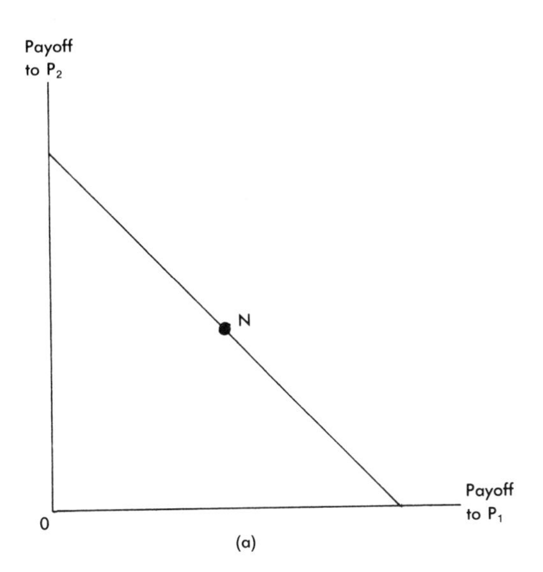

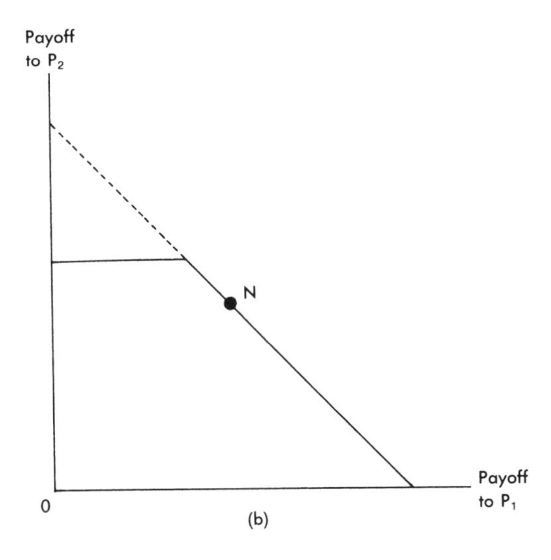

Figure 7.8
Illustrating the axiom of irrelevant alternatives.

applied to an important class of simple games (see chapter 6). This "main simple solution" has a direct interpretation in terms of prices in a vote-selling situation where the imputations describe how the spoils are divided among the winners.

A new and somewhat different index was suggested by Banzhaf (1965). His motivation was to help solve certain legal battles concerning the standards of constitutional fairness for systems of electoral representation at both the state and local levels of government. The *Banzhaf power index* is sometimes called the Banzhaf value, and we shall discuss below its interpretation as a variant of a value applied to a general characteristic function; but in both its intended and its actual usage it is more exactly a power index, and it is naturally contrasted with the Shapley–Shubik index.

The key element in the construction of the Banzhaf index is a *swing*. A swing for player i is a pair of sets $(S, S - \{i\})$ such that S is a winning set and $S - \{i\}$ is not. For each player i in the set of all players N, we may calculate a number $\eta_i(v)$ that represents the number of swings for i in the simple game with characteristic function v. Following the notation suggested by Dubey and Shapley (1979) we let $\bar{\eta}(v)$ be the total number of swings in the game

$$\bar{\eta}(v) = \sum_{i \in N} \eta_i(v).$$

A player with $\eta_i(v) = 0$ is called a *dummy* since he is never needed to help a coalition win. A player with $\eta_i(v) = \bar{\eta}(v)$ is called a *dictator*.

The swing numbers form the nonnormalized Banzhaf index. Since our usual concern is with relative voter power, it is useful to define a *normalized Banzhaf index* such that the weights add to one:

$$\beta_i(v) = \eta_i(v)/\bar{\eta}(v) \quad (i = 1, 2, \ldots, n).$$

Dubey and Shapley (1979) point out some problems with this normalized form and suggest an alternative normalization,

$$\beta'_i(v) = \eta_i(v)/2^{n-1} \quad (i = 1, 2, \ldots, n),$$

which may be deemed more natural. Suppose that each individual tosses a coin to decide whether to vote yes or no. The set of yes votes, Y, is then a random variable giving a probability of $1/2^n$ for any outcome. We call i a *swinger* if his vote affects the passage of the bill. Thus i is a swinger if $(Y \cup \{i\}, Y - \{i\})$ is a swing for i. If S is a set containing i, there are exactly two sets Y such that $Y \cup i = S$; thus the probability of a particular swing for i is twice $1/2^n$ or $1/2^{n-1}$.

Hence the β_i' measure precisely the probability that a player is a swinger.

In contrast, the probability model for the Shapley–Shubik index considers all possible orders in which a vote can take place. For any ordering of the players there will be a unique player who is marginal, that is, who is in a position to provide the coalition with just enough strength to win. He is the *pivot* for the coalition. If a priori all $n!$ orderings are assumed equiprobable, then the Shapley–Shubik index is a measure of the probability that any player is pivotal.

A simple example will clarify the two indices. Suppose that the game has four players with votes of 3, 2, 1, and 1. Five votes are needed to carry a motion. The game is denoted (5; 3, 2, 1, 1), where 5 is the quota and the votes are the weights of the players. For ease in the simple enumeration that follows we use 3 and 2 as the names of the more powerful players but A and B as the names of the two single-vote players.

Table 7.2 illustrates all of the 2^4 possible swings that are needed to calculate the Banzhaf index. The straight normalization is $\beta = (0.5, 0.3, 0.1, 0.1)$. The second normalization is $\beta' = (1.25, 0.75, 0.25, 0.25)$.

In contrast, we may calculate the Shapley–Shubik index as follows:

$3\dot{2}AB$ $2\dot{3}AB$ $2A\dot{3}B$ $2AB\dot{3}$

$3\dot{2}BA$ $2\dot{3}BA$ $2B\dot{3}A$ $2BA\dot{3}$

$3A\dot{2}B$ $A3\dot{2}B$ $A2\dot{3}B$ $A2B\dot{3}$

$3A\dot{B}2$ $A3\dot{B}2$ $AB\dot{3}2$ $AB2\dot{3}$

$3B\dot{2}A$ $B3\dot{2}A$ $B2\dot{3}A$ $B2A\dot{3}$

$3B\dot{A}2$ $B3\dot{A}2$ $BA\dot{3}2$ $BA2\dot{3}$

In each of these 24 orderings the pivot is represented by an overdot above the critical player; thus the Shapley–Shubik index is $(\frac{14}{24}, \frac{6}{24}, \frac{2}{24}, \frac{2}{24})$. We note that the game (5; 3, 2, 1, 1) is not *decisive*; that is, the complement of a losing coalition is not necessarily winning (see chapter 6).

EXERCISE

7.2. *For the decisive game (3; 2, 1, 1, 1), verify that the normalized Banzhaf and Shapley–Shubik indices are the same.*

We note from the example that the Shapley–Shubik index depends on equiprobable permutations of N, whereas the Banzhaf index de-

Table 7.2
A four-person weighted voting scheme

Voter					Swinger			
3	2	*A*	*B*	Outcome[1]	3	2	*A*	*B*
Y	Y	Y	Y	P	X			
Y	Y	Y	N	P	X	X		
Y	Y	N	Y	P	X	X		
Y	Y	N	N	P	X	X		
Y	N	Y	Y	P	X		X	X
Y	N	Y	N	F		X		X
Y	N	N	Y	F		X	X	
Y	N	N	N	F		X		
N	Y	Y	Y	F	X			
N	Y	Y	N	F	X			
N	Y	N	Y	F	X			
N	Y	N	N	F	X			
N	N	Y	Y	F	X			
N	N	Y	N	F				
N	N	N	Y	F				
N	N	N	N	F				
Total				5P, 11F	10	6	2	2

1. P = pass; F = fail.

pends on equiprobable combinations of N. Each permutation produces exactly one pivot, but a single combination rarely produces a single swinger. Thus there is an inherent additive measure in the Shapley–Shubik index which is not present in the Banzhaf index.

Dubey (1975) has provided an axiomatic treatment of the Banzhaf index.

7.7.2 Further considerations of power indices

Among the earliest attempts to define power in a manner congenial to game-theoretic analysis was that of Dahl (1957).[16] He suggested as a measure of the power of individual i over j the extent which i can get j to do something he would not otherwise do minus the extent to which j can get i to do something he would not otherwise do. Dahl mathematizes this measure using conditional probabilities. Allingham (1975), in his application of the Dahl notion to voting systems in general, observes that it may be viewed as a variant formulation of the Banzhaf index.

Coleman (1971), in his consideration of power, distinguishes the ability to prevent and the ability to initiate action. Thus he calculates two indices,

$$\gamma_i = \eta_i/\omega \quad \text{and} \quad \gamma_i^* = \eta_i/\lambda,$$

where ω is the number of winning and λ the number of losing coalitions in N. These are directly related to the Banzhaf index by

$$\frac{1}{\beta_i'} = \frac{1}{2}\left(\frac{1}{\gamma_i} + \frac{1}{\gamma_i^*}\right).$$

Rae (1969) approaches power by examining the responsiveness of a voting system to the electorate. His essential idea is to count the number of ways in which a voter can find his vote in agreement with the outcome of the vote. Denoting Rae's index of agreement by ρ_i, it can be shown that

$$\rho_i = 2^{n-1} + \eta_i, \quad \text{where } n = |N|.$$

In a completely different context, a relationship between the Banzhaf index and the literature in threshold logic and switching functions has been observed (Winder, 1968, 1969). In particular a simple game may be viewed as corresponding formally to a positive switching function in which voting yes or no is reinterpreted as a switch being on or off and a winning coalition is interpreted as the system functioning. Chow (1961) has calculated a set of parameters for electrical-engineering purposes which turn out be directly related to the Banzhaf index for many games; these are known as *Chow parameters*.

Before leaving the subject of power indices, a few general remarks are in order. The power to initiate and pass and the power to block are clearly different phenomena, but the difference possibly lies more in the domain of dynamics than has been indicated by the models in this chapter. The reasons why voting schemes for the election of candidates tend to give rise to decisive simple games, such as simple majority rule, whereas constitutional amendments call for nondecisive simple games with two-thirds or three-fourths or other large majorities, appear to be closely related to the desire to give stability to political systems. In certain systems such as the British House of Lords the structure does nothing more than impose a delay forcing the other house to reconsider its action.

In modeling political or social phenomena the use of the word "power" in the discussion of indices may be unfortunate. Political or social power is clearly a multidimensional concept encompassing items such as "the power to persuade." Furthermore, political power, social power, and economic power appear to be somewhat different and partly, but by no means completely, interchangeable.

7.7.3 Other values

The Shapley value treats all the permutations of coalition formation as equiprobable. It may be argued that this treatment is "sociologically neutral," that it implies no biases whatsoever in the coalitional structure. We may contemplate a modification of this treatment based upon the idea of a fuzzy game (Aubin, 1977) and a multilinear extension of the value (Owen, 1972, 1975).

Suppose we represent the set $S \subset N$ by a vector $(s_1, s_2, \ldots, s_n)$, where

$$s_i = \begin{cases} 1 & \text{if } i \in S, \\ 0 & \text{if } i \notin S. \end{cases}$$

We may picture the 2^n subsets of N as the vertices of a cube. This is shown in figure 7.9 for $n = 3$. Any point other than a vertex may be regarded as representing a fuzzy coalition, to which each player belongs with an intensity between 0 and 1. Thus a point in the cube may be viewed as a random variable giving rise to a coalition formed by including each player i independently with probability p_i.

A real-valued function defined on the whole cube and having the value 0 at 0 is called a *fuzzy game*. Owen's multilinear extension may be viewed as a type of fuzzy game that is useful in studying generalizations of the value. The fuzzy game can be considered a linear interpolation that extends v first to the edges of the n-cube, then to the two-dimensional faces, and so on. All of the interpolation uses lines parallel to the coordinate axes, as shown for the point (x, y) whose coordinates may be expressed as a linear combination of H and J, which in turn are linear combinations of AO and BC.

The general formula for a multilinear extension is

$$\tilde{v}(s_1, s_2, \ldots, s_n) = \sum_{T \subset N} \left\{ \prod_{i \in T} p_i \prod_{i \in T} (1 - p_i) v(T) \right\}.$$

This yields the *expected worth* of the random coalition. We may denote the extension as

$$v(p_1, p_2, \ldots, p_n) = E_p v(\tilde{s}),$$

where $\tilde{s}$ is a fuzzy coalition.

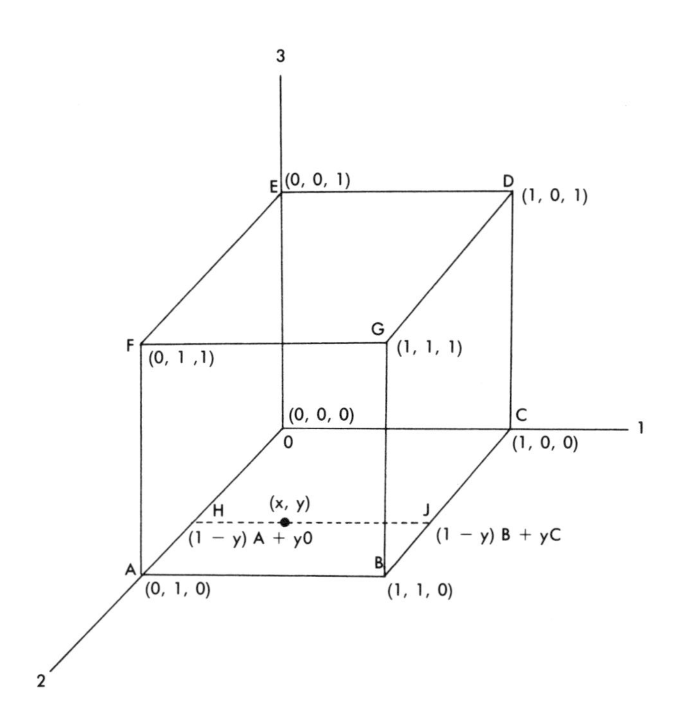

Figure 7.9
Owen's multilinear extension of the value.

A special class of extensions lie on the main diagonal of the cube, denoted by $(t, t, \ldots, t)$, $0 \leq t \leq 1$. The marginal contribution of player i to this fuzzy coalition is

$$\frac{\partial}{\partial x_i} \tilde{v}(t, t, \ldots, x_i, \ldots, t)$$

evaluated at $x_i = t$, but by the multilinearity of $\tilde{v}$ this is

$$\tilde{v}(t, t, \ldots, 1, \ldots, t) - \tilde{v}(t, t, \ldots, 0, \ldots, t).$$

If $T(t)$ denotes the random subset $N \backslash i$ (the set N excluding i), then the probability that $T(t) = s$ is $t^s(1 - t)^{n-s-1}$. For any t we obtain a member of a class of values

$$\phi_i = \sum_{S \subset N|i} t^s(1 - t)^{n-s-1}[v(S \cup i) - v(S)].$$

For $t = 1/2$ this gives the (unweighted) Banzhaf value

$$\beta_i = \sum_{S \subset N \setminus i} \frac{1}{2^{n-1}}[v(S \cup i) - v(S)].$$

The Shapley value may be regarded as an average over all the t-values; that is, if we integrate with respect to t, then the probability of S is $s!(n - s - 1)!/n!$, which is the coefficient in the Shapley value.

The Shapley value is characterized by axioms of symmetry, linearity, efficiency, and treatment of dummy players. Variations on these axioms have been considered (Dubey, 1975; Dubey, Neyman, and Weber, 1981). The linearity and dummy axioms together with a monotonicity axiom give the class of *probabilistic values* in which payments do not necessarily sum to an efficient outcome and individuals may have different evaluations of the chances of coalition formation. If an efficiency axiom is included (but symmetry is still left out), then the family of *random-order values* is obtained.

This type of value, with different probability weightings, has also been studied for games with a continuum of players, and a class of *semivalues,* which are probabilistic values that also satisfy a symmetry condition, has been defined and explained (Dubey, Neyman, and Weber, 1981).

Although the Banzhaf index has been suggested as an alternative to the Shapley–Shubik index and has been considered by some political scientists (e.g., Allingham, 1975) and the courts as the more naturally applicable index for voting problems, several considerations suggest that the Shapley–Shubik index has desirable properties not shared by the Banzhaf index. Weber (1979) has shown that the Shapley value is the unique semivalue arising from consistent expectations of all of the players. Furthermore, as Dubey and Shapley (1979) have shown for oceanic games with one or two major (atomic) players, a parametric exploration yields a much less erratic and intuitively more satisfactory change for the Shapley–Shubik index than for the Banzhaf power index.

It should be stressed that the selection of weightings in the construction of a specific probabilistic value is more a matter of modeling than of mathematics. The more axioms satisfied by the ad hoc selection of probabilities, the more a priori confidence we may attach to it. But there may be a tradeoff between using specific outside knowledge of the phenomenon being studied and using the axiom system selected.

In section 10.2.3 it is shown that t-values may be derived in a strikingly different way through use of a noncooperative model of the game.

7.7.4 A comment on no-side-payment values

The attitude adopted in this volume is eclectic. We do not propose a single solution theory to be applied under all circumstances. In particular we believe it worthwhile to explore specific examples in order to check whether the outcomes suggested by theory and intuition match.

A. E. Roth (1980) has suggested a class of three-person no-side-payment games having a no-side-payment solution that may seem unsatisfactory. Consider a three-person game denoted by $\Gamma(p)$ in which all players acting alone obtain nothing; P_1 and P_2 acting together obtain (1/2, 1/2, 0); P_1 and P_3 obtain $(p, 0, 1 - p)$, $0 < p < 1$; and P_2 and P_3 obtain $(0, p, 1 - p)$. Together all three can obtain any point in the convex hull of $\{(0, 0, 0), (1/2, 1/2, 0), (0, p, 1 - p), (p, 0, 1 - p)\}$, where $0 \leq p \leq 1$. For $0 \leq p \leq 1/2$ Roth argues that the unique outcome acceptable as a solution for this game is (1/2, 1/2, 0), because P_1 and P_2 can agree on that outcome as preferable to any other and can obtain it. Yet when we consider the λ-transfer value for this game, there is a value $\lambda = (\lambda_1, \lambda_2, \lambda_3) = (1, 1, 1)$ which gives an associated (side-payment) characteristic function of

$$v(i) = 0 \quad (i = 1, 2, 3),$$

$$v(12) = 1/2 + 1/2 = 1,$$

$$v(13) = p + (1 - p) = 1,$$

$$v(23) = p + (1 - p) = 1,$$

$$v(123) = 1,$$

which has a (side-payment) value of (1/3, 1/3, 1/3). But this can be obtained without side payments; hence, over the range $0 \leq p \leq 1/2$ this λ-transfer value is insensitive to changes in p.

If we choose weights of $\lambda = (1, 1, 0)$, then the λ-transfer value is (1/2, 1/2, 0).

When $p = 1/2$, Roth notes that the game is completely symmetric, hence the solution (1/3, 1/3, 1/3) is intuitively acceptable. As soon as $p = 1/2 - \epsilon$, Roth opts for (1/2, 1/2, 0). Depending on the λ-values chosen, the λ-transfer value suggests either of these outcomes. Yet one might feel that as p varies over a small range, the λ-value should move smoothly or not at all rather than discontinuously.

7.8 Applications of Values and Power Indices

Value and power-index theory have been applied to political science, to economics and accounting, and to military problems. Here we shall note but not develop some of these applications. In particular the Shapley axioms can be reinterpreted in terms of principles of accounting in the design of joint-cost allocations and incentive systems (Shubik, 1962a).

A natural question for the economist to ask is how the value compares with the core. We have seen from the example in figure 7.1 that the value is not necessarily in the core; a simple example serves to contrast the two solutions in an economic context. Consider three individuals 1, 2, 3 who by cooperation can earn the following amounts:

$$v(1) = v(2) = v(3) = 0,$$
$$v(1, 2) = 0, \quad v(1, 3) = v(2, 3) = 100,$$
$$v(1, 2, 3) = 100.$$

It is easy to check that the core consists of a single point (0, 0, 100), whereas the value is (100/6, 100/6, 400/6). This seems to be intuitively reasonable as an economic imputation of wealth inasmuch as it attaches some value to the contributions of players 1 and 2 in the two-person coalitions (1, 3) and (2, 3), whereas the core allocates zero to them; and if we cast this game in terms of a market where 1 and 2 are selling labor to 3, we find that the price of labor is zero and the imputation selected by the market is precisely that of the core.

Two classes of application of the value depend essentially upon competitive forces determining the probability that certain coalitions form. One is a direct application to voting in which position along an ideological spectrum influences the behavior of the players (Shapley, 1977). The other is a military application in which the n players are interpreted as nodes in a connected military network and the probability that they will be in the final coalition is interpreted as a probability that a particular set of nodes survive after they have been attacked and defended by the opposing forces. The model of the attack and defense, which can be cast as a two-person noncooperative game, will be treated explicitly in section 10.2.3.

The application to voting with ideological differences accounted for was originally suggested by Owen (1971) and extended by Shapley

(1977). The political dimensions are represented by an m-dimensional "issue" or "ideology" space R^m. A one-dimensional space, R^1, might represent a simple political spectrum stretching from an extreme right to a left wing.

We consider political issues ξ and political profiles x^i each as m-dimensional vectors. We assume that each issue is broken down into its component contents by the voters. Thus in figure 7.10, suppose that the two dimensions of concern are attitude toward foreign policy and attitude toward budget policy. We can describe the intensity of each voter's position by a point in this space.

Consider issue 1: we line up the votes by drawing in the perpendiculars from each voter's "ideology point." Thus the lineup is given by 1, 2, 5, 3, 4. For issue 2 the lineup is 1, 2, 3, 4, 5. In an important way we see here that ideological strength weakens one's chances for being pivotal. The stronger one feels about an issue, the higher the probability that one is early in the voting lineup.

For a three-person voting game with all individuals i having distinct positions x^i in a two-dimensional ideology space, we can describe the different voting orders by a relatively simple circular diagram in

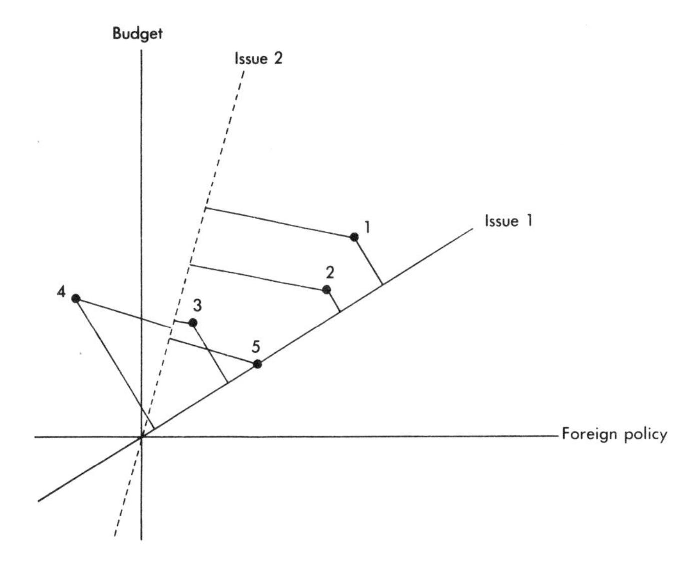

Figure 7.10
An ideology space.

which we imagine a pointer being spun at random, with the point where it comes to rest giving the direction of the ideological mix of the issue selected. Consider the three instances shown in figure 7.11. In the first instance the three voters' ideological views lie on a line; in the second on a triangle with angles α, β, γ; and in the third on an equilateral triangle. No matter in which direction the issue points, if the voters' views are on a line as in figure 7.11a, then the lineup will always be *ABC*, and for a simple majority vote game (2; 1, 1, 1) *B* will *always* be pivotal. Here the center controls the outcome between the balanced extremes. For the other two games the changes in the orderings are proportional to the angles in the triangle. Thus the modified value for the game shown in figure 7.11b is $(\alpha/\pi, \beta/\pi, \gamma/\pi)$. For the equilateral triangle in figure 7.11c the value is the Shapley value (1/3, 1/3, 1/3). This device thus generates all of the semivalues noted in section 7.7.3.

Another natural application has been to corporate voting, in which there are often few major voters and many minority voters. The theory of *oceanic games* (Milnor and Shapley, 1961; Shapley, 1961a) was designed specifically for this sort of situation. In an oceanic game there are several players of finite size, and the remainder constitute an ocean of players of infinitesimal size.

An example that has been worked out in detail elsewhere (Shapley, 1961a) will be sketched here. Consider a struggle for corporate control in which a simple majority of votes is needed, there are two large stockholders, and the rest are small. We denote this by $[1/2; w_1, w_2; \alpha]$, where $\alpha = 1 - w_1 - w_2$ and w_1, w_2 are the percentages of votes controlled by the large stockholders.

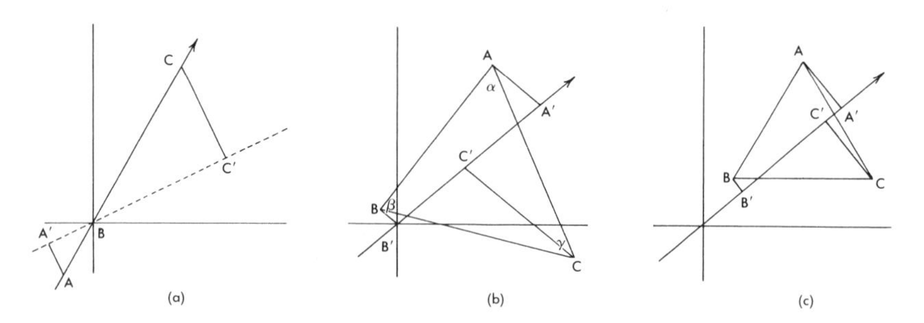

Figure 7.11
A three-person voting game in an ideology space.

A whole class of control situations can be examined by varying w_1 and w_2 over the ranges $0 \leq w_1 \leq 1$, $0 \leq w_2 \leq 1$, and $w_1 + w_2 \leq 1$. Four broad ranges can be identified:

1. $w_1 + w_2 \leq 1/2$ (the "interior" region);
2. $w_1 \leq 1/2, \quad w_2 \leq 1/2, \quad w_1 + w_2 \geq 1/2$ (the balance-of-power region);
3. $w_1 \geq 1/2$ (first player controls);
4. $w_2 \geq 1/2$ (second player controls).

The power index for the first player is

$$\phi_1 = \begin{cases} \dfrac{w_1}{\alpha} - \dfrac{w_1 w_2}{\alpha^2} & \text{in region 1,} \\ \dfrac{(1 - 2w_2)^2}{4\alpha^2} & \text{in region 2,} \\ 1 & \text{in region 3,} \\ 0 & \text{in region 4.} \end{cases}$$

This formula (and a similar one for ϕ_2) are derived by Milnor and Shapley (1961). The power of the oceanic players in toto is

$$\phi = 1 - \phi_1 - \phi_2.$$

A useful measure is the relative power per share. This is given by $R_1 = \phi_1/w_1$, $R_2 = \phi_2/w_2$, and $R = \phi/\alpha$. For example if $w_1 = 0.10$ and $w_2 = 0.40$, then $\phi_1 = 0.04$ and $\phi_2 = 0.64$. Their power ratios are $R_1 = 0.04/0.10 = 0.4$, $R_2 = 0.64/0.40 = 1.6$, and $R_3 = 0.32/0.50 = 0.64$.

This model is highly suggestive as a starting point for a more detailed and somewhat dynamical model of corporate control in which the shares of the ocean can be bought on the open market by the control player or some other group.

The Shapley–Shubik and Banzhaf indices can be applied to multicameral structures. An example is a legislative system with three groups: a president, a senate, and a house (Shapley and Shubik, 1954). Suppose that the president has veto power and that for a bill to pass a majority must be obtained in the senate and house. The senate has 3 members, the house 5 members. The Shapley–Shubik and Banzhaf indices applied to this voting system yield the percentage shares of power shown in table 7.3.

Table 7.3
Power indices for a tricameral legislative system

	Shapley–Shubik	Banzhaf
President	0.381	0.250
Each senator	0.107	0.125
Each congressman	0.060	0.094

Table 7.4
Power indices for the proposed Canadian constitutional amendment scheme

Province	Shapley–Shubik index (ϕ)	Banzhaf index (β)	Percentage of population (1970)	
Ontario	31.55	21.78	34.85	Average 31.90
Quebec	31.55	21.78	28.94	
British Columbia	12.50	16.34	9.38	
Alberta	4.17	5.45	7.33	Average 5.65
Saskatchewan	4.17	5.45	4.79	
Manitoba	4.17	5.45	4.82	
New Brunswick	2.98	5.94	3.09	Average 2.47
Nova Scotia	2.98	5.94	3.79	
Prince Edward Island	2.98	5.94	0.54	
Newfoundland	2.98	5.94	2.47	

We close with one further applied example which arose in relation to a proposed amendment to the Canadian constitution (Miller, 1973; Straffin, 1976). The scheme requires that any constitutional amendment obtain approval by (1) Quebec, (2) Ontario, (3) two of the four Atlantic provinces, and (4) British Columbia and one central province, or all three central provinces. The values of the two indices are shown in table 7.4.

EXERCISE

7.3. *For the game (8; 4, 4, 2, 2, 2, 1) calculate and compare the Banzhaf and Shapley–Shubik indices.*

Notes to chapter 7

1. The confusion is not diminished by the custom of naming values after their authors. This would not be so bad were it not that several authors have participated in the formulation of two or more values apiece.

2. See Shapley (1951b, 1952b, 1953c). At least three other investigators came upon the formula independently at about the same time; the idea was "in the air" and ripe for discovery.

3. Among subsequent contributions to the development of the value concept are those of Selten (1960, 1964), Isbell (1960b), Harsanyi (1963), Vilkas (1963), Miyasawa (1964), Shapley (1962c, 1964c, 1969), Kannai (1966), Owen (1972), and Aumann and Shapley (1968, 1969, 1970a,b, 1974).

4. The political applications came first, via the notion of "power index" (Shapley and Shubik, 1954); see, for example, Riker (1959), David, Goldman, and Bain (1960), Mann and Shapley (1964), Banzhaf (1965, 1968), Riker and Shapley (1968), Nozick (1968), Leiserson (1968), Owen (1971), Riker and Ordeshook (1973), Brams (1975, 1978), and Ordeshook (1978). For some economic applications see Shapley (1961a,b, 1964c,d) and Shapley and Shubik (1969d).

5. More precisely, the value is *relatively invariant* (or *covariant*), since it is expressed in terms of the selfsame utility scales.

6. There has been some investigation of value functions that weight this average differently (see Banzhaf, 1965; Starr, 1966; Owen, 1971; Weber, 1979; Dubey, Neyman, and Weber, 1981; Dubey and Shapley, 1979; Straffin, 1977).

7. Exact proportionality would occur at the imputation ($12, $9, $6).

8. The postulates for a simple game require that the empty set be losing, the all-player set be winning, and every superset of a winning set be winning (see Shapley, 1962b).

9. When there are two big players, the relative power positions may be reversed (see Shapley, 1961a, and section 7.8 below). This may be attributed to the fact that the small players hold the balance of power.

10. This is not unreasonable; for example, if a wealthy woman marries a man who is subject to blackmail, both may suffer.

11. Harsanyi (1959 et seq.) defined the modified characteristic function somewhat differently from our h, in such a way that it coincides with v on non-constant-sum, fixed-threat games as well as on constant-sum games. His definition is equivalent to the present one when used in connection with (7.1).

12. We are temporarily assuming that utilities are extrinsically comparable, the players' units of measurement being equal. Both "efficiency" and "equity," as used here, are relative to this assumption. With the aid of the equivalence principle we seek to discover an intrinsic comparability, arising out of the game itself.

13. Much of the conceptual groundwork for the extension outlined here was laid by Harsanyi (1963), but there are several significant differences

between his value and the present one, which was apparently first proposed by E. L. Kaplan (unpublished). For a discussion of the differences see Shapley (1964c, 1969) or Shapley and Shubik (1969d).

14. For the continuous version we may summarize this condition by the equality of the two boundary integrals:

$$\int_{A_1}^{s} \sqrt{-du_1\, du_2} = \int_{A_2}^{s} \sqrt{-du_1\, du_2}.$$

15. Bloomfield and Wilson (1972), however, have considered no-side-payment simple games for generating an appropriate set of models for the study of social choice.

16. See also Nagel (1975) for a survey of the use of the concept of power.

8 Two-Person Zero-Sum Games

8.1 A Shift in Viewpoint

The solution concepts discussed thus far all belong to what is usually called cooperative game theory. In these solution concepts, as well as in others to be described in chapter 11, the players are generally assumed able to talk to each other, to coordinate plans (in coalition), to transmit threats, and to make trustworthy promises—in short, to negotiate outside the formal structure of the game rules. Some solutions lean more heavily on these cooperative devices than others, or use them in different ways. (The core, for example, can be regarded as the set of outcomes that are proof against collusion.) But all involve in some way the notion of coalition and joint action.

In this and the next two chapters we offer a survey of the *noncooperative* branch of game theory. Its solution concepts contrast sharply with those we have been considering up to now, in that they make no attempt to account for coalition formation or any other mode of collusion among the players. In order to apply a noncooperative solution one must stipulate that the players of the game—for whatever reason—do not, cannot, or will not make any attempt at coordinating their strategic decisions.

The characteristic function is now useless, of course, and we must fall back on the strategic or the extensive form of the game. Since the players do not bargain, interpersonal utility comparisons are meaningless. Cardinal utility, however, may still be required because of mixed strategies or chance moves.

The central idea of the noncooperative approach is, in one word, "selfishness"—not misanthropic or vindictive selfishness, but simple indifference to the desires of others.[1] The n-person problem breaks up into n one-person problems to be solved simultaneously. Each player exercises free rational choice within his splendid isolation. He treats the other players as part of the environment, as behavioral mechanisms without free choice or motivation. He is not influenced by the payoffs they receive or by the strategies they might have chosen but did not.

This solipsistic attitude, which the noncooperative theory in its strictest form attributes to the players, entails a peculiar sort of naiveté. The players are, in effect, assumed to be incapable of understanding the theory. The theory, being symmetric, recognizes all individuals as free decision makers, but it does not permit them this insight. This may be a weakness in applications. Sophisticated players, when they can perceive a clear advantage to be gained from exploiting a commonality (or divergence) between their interests and those of other players, can usually find loopholes in any law against collusion, or ways to circumvent any social or ethical taboos that may superficially seem to prevent cooperation.[2] For this reason we shall find that the noncooperative solutions are most pervasive, and most likely to be useful in application, when they also exhibit some form of multilateral stability, such as Pareto optimality or the core property.

The foregoing discussion emphasizes (some would say exaggerates) the conceptual gap that separates the philosophy of the noncooperative branch of game theory from the resolutely nonbehavioral viewpoint that we have espoused for game theory in general. The gap is not unbridgeable, though. Indeed a major goal of this work is bridge-building in just this area, by means of systematic multisolutional analyses of the key economic models. The noncooperative solutions will have a special value here just because they so often provide a point of contact with classical economics. They will also serve to keep some of our models within hailing distance of the modern disciplines of linear programming and decision theory, which are also essentially noncooperative in their outlook.

In order to divide difficulties we begin by considering an extremely special class of games, to wit, two-person zero-sum games. It should be noted at the start, though, that these games do not provide many useful models for the behavioral sciences.

8.2 The Minimax Solution for Two-Person Zero-Sum Games

In the Dart Duel of chapter 2 we had an opportunity to taste the persuasive force of minimax reasoning applied to a context in which two interests are in direct, head-to-head conflict. Actually, when applied to this narrow class of games, virtually all solution theories, cooperative and noncooperative alike, come down to this same minimax solution. But minimax theory is discussed most naturally under the "noncooperative" rubric because no significant cooperation can be expected to occur, whether or not it is allowed, when one player must lose exactly what the other wins (Aumann, 1961a).

We recall that the dart throwers chose strategies from infinite sets. In introductory expositions, however, it is easier to work with finite sets of strategies, and we shall do so here. But first we need a notation.

A *finite two-person zero-sum game* in strategic form is often called a *matrix game* because it can be completely specified by a rectangular array of numbers, together with certain conventions on how to read it. The entries a_{ij} represent the payoffs to one of the players—call him P_1. Since the game is zero-sum, the payoffs to the other player, P_2, are $-a_{ij}$. The index i conventionally ranges over P_1's strategies (i.e., rows of the matrix), while j ranges over P_2's strategies (i.e., columns of the matrix).

Consider a sample matrix game:

		P_2			
	$j = 1$	2	3	4	
P_1	$i = 1$	5	-1	2	9
	2	5	8	$4^{\#}$	6

Here P_1 has two strategies and P_2 four. If P_1 chooses the first row and P_2 the fourth column, then P_1 wins 9 and P_2 loses 9, and similarly for the other entries.

8.2.1 Saddlepoints

Throughout game theory the most convincing solution concept, without a doubt, is the *saddlepoint in pure strategies.* It is illustrated in the matrix above by the marked entry $4^{\#}$. The first player, seeking to maximize a_{ij}, makes it at least 4 by choosing $i = 2$. The second player, seeking to minimize a_{ij}, makes it at most 4 by choosing $j = 3$. Accordingly 2 and 3 are optimal strategies for P_1 and P_2, respectively, and the number $a_{23} = 4$ is the *saddlepoint value.*

It is easily seen that while there may be many optimal strategy pairs, the saddlepoint value, if it exists at all, is unique.

This solution can be characterized in two equivalent yet intuitively rather different ways, and the distinction will be important to us later. First, let us say that a strategy pair $(i^\#, j^\#)$ has the *strategic equilibrium property* if P_1's choice of $i^\#$ is his best response to P_2's choice of $j^\#$, and vice versa. Formally,

$$\begin{aligned} a_{i^\# j^\#} &= \max_i a_{ij^\#}, \\ -a_{i^\# j^\#} &= \max_j (-a_{i^\# j}). \end{aligned} \tag{8.1}$$

Since the second line could be rewritten

$$a_{i^\# j^\#} = \min_j a_{i^\# j},$$

the strategic equilibrium property does express the idea of a saddlepoint in a matrix, that is, an entry that is simultaneously the minimum of its row and the maximum of its column.

For the second approach, let us say that a strategy has the *maxmin property* if its "safety level"—the worst that could happen to its user if it is used—is as high as that of any other strategy available to him.[3] Formally,

$$\begin{aligned} \min_j a_{i^* j} &= \max_i \min_j a_{ij}, \\ \min_i (-a_{ij^*}) &= \max_j \min_i (-a_{ij}). \end{aligned} \tag{8.2}$$

It is plain that maxmin strategies always exist, but for them to form a saddlepoint we need more: their safety levels must agree. That is, we need

$$\min_j a_{i^* j} = -\min_i (-a_{ij^*}).$$

This equality is only possible if the two "mins" are attained at $j = j^*$ and $i = i^*$, respectively, making both sides equal to $a_{i^* j^*}$. But in that case (8.2) becomes

$$\begin{aligned} a_{i^* j^*} &= \min_j a_{i^* j}, \\ -a_{i^* j^*} &= \min_j (-a_{ij^*}), \end{aligned} \tag{8.3}$$

which is just a simple transposition of (8.1).

These two rationales, "equilibrium" and "safety," reflect opposite principles of competition. They coincide here only because we are dealing with a special kind of game. Following the first principle, one seeks to take as much as possible from a passive, predictable opponent. Following the second, one seeks to save as much as possible in the face of an active, unpredictable opponent. The first confidently assumes that one can make the final adjustment in strategy after the other player has committed himself; the second pessimistically concedes the last word to the enemy. When the two principles lead to the same result, as here, we have a powerful argument indeed for the "rationality" of the solution.[4]

EXERCISE

8.1. *Find the saddlepoints of the following matrix:*

	1	2	3	4
1	1	2	1	4
2	−3	2	−2	2
3	1	1	1	3
4	0	1	0	1

8.2.2 The minimax theorem

Unfortunately, not all matrix games have saddlepoints, if only pure strategies are considered.[5] The following simple example shows why.

PICK A NUMBER. *P_2 thinks of a number between 1 and n. If P_1 guesses it, he wins a dollar from P_2.*

The payoff matrix is the familiar "identity matrix"—a diagonal of 1s on a field of 0s.

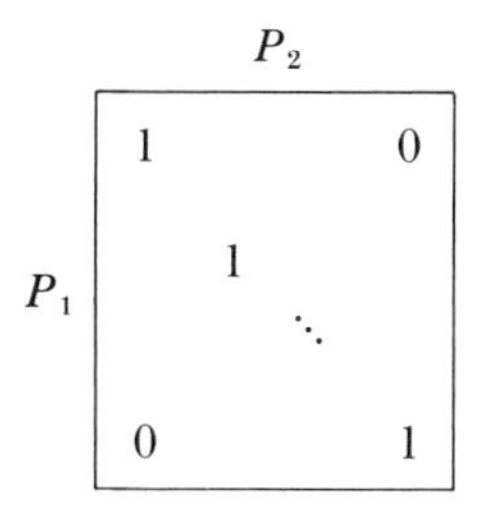

Since P_1 wants $i = j$ and P_2 wants $i \neq j$, no pure-strategy equilibrium is possible, and there is no saddlepoint.[6]

The saddlepoint concept is saved, however, by the introduction of *mixed strategies* whereby, in effect, the players use randomizing robots to choose their strategies and content themselves with merely controlling the probabilities. Mathematically this amounts to enlarging the strategy sets of the players, introducing as additional strategies all possible convex linear combinations of the original rows and columns.

This striking concept was first put forward in mathematical generality by Emile Borel in 1921 (see also Borel, 1924, 1938, 1953; Frechet, 1953; von Neumann, 1953b), but its full significance was not appreciated until some years later when John von Neumann (1928), working independently, discovered the famous "minimax theorem": *if mixed as well as pure strategies are considered, then every matrix game has a saddlepoint and hence a saddlepoint value.* This theorem is generally regarded as the historical cornerstone of game theory. It has since been generalized in many ways, both for game-theoretic purposes and for applications in other fields of mathematics.[7]

The minimax theorem is easily verified for Pick a Number. Simply let each player use the "uniform mix," selecting each number from 1 to n with probability $1/n$. P_1 thereby assures himself $1/n$ dollars (in expected value), regardless of what P_2 might do. Moreover, P_1 can do no better, even if he guesses that P_2 is also using the uniform mix. Similarly P_2 can hold his loss to $1/n$ dollars (in expected value) and can do no better. The saddlepoint value of the game is therefore $1/n$.

Without denying the great importance of the minimax theorem, we must admit that it has very little to do with the aspects of game theory that will engage us here. Indeed one of our objectives is to try to move game theory—for social scientists—out of the shadow cast by von Neumann's brilliant discovery. The impression is too widely held, even today, that game theory begins and ends with minimax theory.[8]

8.2.3 Equalization

One further concept has already been illustrated without comment: the optimal mixed strategies in Pick a Number have the property of making the opponent's payoff independent of his choice of strategy. Given that one player uses the uniform mix, the other can neither help nor hurt himself by varying his strategy. We call these strategies *equalizers.*

If equalizers exist for both players, then they constitute a saddlepoint, that is, they are optimal, since the equilibrium property (8.1) is assured. This does not preclude the existence of other optimal strategies that are not equalizing. Any optimal strategy has a certain tendency toward equalizing, however, since it must at least make the opponent indifferent among the pure strategies that enter into his optimal mixes. A game like Pick a Number is in a sense an extreme case, since all pure strategies enter into the solution; in such "completely mixed games" (see Kaplansky, 1945) the optimal strategies are necessarily equalizers.

An equalizing strategy need not be optimal if only one player has one. The following 3 × 3 matrix game shows how this can happen:

	P_2		
	3	−1	2
P_1	−6	8	0
	6	−14	−4

If P_1 uses his equalizer (2/5, 2/5, 1/5), the expected payoffs are (0, 0, 0). His optimal strategy (7/9, 2/9, 0), however, yields (1, 1, 14/9), assuring him 1 if P_2 plays the first or second column and more if P_2 makes the mistake of playing the third column.

In a non-constant-sum game, equalizers can be even more strikingly nonoptimal, an extreme example being the so-called doomsday machine, which enables any player to assure the worst possible outcome for all players (see section 9.2.2).

EXERCISES

8.2. *What happens in the above 3 × 3 example if one attempts to construct an equalizer for P_2 by solving the appropriate set of simultaneous linear equations, namely $\Sigma_j a_{ij} y_j = v$ (i = 1, 2, 3) and $\Sigma_j y_j = 1$?*

8.3. *Calculate the optimal strategy pairs and the saddlepoint value for the following games:*

	1	2	3
1	2	3	−2
2	0	4	0
3	−2	3	2

	1	2	3
1	16	−8	4
2	−24	16	3
3	1	1	2

Given that one player has selected his strategy, we may ask what would be the best response of the other player if he had that information. Reconsidering the last example, if P_1 uses (7/9, 2/9, 0), then the best response of P_2 is either his first or his second pure strategy. Both yield a payoff of -1. Any mix of them will also yield P_2 a payoff of -1. In particular, if he uses (1/2, 1/2, 0), then P_1's best response would be either of his first two pure strategies.

If P_1 announces a pure strategy and P_2 announces a pure-strategy best response, then if P_1's best response is his original strategy, the zero-sum game has a saddlepoint. There may be more than one saddlepoint, though. In exercise 8.1 it is easy to verify that there are four saddlepoints. If P_1 uses his first or third pure strategy, the best response by P_2 is his first or third pure strategy. Here there is no need for either player to mix his strategy; either can use any mixture of $(p, 0, 1 - p, 0)$, $0 \leq p \leq 1$, since all are optimal. If (i, j) and (g, k) are each saddlepoints, then so are (i, k) and (g, j). The strategy pairs that make up the equilibrium points in a non-constant-sum game do not necessarily have this property of interchangeability.

8.2.4 Interpretation and application

Almost no situations of major importance encountered in society are best considered as strict two-person zero-sum games. Even two gladiators in combat would both like to survive, and societies frequently provide face-saving mechanisms so that individuals or groups engaged in struggles that appear to involve pure opposition are offered a way out.

About the only direct applications of two-person zero-sum game theory involve duels or situations that can be modeled as duels. Two-person parlor games such as Tic-Tac-Toe, Chess, Two-Handed Poker, and Nim provide classic examples. But actual duels and search problems, in which, for example, a destroyer tries to find a submarine while the submarine tries to avoid being found, or two tanks engage in combat, can also be studied as two-person zero-sum games. Thus there are direct military applications to tactical and weapons analysis problems.

Our main interest is in the application of game theory to society, the polity, and the economy, and two-person zero-sum game theory does not provide a useful model of any of these institutions. The desires of the players in such a game are diametrically opposed.

There is no need for communication, language, or side payments. The only actions are deeds; words play no role. There is no cooperation because there is nothing to cooperate about.

Unless explicitly stated otherwise, the assumption of external symmetry essentially rules out all personality traits in the players. Yet we know that even in Chess elaborate precautions must be taken to avoid having personality and psychological tricks disturb the play.

The 2×2 and even the 3×3 matrix games are convenient and easy to use for experiments (Rapoport, Guyer, and Gordon, 1976). Even saddlepoints are not always picked by naive players, and mixed strategies (except possibly with the same odds on all events) are not natural to untrained individuals. The justification for the theory is *normative*: if you follow it, you can best protect yourself against any opponent.

Two simple 2×2 games illustrate the power of the normative prescription:

10	−10
−10	10

10	−5
−15	10

The first matrix is the familiar Matching Pennies game. If the outcome is HH or TT, then P_1 wins \$10; otherwise he loses \$10. The way most people play this game in practice is by flipping coins, and this is what the theory prescribes.

The second matrix may appear to be fair to a naive player, and if both use (1/2, 1/2) as a mixed strategy, then the expected payoff is

$$\frac{1}{4}(10) + \frac{1}{4}(-5) + \frac{1}{4}(-15) + \frac{1}{4}(10) = 0.$$

Suppose that in general P_1 uses p and $1 - p$ and P_2 uses q and $1 - q$ as the probabilities associated with their mixed strategies. Let V be the value of the game. Then

$$\begin{aligned} &10q - 5(1 - q) \geq V, && 10p - 15(1 - p) \leq V, \\ &-15q + 10(1 - q) \geq V, && -5p + 10(1 - p) \leq V, \\ &0 \leq q \leq 1, && 0 \leq p \leq 1. \end{aligned}$$

It is easy to show that the optimal mixed strategy for P_1 is (5/8, 3/8), and that for P_2 is (3/8, 5/8). The value is

$$V = \frac{3}{8}(10) + \frac{5}{8}(-5) = \frac{3}{8}(-15) + \frac{5}{8}(10) = \frac{5}{8}.$$

Thus a distinct bias in favor of P_1 can be obtained by following the advice from the theory of two-person zero-sum games.

Game theory provides the same solution for two-person constant-sum games as for zero-sum games. If a constant payment is made to one or both players regardless of the outcome of the game, this payment should not affect their behavior concerning the payoff that does depend upon strategies. Thus the constant-sum game shown below is strategically equivalent to the second one noted above. Here the first entry in each cell of the matrix is the payoff to P_1 and the second the payoff to P_2:

20, −10	5, 5
−5, 15	20, −10

An allocation game Colonel Blotto has three divisions to defend two mountain passes. He will defend successfully against equal or smaller strength but lose against superior forces. The enemy has two divisions. The battle is lost if either pass is captured. Neither side has advance information on the disposition of the opponent's divisions. What are the optimal dispositions?

Assume that the worth of overall victory is 1 and that that of defeat is −1. Then we may describe the game as follows:

	(2, 0)	(1, 1)	(0, 2)
(3, 0)	1	−1	−1
(2, 1)	1	1	−1
(1, 2)	−1	1	1
(0, 3)	−1	−1	1

Here (3, 0) means that three divisions are sent to defend the first pass and zero to the second.

The optimal strategies are for Blotto to divide his divisions. He plays the mixed strategy (0, 1/2, 1/2, 0) while the enemy hits one pass with all of his force, playing (1/2, 0, 1/2).

Games with vector payoffs In military applications it may be desirable to consider multiple goals for the opponents rather than try to represent an outcome by a single number. If the payoff vectors have $k + 1$ components, which might represent men, money, or supplies, then Shapley (1959c) has shown that a $2k$ parameter set of equilibrium points exists. If weights are introduced on the different outcomes so that they can be compared, then the zero-sum game with vector payoffs can be translated into a non-constant-sum game whose equilibria are related to the equilibria in the vector payoff game.

Consider the following game with two-component vector payoffs:

	1	2
1	(1, 0)	(−2, 1)
2	(−2, 3)	(1, 2)

Against an arbitrary mixed strategy of q and $1 - q$ used by P_2, if P_1 uses his first strategy he obtains $(-2 + 3q, 1 - q)$. If he uses his second strategy he obtains $(1 - 3q, 2 + q)$. For $0 \leq q \leq 1/2$ the second strategy is always the best reply, since each component of the payoff dominates the component of the payoff associated with the first strategy. For $1/2 < q \leq 1$ this is not so.

Against an arbitrary mix of p and $1 - p$ used by P_1, P_2 obtains $(2 - 3p, -3 + 3p)$ with his first strategy and $(-1 + 3p, -2 + p)$ with his second. For $0 \leq p < 1/2$ the second strategy is the best response, giving higher payoffs in both components of the payoff function; this is not so for $1/2 \leq p \leq 1$.

If P_1 weights his goals by (1/2, 1/2), that is, equally, and if P_2 places all his weight on the first component, the game with vector payoffs becomes the following non-constant-sum game:

	1	2
1	1/2, −1	−1/2, 2
2	1/2, 2	3/2, −1

Here the first component in each cell is the payoff to P_1 and the second the payoff to P_2. There is an equilibrium point at the strategy pair (2, 1).

A simple solution method There has been considerable progress in deriving algorithms for the solution of both two-person zero-sum and non-constant-sum matrix games (Lemke and Howson, 1964). Efficient algorithms exist which can handle zero-sum games of considerable size, but these will not be dealt with here. However, some geometrical insight can be obtained from a simple diagram that illustrates the solutions to a matrix game of size $2 \times m$ (or $m \times 2$). We use the nonsymmetric Matching Pennies game as an illustration. In figure 8.1 the points (1, 0) and (0, 1) represent the two pure strategies of P_1. Any point on the line connecting them represents a mixed strategy with weights p on (1, 0) and $1 - p$ on (0, 1).

Suppose that P_1 uses his pure strategy (1, 0), so that $p = 1$. If P_2 then uses his pure strategy (1, 0), the payoff is 10, as denoted by A; if he uses (0, 1), the payoff is -5, as denoted by B. Similarly C and D show payoffs for P_1 using (0, 1) against P_2 using (0, 1) and (1, 0), respectively.

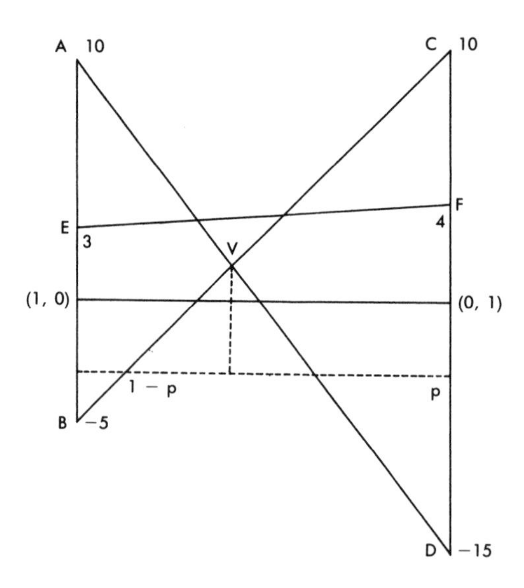

Figure 8.1
Solution of a 2×2 zero-sum game (method 1).

Given that P_2 uses strategy (0, 1), then the payoff to P_1 from any mixed strategy is represented by the line BC. The payoff is $-5p + 10(1 - p)$, which for $p = 1$ gives -5 at B and for $p = 0$ gives 10 at C. Similarly if P_2 uses (1, 0), then the payoff from any mixed strategy by P_1 is $10p - 15(1 - p)$.

The intersection at V indicates the mixed-strategy weights at which the same expected payoff will be obtained regardless of whether P_2 plays (0, 1) or (1, 0). The segments BVD show how P_1's security level changes as he varies his mixed strategy. The most that P_1 can guarantee under all circumstances is $V = 5/8$, where $p = 5/8$ and $1 - p = 3/8$.

If P_2 has more than two strategies, the payoffs resulting to P_1 from their employment can be illustrated by extra lines on figure 8.1. An example is the line EF for the 2×3 matrix game:

10	−5	3
−15	10	4

As can be seen from both the matrix and the diagram, P_2 would never employ his third strategy.

An alternative diagrammatic treatment In figure 8.2 the two axes measure the payoffs to P_1 on the assumption that he uses his first or second pure strategy, respectively. Each point represents the payoffs if P_2 uses a pure strategy. The triangle ABC is the boundary of the convex set of payoffs to which P_2 can restrict P_1. P_2's goal is to select a point in this set that minimizes P_1's payoff under all circumstances. This will be the point V on the 45° line. It yields the same payoff to P_1 whether he uses his first or his second strategy, and it keeps that payoff as small as possible. P_2's mixed strategy involves only his first two strategies; any utilization of his third strategy harms him.

Fictitious play Another method of solving games, different from either an analytical approach or a diagrammatic treatment, is by simulation. Although this method, known as *fictitious play* (G. W. Brown, 1951), may be regarded purely as a means of obtaining a solution, it can also be regarded as a simulation of the manner in which individuals actually play games set in a dynamic context.

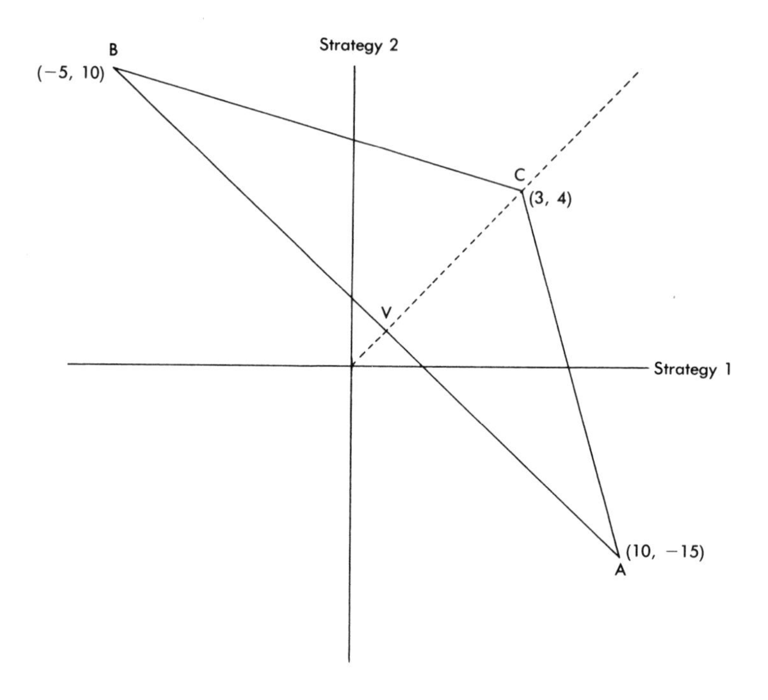

Figure 8.2
Solution of a 2 × 2 zero-sum game (method 2).

Let each player assume that the best prediction of his opponent's behavior is given by the frequency with which he has played his pure strategies previously. The rule each player follows is then: Select a pure strategy that maximizes expected payoffs under this assumption; in case of indeterminacy, randomize.

Consider an example for Matching Pennies:

−1	1
1	−1

Suppose that each player starts with his first move. Table 8.1 records the plays and the statistics on their frequency of play. On play 2, P_1 faces (1, 0) and hence chooses 2; P_2 faces (1, 0) and hence selects 1. On play 3, P_1 faces (1, 0) and P_2 faces (1/2, 1/2). Robinson (1951)

Table 8.1
A simple Matching Pennies game

	P_1's move			P_2's move		
Play	1	2	Frequency	1	2	Frequency
1	1		1, 0	1		1, 0
2		1	$\frac{1}{2}$, $\frac{1}{2}$	1		1, 0
3		1	$\frac{1}{3}$, $\frac{2}{3}$	1		1, 0
4		1	$\frac{1}{4}$, $\frac{3}{4}$		1	$\frac{3}{4}$, $\frac{1}{4}$
5		1	$\frac{1}{5}$, $\frac{4}{5}$		1	$\frac{3}{5}$, $\frac{2}{5}$
6		1	$\frac{1}{6}$, $\frac{5}{6}$		1	$\frac{1}{2}$, $\frac{1}{2}$
7		1	$\frac{1}{7}$, $\frac{6}{7}$		1	$\frac{3}{7}$, $\frac{4}{7}$
8	1		$\frac{1}{4}$, $\frac{3}{4}$		1	$\frac{3}{8}$, $\frac{5}{8}$
9	1		$\frac{1}{3}$, $\frac{2}{3}$		1	$\frac{1}{3}$, $\frac{2}{3}$
10	1		$\frac{2}{5}$, $\frac{3}{5}$		1	$\frac{3}{10}$, $\frac{7}{10}$
11	1		$\frac{5}{11}$, $\frac{6}{11}$		1	$\frac{3}{11}$, $\frac{8}{11}$

proved for the two-person zero-sum game that the frequencies eventually approach the mixed-strategy equilibrium of the game—in this instance (1/2, 1/2) and (1/2, 1/2). Unfortunately this does not generalize for non-constant-sum games. An example of a non-constant-sum game for which fictitious play fails is given by Shapley (1962a):

1, 0	0, 0	0, 1
0, 1	1, 0	0, 0
0, 0	0, 1	1, 0

8.3 Two-Person Zero-Sum Games in Extensive Form

Chess is a zero-sum game with perfect information. Individuals play this and other such games in the extensive form. In spite of its size we can conceive of a full game tree for Chess or any other finite-move game. Von Neumann and Morgenstern (1944) showed that perfect information was a sufficient condition for any two-person constant-sum game to have a saddlepoint. This result follows immediately by backward induction. At each terminal choice point the player whose turn it is will choose the move that is best for him. We can delete the final branches of the tree and replace them by the payoffs determined by all ultimate moves. We may then consider all penultimate moves in the old game as ultimate moves in the shortened game. This process will end at the initial node of the tree or the starting point of the game.

Figures 8.3a and 8.3b show the simple game of Matching Pennies, given perfect information, in extensive and strategic form (for HHTT read "If he plays heads, I play heads; if tails, I play tails").

EXERCISE

8.4. *Apply backward induction to the extensive form shown in figure 8.3a. What is the saddlepoint?*

Backward induction shows that Chess has a saddlepoint, but to date no one has been able to prove that it is a win or draw for White.

The normative theory for the play of two-person zero-sum games as complex as Chess calls for strategic planning that is beyond most individuals' capabilities. Yet except for computational difficulties the minimax principle appears to be satisfactory. Even setting aside computational problems, though, the adequacy of the principle has been challenged (Aumann and Maschler, 1972).

In a two-person zero-sum game with perfect recall (see chapter 3) Kuhn (1953) has shown that all mixed strategies are equivalent to

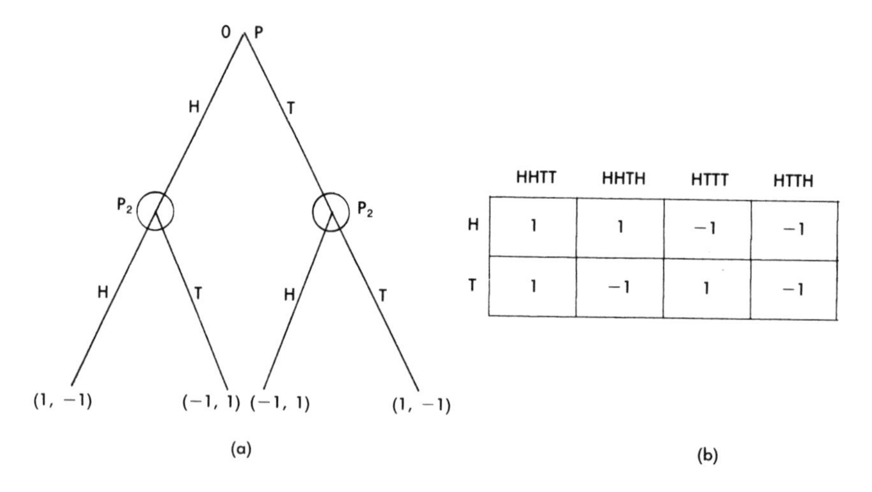

	HHTT	HHTH	HTTT	HTTH
H	1	1	−1	−1
T	1	−1	1	−1

Figure 8.3
The extensive and strategic forms of a Matching Pennies game with perfect information.

behavior strategies. From this it follows that in a game with perfect recall the players should use optimal behavior strategies.

Consider the game illustrated in figures 8.4a and 8.4b. Nature (P_0) moves first and gives the move directly either to P_2 with probability 1/3 or to P_1 with probability 2/3. P_2 is not informed whether Nature has called upon him directly or whether P_1 has moved. Each player has two pure strategies, and the strategic form of the game is shown in figure 8.4. This game has unique optimal mixed strategies of (1/8, 7/8) for P_1 and (1/2, 1/2) for P_2. The value is $3\frac{1}{2}$.

It might be argued that, given the information that Nature has given him the move, P_1 would be better off switching to (1/2, 1/2). But if P_2 were given the information that P_1 is employing (1/2, 1/2), his best reply would be (0, 1), which yields $\frac{1}{3}(0) + \frac{2}{3}(1/2)(6) = 2$. And if P_1 knew that P_2 used (0, 1), his best reply would be (0, 1), which yields $\frac{1}{3}(0) + \frac{2}{3}(6) = 4$.

Aumann and Maschler argue that the minimax solution for a two-person zero-sum game can be defined by both equilibrium-point and maximum-security-level arguments, but that here, given the information that Nature has called on P_1 to move, the strategy (1/8, 7/8) no longer provides maximum security against any strategy of P_2.

If P_1 makes a binding precommitment to (1/8, 7/8) before P_0 moves, then the extensive form is being essentially ignored. P_1 is throwing

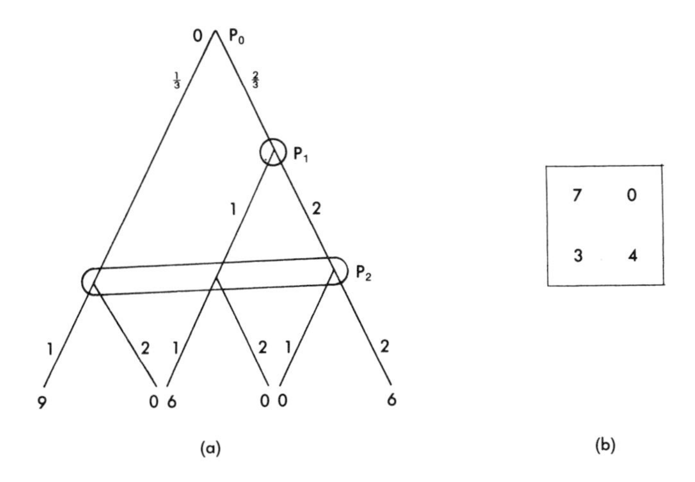

Figure 8.4
Different extensive forms for the same strategic form (1).

away his chance to use his knowledge of Nature's move. Aumann and Maschler stress that they are considering the game being played only once without precommitment. The following exercises may help the reader to understand their point of view further (see also M. Davis, 1974; Owen, 1974).

EXERCISES

8.5. *Calculate the strategic form of the game shown in extensive form in figure 8.4a. If you had the choice to play as P_1 in the game shown in figure 8.4a, or in figure 8.5, which would you prefer, or would you be indifferent?*

8.6. *Calculate the strategic forms for the two games shown in figure 8.6. Do these games, in your opinion, differ from those in figures 8.4 and 8.5? If so, indicate why.*

Games such as Chess or Poker are played move by move. The strategic and extensive forms of these games are formally linked, but

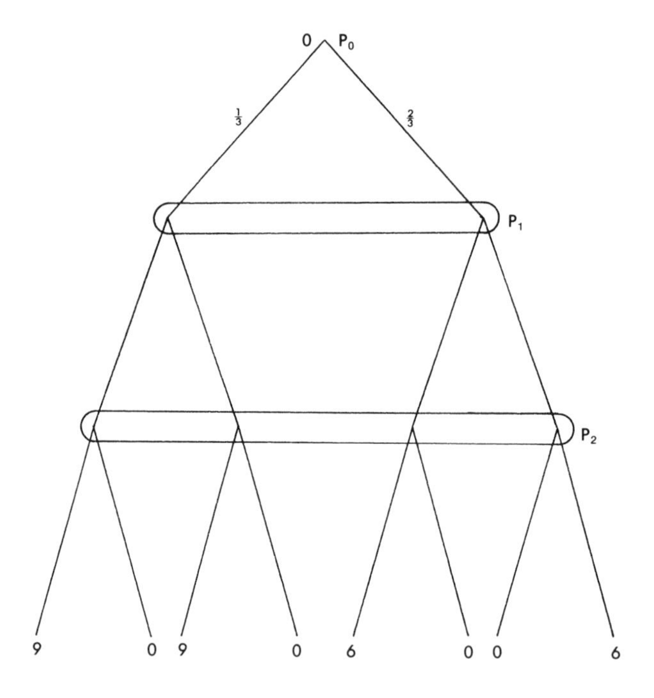

Figure 8.5
Different extensive forms for the same strategic form (2).

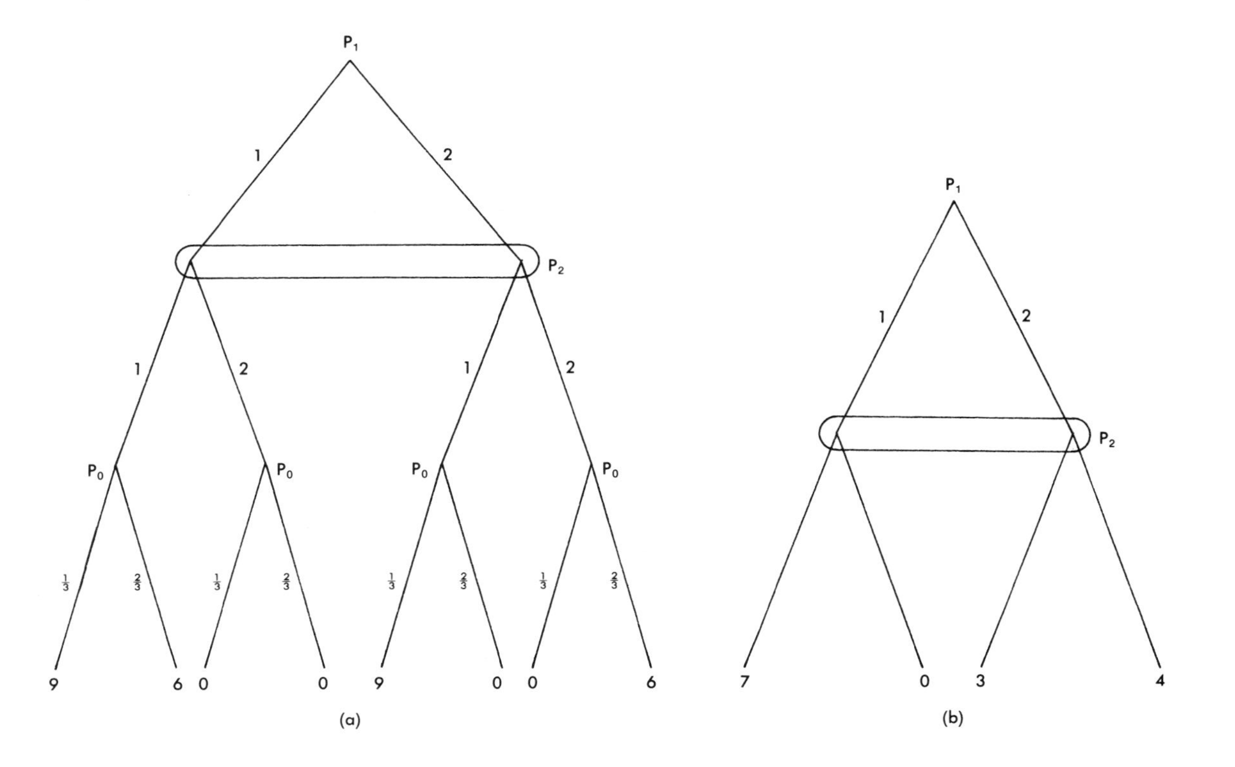

Figure 8.6
Different extensive forms for the same strategic form (3).

human behavior appears to be different in one-shot matrix games and in games played move by move. The solution theory of minimax is prescriptive. It provides advice as to how one should play if one could calculate in sufficient detail. There is no dynamics. The linkage suggested between the extensive and strategic forms demonstrates that the problems being treated are essentially static.

8.4 Games of Indefinite Length

In chapter 3 we discussed briefly the ideas behind stochastic games, recursive games, and games of survival. In this volume we do not propose to deal with dynamics in any depth; a few comments are in order, however, if only to point out new problems as the models are extended.

A two-person zero-sum stochastic game (Shapley, 1953c) is one in which play proceeds from matrix game to matrix game according to transition probabilities influenced by both players. We may consider a finite number of matrix games N with a finite number of pure strategies m_k, n_k for P_1 and P_2 in the kth game. One game is distinguished as the starting state. At each game k, if P_1 and P_2 select alternatives i and j, respectively, there is a probability $s_{ij}^{k} > 0$ that the supergame or overall game ends and a probability s_{ij}^{ke} that it goes to game e. Thus the game may continue indefinitely, but the probability that it ends by time t is at least $1 - (1 - s)^t$ where $s = \min s_{ij}^{k}$. The expected total gain or loss is bounded, and it has been shown that a solution in *stationary strategies,* that is, behavior strategies that call for a player to employ the same probabilities whenever he encounters the same subgame, exists. Shapley (1953) has shown that the game has a value. Hence we have a natural extension of the results from finite two-person zero-sum games.

Bewley and Kohlberg (1978), in a paper which also contains a useful bibliography on stochastic games, investigate what happens if the stop probabilities are zero. This can be done by two methods that avoid the pathologies of the infinite: one can consider a game in which the payoffs are discounted with an interest rate close to zero, or one can consider the average per-period payoff in a game with a finite number of periods T and let T become large. Bewley and Kohlberg provide a description of the asymptotic behavior of the strategies and the value as the interest rate approaches zero in the discounted payoff game and as T goes to infinity in the finite-horizon game. They establish the existence of asymptotically optimal strategies and an asymptotic value.

An important variant of a two-person zero-sum game of indefinite length is the game of survival (Milnor and Shapley, 1957). Two players with initial resources r_0 and $R - r_0$ play a zero-sum matrix game $A = \|a_{ij}\|$ until one of them is ruined. The matrix game, unlike those previously described, does not have final payoffs in its entries but points or "chips" or resources that are transferred between the players after each play. The outcomes are individual ruin or survival or a never-ending game. The payoffs are the values attached to the three outcomes. We may attach a value of 1 to survival, 0 to ruination, and Q to a game that never ends. If both survive, then the payoffs are $Q \geq 0$ and $1 - Q \geq 0$, where Q may be some function of the play. The game is completely specified by $\|a_{ij}\|$, R, Q, and r_0, the initial fortune of P_1.

The following properties have been established. We may express the value V as a function of P_1's resources r; if it exists, $V(r)$ is a monotonic increasing function of r and a solution of the functional equation

$$\phi(r) = \text{val}\,\|\phi(r + a_{ij})\|, \qquad 0 < r < R, \tag{8.4}$$

where "val" is the minimax value of the matrix game with boundary conditions

$$\phi(r) = 0, \qquad r \leq 0,$$

$$\phi(r) = 1, \qquad r \geq R.$$

If $Q \equiv 0$ (or 1), the value exists, as does an optimal strategy for the first (second) player. If there are no zeros in A, the value exists and is independent of Q, and optimal strategies exist for both players.

An example will show some of the features of the stochastic game. Consider the following game:

$$A = \begin{vmatrix} -1 & 0 \\ 2 & -1 \end{vmatrix}, \qquad Q = 0.$$

Here $V(r) = 0$; but if $Q \neq 0$ and $r < R$, P_2 has no optimal strategy independent of Q. This contrasts with the one-shot game, in which P_2 has the optimal strategy (1/4, 3/4).

Notes to chapter 8

1. This is not quite accurate, in the sense that an implicit consideration for the welfare of others may be coded into the individual's utility function.

2. Behavioral scientists sometimes design laboratory experiments in which the noncooperative canon is rigorously enforced, with all loopholes tightly guarded. In this way they can subject the players, and the artificial "society" they comprise, to a variety of instructive stresses. ["Rebellion against the experimenter" may be reported in these experiments; see, e.g., Bonachek (1968).]

While the psychological and sociological insights gained by these experiments are certainly interesting in themselves and may be of some incidental use in game theory, we do not regard them as having much to say on the question of what solution concept to apply outside the laboratory, where noncooperation can hardly ever be so rigorously enforced.

3. Referring again to the example, we see that the safety levels for P_1's strategies are -1 and 4, respectively, making $i^* = 2$ his maxmin strategy. For P_2 they are -5, -8, -4, and -9, respectively, making $j^* = 3$ his maxmin strategy.

4. The terms "minimax," "maxmin," "minmax" are often used interchangeably, denoting a situation, process, or mathematical formula in which maximization and minimization of the same function are taking place with respect to different variables. Sometimes, as in the phrase "maxmin property," the order of the syllables is significant.

"Minimax" and "Min-Max" are popular brand names, appearing on articles ranging from men's suits to fire extinguishers. "Maximin" is the name of a minor saint honored by a fine medieval church near Aix-en-Provence, and also of two Roman emperors who were noteworthy not only for their humble origins and their feats of physical prowess but also for the outpourings of public joy at their demise. Of the earlier we read, "He and his son were murdered in their tent by a body of praetorians. Their heads were cut off and despatched to Rome, where they were burned on the Campus Martius by the exultant crowd." Of the latter, "[His] death was variously ascribed 'to despair, to poison, and to the divine justice.'" (Quotes from the *Encyclopaedia Britannica,* 11th ed.)

The minimax concept has proved surprisingly difficult for modern lexicographers. *Webster's New International Dictionary* (3rd ed., 1961) omits the term, but the "theory of games" entry lays the groundwork for confusion: ". . . to maximize one's gain or to minimize one's loss. . . ." *The Random House Dictionary* (1966) similarly speaks of "maximizing gains and minimizing losses" and then gives us a pair of nouns(!): "*maximin*: a strategy of game theory employed to maximize a player's minimum possible gain"; and "*minimax*: a strategy of game theory employed to minimize a player's maximum possible loss." (Is there is a difference?) According to *The American Heritage Dictionary* (1969), "minimax" refers to the principle by which "a player selects the strategy to minimize an opponent's greatest possible gain and maximize his own." (His own greatest possible gain?) The essential idea—that there is *one* function of *two* independent variables—is repeatedly obscured, if not misstated.

5. One could calculate the probability that a "random" $m \times n$ matrix game will have a saddlepoint if the payoff entries are assumed to be identically distributed independent random variables. If the distribution is continuous, so that ties have probability 0, then the probability of a saddlepoint is $m!n!/(m + n - 1)!$. Thus for a 2×4 matrix the probability is 40 percent; for a 10×10 matrix it is about 0.01 percent. (See Goldman, 1957; Dresher, 1970.)

In practice, the payoff entries are rarely independent of each other. Indeed the number of different possible outcomes in a multimove game is often distinctly less than the number of pure strategies available even to one player (see the comments on Chess in section 3.2.1), and when this happens ties are inevitable in every row and column of the matrix.

6. Alternatively we could observe that the two safety levels do not agree; they are 0 for P_1 and -1 for P_2.

7. See, e.g., von Neumann (1937, 1945), Ville (1938), Wald (1945b), Loomis (1946), Nash (1950c), Karlin (1950, 1953c), Dantzig (1951, 1956), Gale, Kuhn, and Tucker (1951), Glicksberg (1952), Fan (1952), Shapley (1953c), Nikaido (1954), Kemeny, Morgenstern, and Thompson (1956), Parthasarathy (1967), and Parthasarathy and Raghavan (1971).

8. Perhaps a dozen textbooks could be cited, written for economics or operations-research students, that contain a single pro forma chapter on "game theory," touching nothing but minimax theory and sometimes presenting even that subject as merely an interesting impractical offshoot of linear programming.

9 Noncooperative Solutions to Non-Constant-Sum Games

"Love and War are the same thing, and strategems and policy are as allowable in one as in the other."

Miguel de Cervantes, *Don Quixote*

9.1 Equilibrium Points

In this chapter we consider the general n-person noncooperative game, presented in finite strategic form.[1] The fundamental solution concept for such games was formulated by John Nash (1950c, 1951) in his doctoral dissertation, though many instances of this general kind of solution are found in the economic literature dating back to Augustin Cournot (1838). Unlike most of the cooperative solutions, Nash's solution is not couched in terms of payoffs. Instead it is a set of *strategies*—one for each player. It is a direct extension of the strategic equilibrium property (8.1) and is called an *equilibrium point*, or EP for short.[2]

The basic definition is simple and intuitive. Let $H_i(s_1, s_2, \ldots, s_n)$ be the payoff to player P_i as a function of everybody's strategies including his own. An EP is any vector of strategies $(s_1^\#, s_2^\#, \ldots, s_n^\#)$ such that for each $i = 1, 2, \ldots, n$,

$$H_i(s_1^\#, s_2^\#, \ldots, s_i^\#, \ldots, s_n^\#) = \max_{s_i} H_i(s_1^\#, s_2^\#, \ldots, s_i, \ldots, s_n^\#). \quad (9.1)$$

In words, an EP is a vector of strategies such that no one player, regarding the others as committed to their choices, can improve his lot.

A glance back at (8.1) will confirm that the EP is indeed a generalization of the minimax or saddlepoint solution for the two-person zero-sum game, although it is not the only possible generalization.

9.1.1 Two examples

THE WINDING ROAD. *Two cars approach a blind curve from opposite directions. To avoid the need for a sudden stop, both must be driving to the right of the road, or both to the left.*

Let us suppose each player has just two strategies, L and R. Then the outcome matrix is

	L	R
L	Pass on left	Sudden stop
R	Sudden stop	Pass on right

For a payoff matrix we may take

	L	R
L	0, 0	−1, −1
R	−1, −1	0, 0

There are two obvious EPs here, namely (L, L) and (R, R), or "drive left" and "drive right."

This is a simple prototype of the *game of coordination*: if the players can coordinate their strategies, both will benefit (Schelling, 1960a). In this case the two ways to do this yield equal benefits. More generally, the benefits might be unequal, as in many bargaining models; an element of competition is then added to the coordination problem.

EXERCISE

9.1. *Suppose that both players prefer passing on the right to passing on the left. Is (L, L) still an EP?*

Aumann (1961a) has suggested a class of non-constant-sum games that are closely related to strictly competitive or two-person constant-sum games. At an equilibrium point in a two-person game neither player can increase his payoff by a unilateral change of strategy. A *twisted equilibrium point* is a pair of strategies such that neither player can decrease the other's payoff unilaterally. When these two phenomena coexist, Aumann calls the game *almost strictly competitive*. An example is

4, 4	0, 5
5, 0	1, 1

All such games have a unique equilibrium payoff, and thus values (V_1, V_2), which each player can guarantee for himself, can be attached to the game.

9.1.2 The existence theorem

The equilibrium-point concept owes much of its appeal, at least for mathematicians, to the beautiful existence theorem proved by Nash (1950c, 1951), which states that if mixed as well as pure strategies are considered, *every finite n-person game has at least one EP* (Glicksberg, 1952; Nikaido and Isoda, 1955). This is, to be sure, a generalization of von Neumann's theorem, but it has a rather different mathematical flavor, since it lacks the latter's intimate connection with linear algebra. Furthermore, there is an important conceptual difference: the minimax theorem provides a *value* for the game; Nash's theorem does not. If, as frequently happens, there are many EPs in a game, there is no a priori reason to expect their payoff vectors to be equal.

Two-person games with finite strategy sets are called *bimatrix games,* since they can be represented in much the same manner as matrix games except that a pair of matrices (or a single matrix with number pairs as entries) is now required. Bimatrix games have been widely studied (Vorobyev, 1958; Kuhn, 1961; Shapley, 1964b; Mangasarian, 1964; Lemke and Howson, 1964). They have some of the algebraic tractability of matrix games, since their EPs can be determined by solving systems of simultaneous linear equations based on submatrices.[3] When there are more than two players, however, the corresponding equations are multilinear rather than linear, and relatively less is known about the nature of their noncooperative equilibria (see, however, Nash, 1950c, 1951; H. Mills, 1960; Rosenmüller, 1971c; Wilson, 1971b; Sobel, 1970; Garcia, Lemke, and Leuthi, 1973).

EXERCISES

9.2. *A three-person game is illustrated in figure 9.1. Show that there are two "completely mixed" EPs of the form* $(p, 1-p), (p, 1-p), (p, 1-p)$.

9.3. *Determine the other seven EPs of this game.*

9.1.3 Uniqueness, optimality, and interchangeability

The fact that different EPs in the same game may yield different payoffs makes the question of *uniqueness* especially urgent. Unfortunately it is only for restricted classes of games—most notably for the

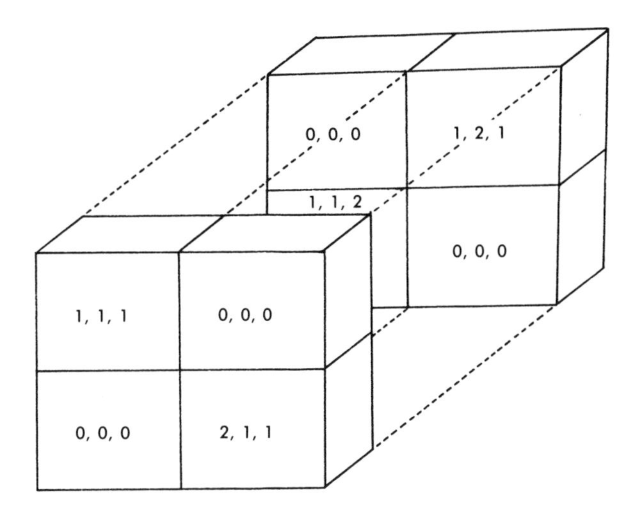

Figure 9.1
A three-person game.

two-person zero-sum (and constant-sum) games—that uniqueness is the rule and multiplicity the exception. Thus almost all matrix games have unique saddlepoints (in pure or mixed strategies), in the same sense and for almost the same reason that almost all square matrices are nonsingular.[4] When we move on to more general games, multiple EPs become more and more common.[5]

We have already identified two EPs in The Winding Road, and since both seem to make sense, the fact of nonuniqueness is not particularly unsettling. There is, though, a third EP to this game that is not so easy to accept. Let M denote the mixed strategy that causes both L and R to be played with probability 1/2. Then it is easily shown that the strategy pair (M, M) is an EP. Indeed M is an equalizer for both players, yielding the fixed payoff $-1/2$ to each. It makes no difference which side of the road you choose if the other car is assumed to be driving at random on either side.

This example illustrates rather vividly that there is no guarantee of Pareto optimality in an EP and, moreover, that even in extremely simple non-zero-sum games an EP can prescribe behavior that most people would consider irrational.[6]

Still other EPs can be found in this little vehicular contretemps if we are willing to adjust the model to include the possibility of driving down the center of the road, blocking both lanes. Call this new strategy C. Then we have

	L	C	R
L	0, 0	−1, −1	−1, −1
C	−1, −1	−1, −1	−1, −1
R	−1, −1	−1, −1	0, 0

To our dismay we find that C is the new equalizer, so that the pair (C, C) is an EP, yielding the payoffs (−1, −1).[7]

Two morals can be drawn from this new solution, depending on whether we regard it as plausible or absurd. To the extent that (C, C) is realistic ("Why should I drive on the side of this narrow road, when I know full well that the other car will be in the center?"), the existence of this solution should reinforce our earlier warning against too hastily eliminating strategies like C from the model merely because they seem intuitively "bad" or "irrational." Better to let our solution concept do the eliminating and, if it fails to do so, be prepared to justify explicitly any suppression of unattractive solutions that we feel compelled to carry out. At the same time, it is a sobering observation that the outcome that is the very worst for everyone can result from the use of an EP.

It is not possible for a unique EP in a bimatrix game to produce the worst possible result for both players. However, in the symmetric game

1, 1	0, 2
2, 0	0, 0

there is only one symmetric EP, and it is "worst possible." Moreover, by considering games of the Prisoner's Dilemma form, such as

1, 1	0, 2
2, 0	ϵ, ϵ

we can find unique EPs that are arbitrarily close to being worst possible.

Similar examples are easily found in the context of infinite games (including differential games) and n-person games. They should warn us against carelessly labeling noncooperative equilibria as "optimal" or "rational" merely because the definition of EP is a generalization of the one-sided utility-maximizing principle of decision theory or control theory.[8]

EXERCISES

9.4. *Consider a noncooperative form of pure bargaining in which each player independently selects one of r possible agreements. If all select the same, then the outcome associated with that agreement occurs; otherwise the disagreement point occurs. Assuming that everyone prefers every agreement to the disagreement point, show the following:*

a. *In the case of two players, there are* $2^r - 1$ *different EPs.*

b. *In the case of n players, with* $2 < n < r$, *there are infinitely many EPs.*

9.5. *Consider the* 3×3 *bimatrix game just discussed, with the payoff for (C, C) changed to* $(-1 + x, -1 + x)$. *Using the method of square submatrices described in section 9.1.2, show that if x is slightly negative, there are only three EPs, but if x is slightly positive, there are seven.*

The general property of interchangeability of strategy pairs giving the value of a zero-sum game—as illustrated in exercise 9.1 where (1, 1), (3, 3), (3, 1), and (1, 3) are all equilibria—does not carry over to non-constant-sum games. Consider two examples:

2, 2	0, 0
0, 0	1, 1

3, 4	1, 4
3, 2	1, 2

In the first matrix the strategy pairs (1, 1) and (2, 2) form equilibrium points, but (1, 2) and (2, 1) clearly do not. The second example displays interchangeability.[9] Here note that P_1 controls the fate of P_2 and vice versa, and the values of the EPs are different.

A *symmetric* two-person zero-sum matrix game has a symmetric solution. So does a symmetric non-constant-sum matrix game, though it may also have nonsymmetric solutions. Consider the following example of a symmetric two-person non-constant-sum game with two pure-strategy nonsymmetric equilibrium points:

2, 1	0, 0
0, 0	1, 2

Figure 9.2 shows a different representation of the game and the value of the symmetric mixed-strategy equilibrium. Here X_1 and X_2 are the payoffs to P_1 and P_2, respectively. The two pure-strategy equilibria using strategies (1, 1) and (2, 2) have payoffs of (2, 1) and (1, 2), respectively. The mixed-strategy equilibrium, ((2/3, 1/3), (1/3, 2/3)), has the payoff (2/3, 2/3). The shaded area is the feasible set of outcomes arising from all mixed strategies, and the bounding curve shows the payoffs when the players employ the same (noncorrelated) mixed strategy. We note that the symmetric equilibrium is not Pareto-optimal, lying as it does in the interior of the feasible set, whereas the other equilibria are optimal.

If both players were myopic enough, in the example above, to play their maxmin strategies, they would obtain a payoff of (2/3, 2/3), but the maxmin strategies are not in equilibrium. The maxmin strategy for P_1 is (1/3, 2/3), but P_2's best response is not his maxmin strategy of (2/3, 1/3) but (0, 1).

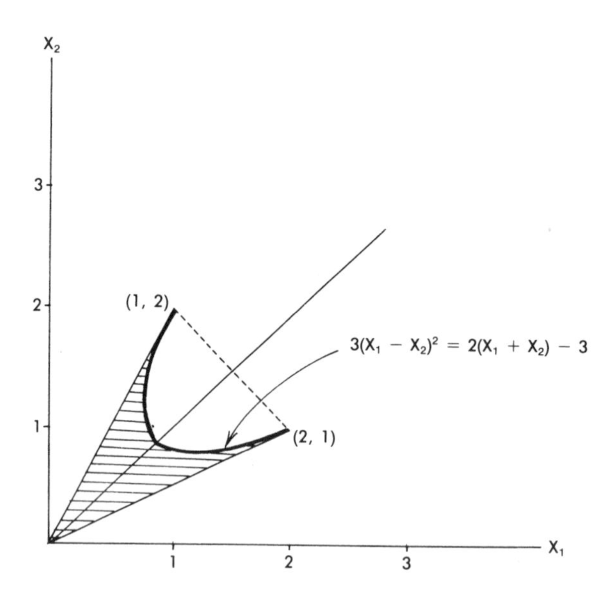

Figure 9.2
A symmetric mixed strategy.

Optimality, conservatism, and equilibrium may or may not coincide. Consider the following matrices:

4, 4	3, 2
2, 3	1, 1

4, 4	1, 3
3, 1	2, 2

In the first there is a unique equilibrium point at (1, 1) which is optimal and would be obtained if both players played their maxmin strategies. In the second there are two equilibria, at (1, 1) and (2, 2); the maxmin strategies yield the second equilibrium, which is strictly dominated by the first.

The more other "nice" properties an equilibrium point has, the more plausible or reasonable as a solution it appears to be (for experimental evidence see Shubik, 1962b). Uniqueness and optimality are appealing as extra conditions. Unfortunately an optimal noncooperative equilibrium is a rare phenomenon. Dubey (1978) has shown that noncooperative equilibria are almost always inefficient. It is easy to select whole classes of games for which this is not so. Indeed a key element in social engineering is to frame the rules of the game so that there will be optimal noncooperative equilibria. This is the problem of designing *self-policing systems.*

Nash regarded a game as solvable if every equilibrium pair were interchangeable. Luce and Raiffa (1957, pp. 106–107) added the condition of *joint admissibility,* that is, Pareto optimality. Thus the Prisoner's Dilemma is solvable in the sense of Nash, but in the terminology of Luce and Raiffa it does not have a *solution in the strict sense.*

9.1.4 Correlated strategies

In figure 9.2 the feasible set of outcomes obtainable from all noncorrelated mixed strategies is not convex. In this game there is a possibility for collusion without explicit side payments. If side payments are permitted, then any point on the line joining (2, 1) to (1, 2) can be achieved. Without side payments, if a correlation of the players' mixed strategies is permitted, they can achieve any outcome in the convex set bounded by (0, 0), (2, 1), and (1, 2).

For strategies to be correlated there must be some mechanism for communicating and contracting between the players. But even here we can make a distinction in the level of enforcement needed for a given agreement. In the given example adherence to the outcome of a joint randomization is self-policing. Even if an individual has the choice to deviate, he does not have the motivation to do so. The players are at one pure-strategy equilibrium or another. This is not so for the game illustrated in figure 9.3 and the following matrix:

8, 4	0, 0
7, 7	4, 8

As in the previous example any outcome on the line connecting (4, 8) to (8, 4) can be obtained from a correlation of strategies involving playing (1, 1) and (2, 2). If the players wish to achieve outcomes in the triangle *ABC*, they must correlate on the strategy pairs (1, 1),

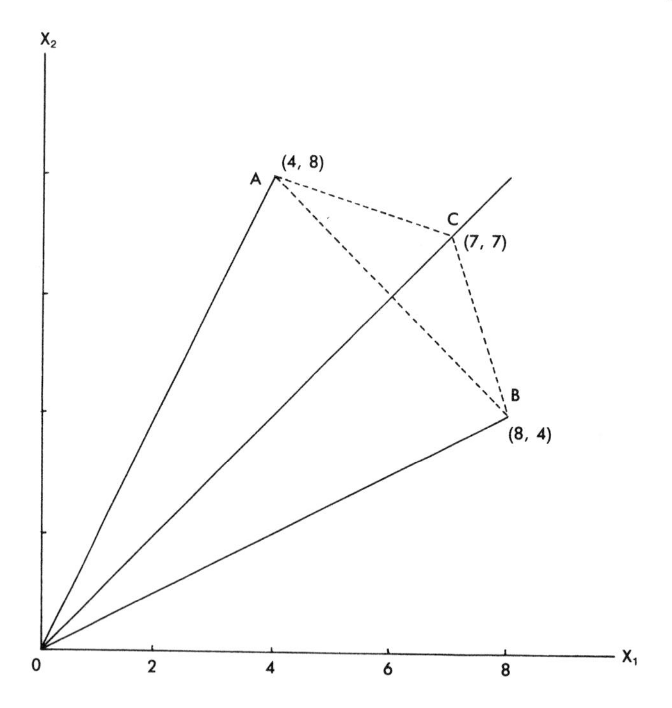

Figure 9.3
The gain from a correlation of strategies.

(2, 2), and (2, 1). If the outcome were a directive for the players to use (2, 1), there would be an incentive for P_1 to deviate if the contract were not binding.

We may interpret conventions such as driving on the left or right of the road as precommitted correlated strategies between any two drivers who immediately know how to play the game and know that the law enforces the societal contract.

Aumann (1974) has provided the following example in which the three players can all gain through a correlated set of strategies:

	1	2
1	0, 0, 3	0, 0, 0
2	1, 0, 0	0, 0, 0

	1	2
1	2, 2, 2	0, 0, 0
2	0, 0, 0	2, 2, 2

	1	2
1	0, 0, 0	0, 0, 0
2	0, 1, 0	0, 0, 3

In this game P_1 picks the row, P_2 the column, and P_3 the matrix. There are three equilibrium payoffs: (0, 0, 0), (1, 0, 0), and (0, 1, 0). Suppose that P_1 and P_2 can correlate so that on the toss of a coin they play (1, 1) with probability 1/2 and (2, 2) with probability 1/2. P_3 plays the second matrix. This is in equilibrium and yields all a non-cooperative-equilibrium payoff of (2, 2, 2).

9.1.5 Instability of mixed-strategy equilibrium points

In the two-person constant-sum case the use of mixed strategies can be justified on defensive grounds. Using your correct, optimal probabilities protects you against any action by your opponent. You do not have to assume that he will play optimally himself; indeed, if his strategy is optimal, it does not matter what probabilities you use, since all of the strategies involved in your optimal mix are then equally good.[10]

In the general case the use of mixed-strategy EPs is harder to justify, since their defensive virtues are no longer clear. Playing a pure-strategy EP may be justified by the fact that it gives (perhaps uniquely) the best possible result on the assumption that the other players will adhere to the EP. But why, in a mixed-strategy EP, should a player want to use the "correct" probabilities?

The defensive argument might again be tried. If you were to use "wrong" probabilities, the other player or players might be motivated to depart from the EP and thereby take advantage of your "mistake."

But an advantage to an opponent need not imply a disadvantage to yourself, since this is no longer a constant-sum situation. In fact the opposite can easily be true, as the following nearly zero-sum example demonstrates:

	P_2	
P_1	$-9, +9$	$+9, -10$
	$+10, -9$	$-10, +10$

The only EP is the pair of mixed strategies (0.5, 0.5), (0.5, 0.5), each of which equalizes the opponent's payoff at zero. Suppose that P_1 deviates to (0.51, 0.49). If P_2 does not anticipate this shift, then P_1's expected payoff will remain zero. But if P_2 does anticipate the shift (or reacts to it, if we imagine the game being replayed a number of times), then his best interests will dictate choosing the first column with an expected payoff of $0.51(+9) + 0.49(-9) = +0.18$ rather than the second column with its expected payoff of $0.51(-10) + 0.49(+10) = -0.2$.

But when P_2 chooses the first column, P_1's expected payoff is also increased to $0.51(-9) + 0.49(+10) = +0.31$. In short, if the deviation is not found out, it doesn't matter; if it is found out, then the natural reaction of the other player benefits both players, the new payoff vector being (0.31, 0.18) rather than (0, 0).

If we consider other deviations in this example, we find that some are self-inhibiting, since they risk a loss with no prospect of gain. Exercise 9.7 gives an example of a game in which every deviation from the EP rewards the deviator, if the other player reacts in his own best interests.[11]

In the face of such evidence it is difficult to make a case for EPs that involve mixed strategies, unless they can be shown to have stronger stability properties than promised by the definition alone. Two other sorts of justification may be possible on occasion: (1) Sometimes one can step outside the usual conceptual framework of game theory and regard the probabilities as subconscious behavioral parameters rather than conscious choices. (2) Sometimes it may be plausible to postulate that the players have some sort of stake in maintaining an equilibrium per se.

Our usual position in applications will be that the EP in pure strategies is the really significant noncooperative solution concept. If it fails to exist, as for example in certain price-cutting duopoly models (Shubik, 1955a; Levitan and Shubik, 1971a), then the fact of its nonexistence may be of central importance to the understanding of the model. But the mixed EP that may exist in the absence of pure EPs may be only of peripheral interest.

EXERCISES

9.6. *In the 2 × 2 example of the text, which mixed strategy is best for P_1 on the assumption that P_2 will react optimally?*

9.7. *Determine the unique EP of the following bimatrix game:*

0, 0	50, 40	40, 50
40, 50	0, 0	50, 40
50, 40	40, 50	0, 0

Show that if either player adopts any other strategy, then the optimal response by the other will result in a superior outcome for both.

9.8. *Determine the unique EP of the following bimatrix game:*

20, 60	40, 20
60, 0	0, 40

Show that each player's maxmin strategy is different from his EP strategy and yet guarantees him as much as he gets in the EP.

9.1.6 Existence of pure-strategy noncooperative equilibria

A result of Dresher (1970) indicates that the existence of pure-strategy noncooperative equilibria is highly probable in non-constant-sum matrix games.[12] Consider an n-person game in which each player i has M_i pure strategies. If all the entries in the n payoff matrices are chosen at random, then for any two players i and j, as the size of their strategy sets is increased, the probability of the existence of a pure-strategy equilibrium approaches $1 - 1/e$, where e is the base of natural logarithms.

Exogenous uncertainty (Nature's move), which is present in many applications, may improve the chances for a pure-strategy equilibrium in a zero-sum or non-constant-sum game. Levitan and Shubik (1971a) provide an example of a class of price-strategy duopoly models without pure-strategy equilibria in which enough uncertainty restores a pure-strategy equilibrium.

As soon as we consider multimove games with information and the possibility of threats (section 9.4), our problem will not be to find pure-strategy equilibria but to cut down on their proliferation.

9.1.7 Another example

The previous four subsections have emphasized weaknesses and limitations in the EP concept. The following example, which is intended to give an idea of the models that will be considered in the next volume, shows the noncooperative solution in a more favorable light.

THE TOMATO MARKET. *Ten farmers have each harvested 150 bushels of tomatoes and must decide independently how much to ship to market and how much to let rot. The market price P in dollars per bushel will depend on the total amount shipped, Q, as*

$$P = 10 - \frac{Q}{100}, \tag{9.2}$$

unless this price is less than the "support" price of 10 cents per bushel.

One EP is immediately apparent: everyone ships his entire crop and collects the support-price payoff of \$15.00 each. Since 1500 bushels are being shipped, no individual, by holding back part of his crop, can even raise the price, let alone raise his profits.

There is another EP, however, which represents a typical oligopolistic practice. If everyone holds back a certain proportion of his crop, not only will the price go up—and the profits—but the strategies will be in equilibrium, so that no individual can gain by shipping more or less than the others.

As it works out, the holdback strategy in question is to ship 91 bushels to market and let 59 bushels rot. (We assume whole bushels and no shipping costs.) It is easy to check this. The price will be

$$P = 10 - \frac{(10)(91)}{100} = 90 \text{ cents per bushel;}$$

the payoff is therefore \$81.90 to each farmer. Suppose one farmer considered shipping x additional bushels. This would depress the price by x cents, according to (9.2), and so his payoff would be

$$(91 + x)\frac{(90 - x)}{100} = 81.90 - \frac{x + x^2}{100}.$$

In short, he would lose money. Likewise, taking x to be a negative integer, the same computation shows that he cannot gain by shipping less than 91 bushels. Hence we have an EP.[13]

The restrained, "tacit" collusion exemplified in this oligopolistic solution may be contrasted with the open collusion that cooperative game theory would envisage. At the value solution, for example, the farmers, in full coalition to maximize total profits, would agree to bring only 500 bushels to market. A price of $5 would result, and each farmer would go home with $250 in his pocket.

EXERCISE

9.9. *Suppose the original harvest of 1500 bushels were evenly distributed among n farmers. Discuss the pure-strategy EP solutions as a function of n.*

9.1.8 Strong equilibrium points

Suppose that in an n-person game in strategic form the set of strategies $(s_1, s_2, \ldots, s_n)$ form an equilibrium point. We might wish to consider higher orders of stability than are given by the definition of a noncooperative equilibrium. Suppose that any subset of players is allowed to coordinate strategies in an effort to overthrow the equilibrium. Aumann (1959) has defined a *strong equilibrium* to be one that is in equilibrium against any subset. He has linked this with multistage games, but here we consider only "one-shot" or matrix games. For an equilibrium to be strong it must be Pareto-optimal, and given all the conditions on its stability against all coalitions, it will be in the beta core of the associated game in cooperative form (see section 6.2.5). Strong equilibria are rare, though they do appear in certain economic models (Dubey, 1980; Dubey and Shubik, 1980; Young, 1978b).

9.2 Non-Constant-Sum Games in Extensive Form

A satisfactory dynamic theory of games is still in its early developmental stages. Difficulties have been encountered both in modeling and in the devising of adequate solution concepts. Many of the failures in obtaining useful applications of game theory to bargaining, international affairs, and economic dynamics stem from inadequate modeling and a confounding of the problems of modeling with those of selecting a solution concept.

This and subsequent sections are aimed at clarifying the differences between the task of model building and that of selecting a solution concept for multimove games. Many dynamic problems are most naturally cast in terms of a game with an indefinite time horizon. For purposes of dividing difficulties and considering the relationships between the extensive and strategic forms as closely as possible, however, we begin by considering only games with a finite number of moves. A simple well-known example—the Prisoner's Dilemma, played twice—will illustrate most of the difficulties both in modeling and in the selection of a solution.

The matrix below provides an example of the Prisoner's Dilemma (Tucker, 1950):

	1	2
1	5, 5	−10, 11
2	11, −10	0, 0

The strategy pair (2, 2) forms the only noncooperative equilibrium. Of the four outcomes in this game, the equilibrium is the only one that is not Pareto-optimal! This game and dynamic variants of it have been the source of well over a thousand experiments. Here we will be concerned with the mathematical model of the one-period game, its relationship to the verbal description usually given, and the problems that arise when the game is played more than once.

The game is called Prisoner's Dilemma because the following verbal scenario is frequently supplied. Two prisoners who are suspected of having committed a crime are interrogated separately by the police. If both maintain silence, at most they can be booked on a minor charge. Each is encouraged to incriminate the other with a promise of leniency if he is not himself incriminated. If they double-cross each other, they are both in trouble.

The story provides a dynamic context for the playing of this "one-shot" game that is more interesting than having unknown individuals come into a laboratory and play a 2 × 2 matrix game anonymously. In particular, the prisoners clearly have a history of prior communication with and knowledge of each other.[14] Random experimental subjects will not have such a history.

Even the laboratory experiments are not devoid of context. The players may not know each other, but unless the experimenter lies they usually know the characteristics of the population from which they are drawn. However, experimental evidence (Wolf and Shubik, 1974) with a single play of this matrix indicates that with anonymous players and no communication an outcome in cell (2, 2) is a reasonably good prediction. Its predictive power is influenced by the relative and absolute sizes of the entries (Rapoport and Chammah, 1965). This lends support to the assumption that individuals not only use utility scales but even make interpersonal comparisons.

When the matrix game is played more than once, even by anonymous players without verbal or face-to-face communication, the stage is set for the development of communication and history in the actual playing of the game. Consider the Prisoner's Dilemma game noted above, played twice. The new game (frequently referred to as a supergame) can be usefully considered in three representations. The extensive form is shown in figure 9.4. Even here care must be taken to distinguish between the case in which the individuals are paid after each play of the 2×2 game and the one in which they are paid at the end of play. We have illustrated the latter. The strategic form gives rise to the single 8×8 matrix shown in figure 9.5. The third representation, which is the one most frequently used in gaming experiments, is usually not fully formulated but consists of the 2×2 matrix presented twice with some sort of accounting or reporting scheme to record the results of the play of the subgame.

A basic assumption of most game theory is that all of these representations are strategically equivalent and hence will not influence any solution or the way the game is played. Empirically this does not appear to be the case (Shubik, Wolf, and Poon, 1974), and furthermore there is no reason why a solution theory should not distinguish among these representations.

If communication did not matter and all aspects of the situation of importance to the game were modeled in the formal structure, then the case for claiming strategic equivalence would be strong. When communication matters, it becomes a problem in modeling to show that it has been accounted for in the formal model.

In the various static solution theories that are called cooperative, the discussion, bargaining, and signaling are all presumed to take

	1	2	3	4	5	6	7	8
(1: 1, 1; 1, 2)	10, 10	10, 10	−5, 16	−5, 16	−5, 16	−5, 16	−20, 22	−20, 22
(1: 1, 1; 2, 2)	10, 10	10, 10	−5, 16	−5, 16	0, 0	0, 0	−10, 11	−10, 11
(1: 2, 1; 1, 2)	16, −5	−20, 22	5, 5	5, 5	−10, 11	−10, 11	−20, 22	−20, 22
(1: 2, 1; 2, 2)	16, −5	−20, 22	5, 5	5, 5	0, 0	0, 0	−20, 22	−20, 22
(2: 1, 1; 1, 2)	16, −5	0, 0	11, −10	0, 0	5, 5	−10, 11	5, 5	−10, 11
(2: 1, 1; 2, 2)	16, −5	0, 0	11, −10	0, 0	11, −10	0, 0*	11, −10	0, 0*
(2: 2, 1; 1, 2)	22, −20	11, −10	22, −20	22, −20	5, 5	−10, 11	5, 5	−10, 11
(2: 2, 1; 2, 2)	22, −20	11, −10	22, −20	22, −20	11, −10	0, 0*	11, −10	0, 0*

Figure 9.5
The strategic form of the twice-iterated Prisoner's Dilemma.

place outside the formal context of the game. In noncooperative equilibrium theory it is assumed that *all* communication takes place within the game. Thus, if discussions are to be held among players, these must be modeled as part of the game. The difficulty is not with the extensive form but with the description or coding of information and communication. It is possible to experiment with a game in which a limited set of messages can be communicated. The selection of a message constitutes a move. Suppose we replace the one-shot game by a two-stage game in which, as a first move, the players simultaneously send one of three messages printed on cards and numbered like mass-produced greeting telegrams. The three messages could be:

1. I intend to select move 1.
2. I intend to select move 2.
3. I do not choose to tell you my move in advance.

The first move for each is thus talk, and the second is action, that is, the actual move.

In general there is no easy way to describe complex sets of messages, verbal inflections, and gestures as sets of moves in a game. Yet all of these may figure importantly in face-to-face communication during bargaining.

When we contemplate a multiperiod game, it is reasonable to consider the possibility of solutions that depend explicitly upon the types of mutual communication, trust, and knowledge that characterize the players. These conditions must be made explicit in the models before a solution is investigated. Otherwise the models will fail to be adequate representations of many political and societal processes.

The first of the two matrices below shows a game that is essentially self-coordinating:

10, 10	0, 0
0, 0	0, 0

10, 10	0, 0
0, 0	10, 10

The players do not need any communication to select their first moves. In the second game the players need to communicate or set up a convention if they are to play in an optimal manner.

9.2.1 Equilibrium points in multimove games

The concept of a saddlepoint and the minimax theorem can be extended to zero-sum games with lengthy but finite periods of repeated play. When we consider non-constant-sum games, even with an extremely simple structure, the extension of the solution concept of the noncooperative equilibrium is neither simple nor satisfactory. For the simplest example, consider Prisoner's Dilemma played twice with each individual required to inform the referee of his first move without prior communication with the other. After this the referee tells both players what moves were selected and what the outcome is. They then each select their second move.

Using backward induction we consider the last play of the matrix game and observe that there is a unique noncooperative equilibrium at (2, 2). Having settled what will be done on the last play, the first play can now be considered, and again the unique equilibrium is (2, 2). We have established that if each player selects his second move each time, this behavior results in a noncooperative equilibrium in the supergame. As we can see from examining figure 9.5, however, this is no longer unique. The outcome (0, 0) is unique, but not the pairs of strategies leading to the outcome. The new game has four equilibrium points, marked with asterisks in the 8 × 8 matrix of figure 9.5. (Note that they are interchangeable.)

The notation on the side of the matrix spells out the overall strategies in detail. For example, the message (1: 1, 1; 1, 2) should be read "I play 1 to start; then on the second trial I play 1 if he has played 1, and I play 1 if he has played 2." The numbers 1–8 above the columns of the matrix are merely a coding or name for each of these eight strategies.

This proliferation of equilibrium points is both quantitative and qualitative. A distinction can be made between state or Markovian strategies and historical strategies.

Consider a modified version of this game in which at each of the two trials the players confront the following 3 × 3 matrix:

	1	2	3
1	5, 5	−10, 11	−19, −20
2	11, −10	0, 0	−19, −20
3	−20, −19	−20, −19	−20, −20

An extra row and column have been added to the original matrix. They contain highly negative entries which serve as threats if the players can communicate.

Using backward induction, we can show that if each player utilizes his second move in each subgame, the result is a noncooperative equilibrium. All four equilibria in the previous multimove game are still equilibria here, but further equilibria with higher payoffs now appear. For example, if each uses the same strategy (1: 2, 1; 3, 2; 3, 3)—read "I play 1 to start; then on the second trial I play 2 if he played 1, otherwise I play 3"—we obtain an equilibrium with payoffs of (5, 5). This is not quite Pareto-optimal but is certainly better than (0, 0). Furthermore, the more times the game is played, the closer an analogous "threat-strategy" equilibrium comes to being Pareto-optimal.

EXERCISES

9.10. *Calculate the number of noncooperative equilibria for the game displayed in the 3×3 matrix above, played twice.*

9.11. *Specify a pair of threat strategies for this game played k times such that there is an equilibrium point that yields each of the players a payoff that differs from the jointly maximal payoff by less than ϵ. How large must $k(\epsilon)$ be?*

It is easy to show that these new strategies satisfy the formal definition of an equilibrium point without explicitly considering the role of communication. We could think of the game in strategic form as consisting of a large one-shot matrix game. These equilibria become more plausible, however, if we formally model communication as part of the actual moves of the game. In this example we could consider the first move for each as the sending of a message to the other concerning his intentions.

9.2.2 An example of equilibrium points in a multimove game

There are two players and N time periods ("rounds"). In each round the players (simultaneously) choose to help (he), hurt (hu), or pass (pa), as follows:

he: Pay \$1 to make the bank give \$2 to the other player.

hu: Pay \$2 to make the bank take \$1 from the other.

pa: Do nothing.

Thus the payoff matrix for each round is:

	he	hu	pa
he	1, 1	−2, 0	−1, 2
hu	0, −2	−3, −3	−2, −1
pa	2, −1	−1, −2	0, 0

Note that pa dominates he and he dominates hu. Figure 9.6 indicates the eight possible changes in the players' fortunes in a single round. Figure 9.7 displays the possible situations after five rounds of play (the numbers indicate the earliest round on which each point can be reached).

For $N = 1$ there is a unique EP, namely pa-pa, because of the strict domination of strategies. (Note that if we eliminate hu, we have a Prisoner's Dilemma matrix.)

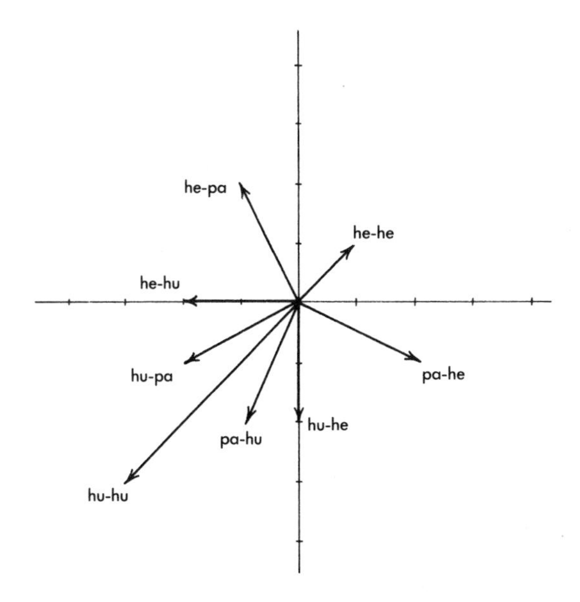

Figure 9.6
The possible changes in each player's fortunes in one round of a multimove game.

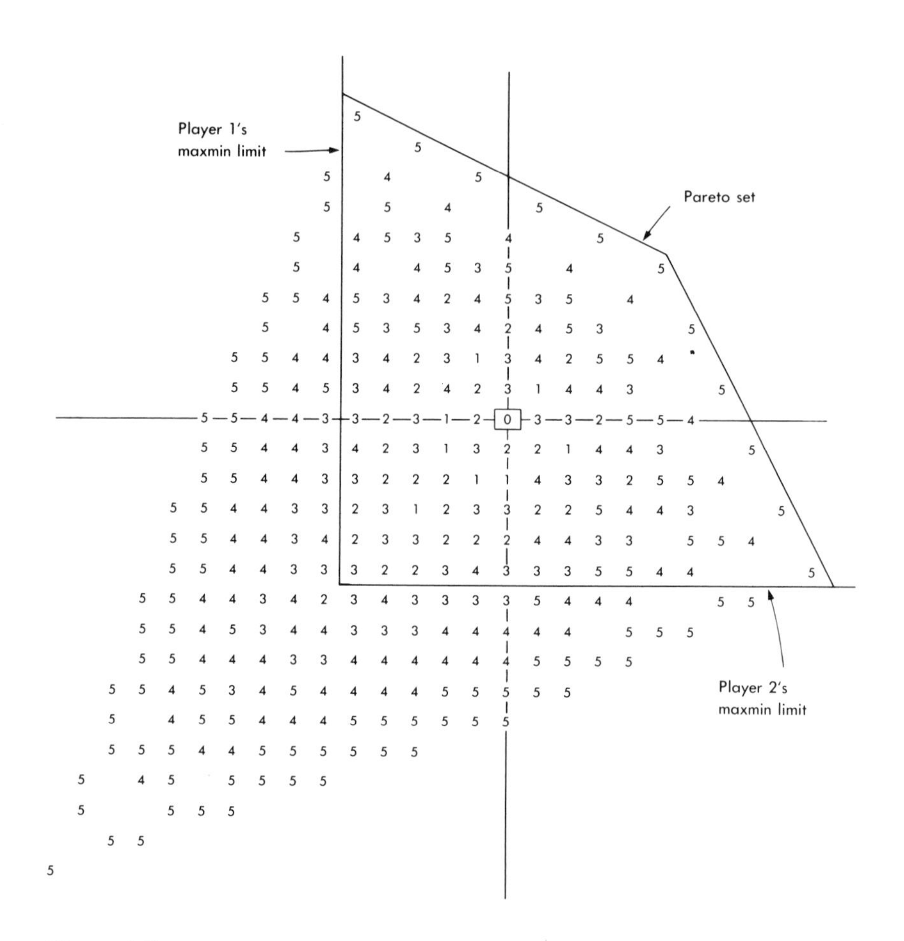

Figure 9.7
Equilibrium points in the iterated Prisoner's Dilemma for $N = 5$.

For any N every EP must cause the last round to be played pa-pa for the same reason. Moreover, there will be a unique EP that is perfect or global, namely the one in which the players' strategies require them to pass under all circumstances. The associated final payoff is of course (0, 0), which may be compared with the cooperative joint maximum at (N, N) and with the maxmin level of $-N$ for each player individually.

The number of operationally different pure strategies can be shown to be

$$3 \cdot 3^3 \cdot 3^{3 \cdot 3} \cdot 3^{3 \cdot 3 \cdot 3} \cdot \ldots \cdot 3^{3^{N-1}} = 3^{(3^N - 1)/2}.$$

We now define a special class of strategies: the *T-strategies,* where T stands for a play of the game, that is, a sequence of N pairs of moves, such as

he-he, he-hu, hu-pa, . . . , pa-he, pa-pa.

The T-strategy for either player is defined by the rule: *Make the choice indicated by T on the first round and on any subsequent round immediately following a T-choice by the other player; otherwise choose hu.* Clearly if both players use their T-strategy, then the move sequence T will result.

Intuitively a T-strategy punishes each departure from T by the other player, by hurting him in the following round.

THEOREM. *If T contains no "hurts" and ends with pa-pa, then each player playing his T-strategy constitutes an equilibrium point.*

Proof. If a player disobeys T on any given round, this will save him at most \$1 on that round, but will lose him at least \$1 on the next round. ■

More generally, a pair of T-strategies is an EP if and only if T ends with pa-pa and each hu in T is preceded by pa for the other player and followed by he for that player. (Proof is similar.)

The interesting point about this example, which is typical of many multimove noncooperative games, is that most of the outcomes that are feasible and individually rational can be achieved by T-strategy EPs, that is, EPs involving only a minimal threat of punishment for violating the behavioral norm. In particular, outcomes approaching the Pareto set along its entire extent can be so attained (see figure 9.8). For example, the outcome (−4, 8) will result from the T-strategy EP with

T = (he-pa, he-pa, he-pa, he-pa, pa-pa).

Figure 9.8
Types of noncooperative equilibria.

With the aid of heavier punishments, EPs can be constructed having any outcome whatever that is individually rational and can be reached in $N - 1$ rounds. These are indicated by the hexagons in figure 9.8. For example, to reach $(-5, -5)$ it suffices to take the move sequence

T = (hu-hu, hu-hu, he-he, pa-pa, pa-pa)

and let the strategy of each player be to obey T as long as the other does, but to play hu continuously to the end of the game if the other player ever departs from T.

9.2.3 Perfect equilibria

Our concern for the existence of pure-strategy equilibria in one-shot matrix games has been replaced by an overwhelming multiplicity of equilibria in multimove games. By imposing extra properties beyond the simple Nash definition, though, we can cut down the number of equilibria eligible as solutions. One general approach to the selection of equilibria has been suggested by Selten (1965, 1975), who has introduced the idea of a *subgame-perfect equilibrium point.*

Consider an n-person game Γ in extensive form. Let Γ' be a subgame of Γ, and let $q = (q_1, q_2, \ldots, q_n)$ be a strategy combination for Γ. The system of probability distributions assigned by q_i to the information sets of player i in Γ' is a strategy q_i' for Γ'. This strategy q_i' is said to be *induced by* q_i *on* Γ', and the strategy combination $q' = (q_1', q_2', \ldots, q_n')$ is said to be *induced by* q *on* Γ'. A subgame-perfect equilibrium point $s = (s_1, s_2, \ldots, s_n)$ is then an equilibrium point in behavior strategies that induces an equilibrium point on every subgame of Γ. In other words, a subgame-perfect equilibrium is an equilibrium point in every subgame of the supergame, given that the supergame can be partitioned into subgames at any point of perfect information for any player as long as the subtree does not have nodes contained in information sets with other nodes not within the subtree.

Figure 9.9 illustrates the concept. We can regard the segment of the tree from P_1's second move onward as a subgame. The rest of the tree cannot be partitioned. In the iterated Prisoner's Dilemma game of figure 9.4 there are four subgames that can be treated separately, one for each node at which P_1 has perfect information (leaving out the initial node). The unique subgame-perfect equilibrium involves each player choosing his second pure (local) strategy in each subgame.

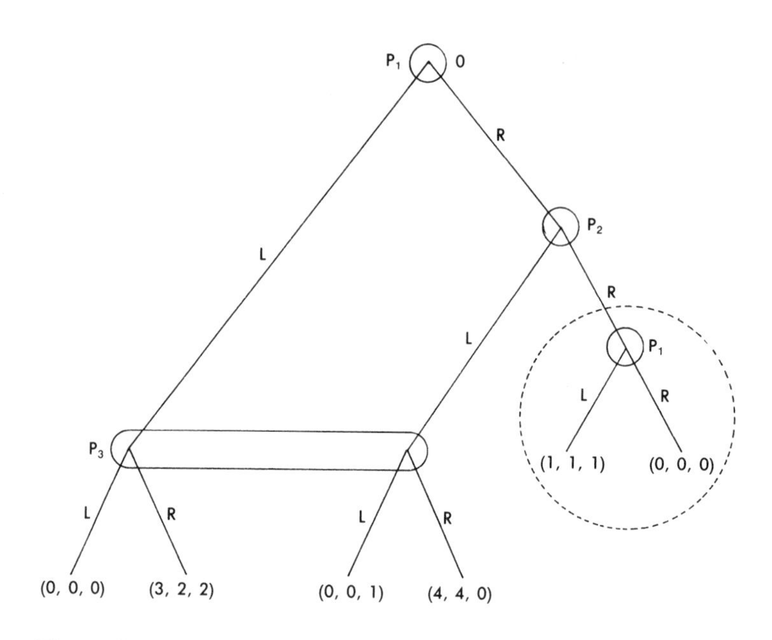

Figure 9.9
Perfect equilibria (example 1).

Returning to the game illustrated in figure 9.9, we can show that in the subgame P_1 will choose L; hence we can replace his second node with the value (1, 1, 1). In the game that remains, each player has one information set and a choice of L or R. Denote the probability that P_i chooses R by p_i. Then it is straightforward to verify that the game has two types of subgame-perfect equilibrium points:

Type 1: $p_1 = 1, \quad p_2 = 1, \quad 0 \leq p_3 \leq 1/4$ with outcome (1, 1, 1).

Type 2: $p_1 = 0, \quad 1/3 \leq p_2 \leq 1, \quad p_3 = 1$ with outcome (3, 2, 2).

For illustration we follow through Selten's analysis of this example. In a type 2 equilibrium point, P_2's information set is not reached, hence his expected payoff does not depend on his strategy. Consider, for example, the equilibrium $p_1 = 0$, $p_2 = p_3 = 1$. If there were even a minute chance of P_1 choosing R instead of L, it would be foolish for P_2 to choose R instead of L, since this would yield a payoff of (4, 4, 0). Furthermore, at the equilibrium, P_2's payoff is uninfluenced by his actions. Thus P_2 can compute his conditional expectation of payoff given that his information set is reached. But P_3 cannot; his information set is reached in two ways, and he is therefore unable to distinguish the actions of P_1 or P_2.

The number of subgame-perfect equilibria can be cut down to the class of perfect equilibria by using a sensitivity analysis, introducing a small perturbation on all of the local moves. This means in particular that there will always be a small chance that all branches in the game tree are reached. We then study the limiting behavior of the equilibria in a sequence of "perturbed games" as the error goes to zero and consider the set of limit equilibrium-point behavior strategies thus obtained.

Applying this idea to the example in figure 9.9 modified for the perturbed game $\hat{\Gamma}^k$, we find that the probability that P_i plays R is p_i, where $1 - \epsilon_k \geq p_i \geq \epsilon_k$ $(i = 1, 2, 3)$. Select the sequence $\epsilon_1, \epsilon_2, \ldots,$ where $\epsilon_1 < 1/4$ and $\epsilon_k \to 0$ as $k \to \infty$; then the perturbed game $\hat{\Gamma}^k$ has a unique equilibrium point of the form

$$p_1^k = 1 - \epsilon_k, \qquad p_2^k = 1 - \frac{2\epsilon_k}{1 - \epsilon_k}, \qquad p_3^k = 1/4. \tag{9.3}$$

We first show that this is an equilibrium by checking best replies. Consider P_3. The probabilities that his information set is reached are given by $1 - p_1$ from P_1 and $p_1(1 - p_2)$ from P_2. If his information set is reached and his choice is R, his expected payoff is $2(1 - p_1)$; if he takes L, then it is $p_1(1 - p_2)$. Hence p_3 is a best reply to the proposed strategies if

$$p_3 = \epsilon_k \quad \text{for} \quad 2(1 - p_1) < p_1(1 - p_2), \tag{9.4}$$

$$\epsilon_k \leq p_3 \leq 1 - \epsilon_k \quad \text{for} \quad 2(1 - p_1) = p_1(1 - p_2), \tag{9.5}$$

$$p_3 = 1 - \epsilon_k \quad \text{for} \quad 2(1 - p_1) > p_1(1 - p_2). \tag{9.6}$$

From (9.3) it follows that (9.5) is satisfied, hence p_3^k is a best reply.

Consider P_2. He has a best reply if

$$p_2 = \epsilon_k \quad \text{for} \quad p_3 > 1/4, \tag{9.7}$$

$$\epsilon_k \leq p_2 \leq 1 - \epsilon_k \quad \text{for} \quad p_3 = 1/4, \tag{9.8}$$

$$p_2 = 1 - \epsilon_k \quad \text{for} \quad p_3 < 1/4. \tag{9.9}$$

Since $p_3 = 1/4$ from (9.8), p_2^k is a best reply.

Finally, P_1 has a best reply if

$$p_1 = \epsilon_k \quad \text{for} \quad 3p_3 > 4(1 - p_2)p_3 + p_2, \tag{9.10}$$

$$\epsilon_k \leq p_1 \leq 1 - \epsilon_k \quad \text{for} \quad 3p_3 = 4(1 - p_2)p_3 + p_2, \tag{9.11}$$

$$p_1 = 1 - \epsilon_k \quad \text{for} \quad 3p_3 < 4(1 - p_2)p_3 + p_2. \tag{9.12}$$

From (9.12) p_1^k is a best reply.

Uniqueness is established as follows. Suppose $p_3 < 1/4$. Then from (9.9) $p_2 = 1 - \epsilon_k$, hence $3p_3 < p_2$. From (9.12) $p_1 = 1 - \epsilon_k$, hence (9.6) applies to p_3. But $p_3 = 1 - \epsilon_k$ is contrary to the assumption that $p_3 \leq 1/4$.

Suppose $p_3 > 1/4$. From (9.7) $p_2 = \epsilon$. But $1 - p_2 > 3/4$, hence, from (9.12), (9.4) applies to p_3 contrary to the assumption that $p_3 > 1/4$.

We have established that an equilibrium point of $\hat{\Gamma}^k$ must have $p_3 = 1/4$. From (9.12) we have $p_1 = 1 - \epsilon_k$, and from (9.4) we have $2(1 - p_1) = p_1(1 - p_2)$, hence $1 - p_2 = 2\epsilon_k/(1 - \epsilon_k)$.

Using a similar argument it can be shown that every type 1 equilibrium point is perfect. Selten establishes that type 2 equilibrium points are not perfect.

Myerson (1978) offers a modification of Selten's perfect equilibria which he calls *proper equilibria.* Proper equilibria are a nonempty subset of perfect equilibria. An example will suffice to differentiate the two concepts. In the game below there are equilibria at (1, 1) and (2, 2):

1, 1	0, 0
0, 0	0, 0

Only (1, 1) is a perfect equilibrium. Any small probability that the first strategies will be used wipes out (2, 2) as an equilibrium.

In the game below there are equilibria at (1, 1), (2, 2), and (3, 3):

1, 1	0, 0	−9, −9
0, 0	0, 0	−7, −7
−9, −9	−7, −7	−7, −7

Here (1, 1) and (2, 2) are both perfect but (3, 3) is not. We may discriminate between (1, 1) and (2, 2), however, by defining an ϵ-

proper equilibrium to be any combination of totally mixed strategies such that every player gives his better responses considerably more probability weight than his worse responses. Specifically if $(\sigma_1, \sigma_2, \ldots, \sigma_n)$ is the n-tuple of totally mixed strategies at an ϵ-proper equilibrium, and if the pure strategy s_i yields a higher expected value to i than a different pure strategy s_i^1, then s_i will be given a weight at least $1/\epsilon$ more than s_i^1. In our example, since strategy 2 dominates strategy 3 for P_1, $\sigma_{13} \leq \epsilon\sigma_{12}$, where σ_{ij} is the probability weight that P_i places on strategy j. This implies that P_2's first strategy will yield more than his third, hence $\sigma_{23} \leq \epsilon\sigma_{21}$. This implies that P_1's first strategy yields more than his second, hence $\sigma_{12} \leq \epsilon\sigma_{11}$. Thus $\sigma_{12} \leq \epsilon$ and $\sigma_{13} \leq \epsilon^2$, and similarly for σ_{22} and σ_{23}. Since probabilities sum to 1, $\sigma_{11} \geq 1 - \epsilon - \epsilon^2$ and $\sigma_{21} \geq 1 - \epsilon - \epsilon^2$. As $\epsilon \rightarrow 0$ the ϵ-proper equilibrium converges to the equilibrium pair (1, 1).

In essence, Myerson has introduced one more level of probabilistic smoothing onto the Selten procedure for selecting perfect equilibria. But in the game shown below both of the equilibrium points are strong and proper:

2, 2	0, 0
0, 0	1, 1

Thus although the idea of a perturbation is intuitively appealing as a way of cutting down the equilibrium points, there may be other criteria we wish to consider.

E. Kohlberg (private communication) has noted that the Selten construct for perfect equilibria is not robust against the addition of completely dominated strategies. Consider the games shown in figures 9.10 and 9.11, where p_i is the probability that P_i will choose L. In figure 9.10 the outcomes (2, 2) and (3, 1) can result as equilibria: the strategy pair $p_1 = 1$, $0 \leq p_2 \leq 2/3$ gives the outcome (2, 2) as an equilibrium; and $p_1 = 0$, $2/3 \leq p_2 \leq 1$ gives (3, 1). Only the latter is a perfect equilibrium point.

The game in figure 9.11 is obtained by enlarging P_2's possibilities but not allowing him extra information. As far as P_1 is concerned, the enlargement of the game is irrelevant since it is dominated by the previous alternatives. Yet the outcome (2, 2) is now a perfect

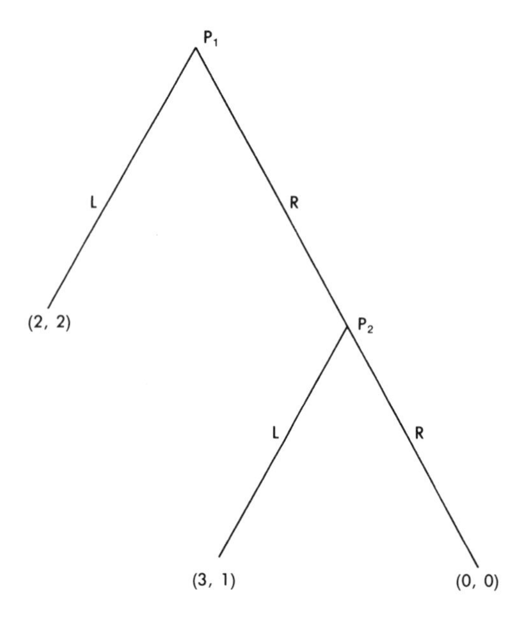

Figure 9.10
Perfect equilibria (example 2).

equilibrium, as can be seen by checking P_2's best reply against $(1 - 2\epsilon, \epsilon, \epsilon)$, which is $(\epsilon, 1 - \epsilon)$, and P_1's best reply against $(\epsilon, 1 - \epsilon)$. The extra branches in figure 9.11 (over 9.10) only provide P_2 with an incentive for an empty threat.

The Selten ϵ-perturbation is one of several perturbations to which we might wish to subject a game in order to test for robustness of the equilibria. For instance, we could consider perturbations in payoffs or in information leaks. None of these seems to be a normative necessity. Hence, in deciding to stress one technique or another for pruning equilibrium points, we may wish to examine the properties of the equilibria that remain and the properties of those removed. In particular, in the example of figure 9.9 the equilibrium points that are removed have optimal payoffs of (3, 2, 2), and the ones that are kept have payoffs of (1, 1, 1), which are strictly dominated by the others. This result is not intuitive and casts some doubt on the fruitfulness of this approach to perfect equilibria.

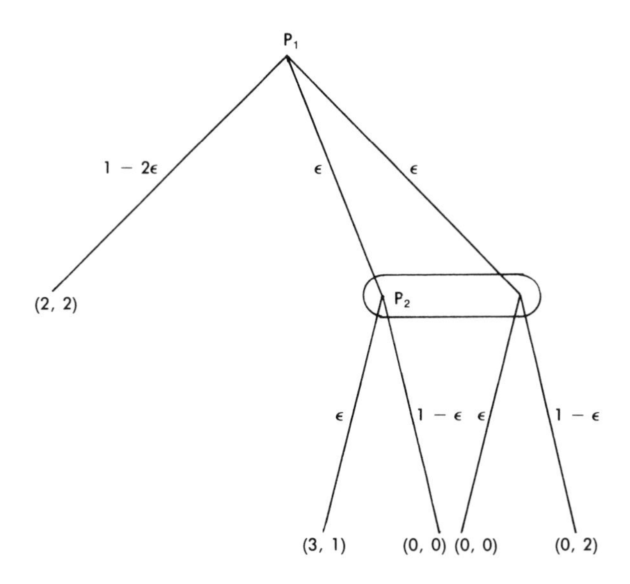

Figure 9.11
Perfect equilibria (example 3).

9.2.4 Equilibria and more information

We now turn to a sensitivity analysis of noncooperative equilibrium points in finite games in extensive form as we vary information conditions.

Consider two games Γ and $\hat{\Gamma}$. We say that $\hat{\Gamma}$ is a *refinement of* Γ (denoted $\Gamma \leq \hat{\Gamma}$) if it is obtainable from Γ only by forming partitions of the information sets in Γ. Calling $\mathscr{C}$ the collection of games in extensive form that arise by imposing all possible information sets on a fixed underlying tree (that is, the nodes, moves, indexing sets, and payoffs are all held fixed, and only the information sets are varied), it is clear that $\lesssim$ gives a partial ordering on $\mathscr{C}$. There is in this ordering a unique maximal element $\Gamma_{\mathrm{R}}^{\mathscr{C}}$ and a unique minimal element $\Gamma_{\mathrm{C}}^{\mathscr{C}}$ in $\mathscr{C}$. In $\Gamma_{\mathrm{R}}^{\mathscr{C}}$ all information sets are singletons. The information sets of $\Gamma_{\mathrm{C}}^{\mathscr{C}}$ are obtained by constructing for each player i the coarsest partition of his nodes. This is illustrated in figure 9.12. For any $\Gamma \in \mathscr{C}$, $\Gamma_{\mathrm{R}}^{\mathscr{C}}$ is sometimes called its *most-refined form* and $\Gamma_{\mathrm{C}}^{\mathscr{C}}$ its *most-coarsened form.* The broken lines describe $\Gamma_{\mathrm{C}}^{\mathscr{C}}$; the unbroken lines, $\Gamma_{\mathrm{R}}^{\mathscr{C}}$.

Let $\Gamma \lesssim \hat{\Gamma}$. Though the strategy sets S^i and $\hat{S}^i$ of i in Γ and $\hat{\Gamma}$ are formally different, there is a natural inclusion $S^i \subset \hat{S}^i$. Every strategy

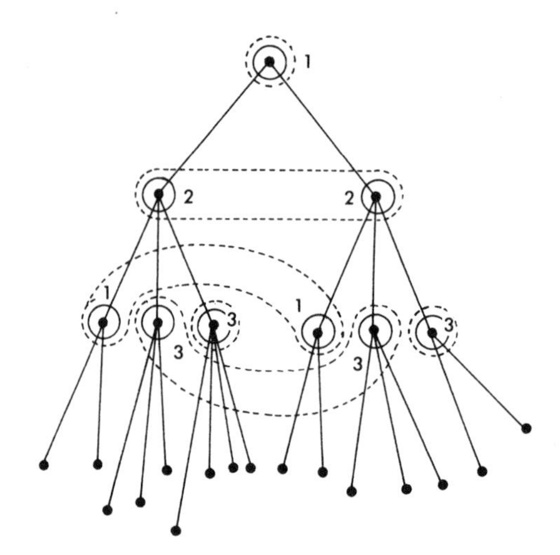

Figure 9.12
Information and pure strategies.

s^i in S^i is identified with a strategy $\hat{s}^i$ in $\hat{S}^i$ as follows: the move chosen by $\hat{s}^i$ at any information set $\hat{I}^i_t$ of i in $\hat{\Gamma}$ is the same as the move chosen by s^i at I^i_j, where I^i_j is the unique information set of i in Γ for which $\hat{I}^i_t \subset I^i_j$.

For any game Γ we denote by $\mathcal{N}(\Gamma)$ the set of all its pure-strategy noncooperative equilibrium points. The following straightforward but striking result has been proved by Dubey and Shubik (1981).

THEOREM. *Suppose* $\Gamma \leq \hat{\Gamma}$. *Then* $\mathcal{N}(\Gamma) \subset \mathcal{N}(\hat{\Gamma})$. *That is, the pure-strategy equilibria of the games with less information are also equilibria of the games with more information.*

We call the set of equilibria in the most-coarsened form of the game tree *information-insensitive equilibria* since they are not destabilized by additional information. They provide an example in which, if all players but one choose to ignore extra information, that individual cannot gain from exploiting this information. This game-theoretic conclusion modifies an old proverb: In the kingdom of the blind, the one-eyed man had better conform.

This proposition, we should note, is not true for mixed strategies. Consider a simple game of Matching Pennies in which player 1 wins if they match. If both players move simultaneously, the payoff matrix is 2 × 2—each player has two moves that coincide with his strategies. The only noncooperative equilibrium comes when each player uses a (1/2, 1/2) mixture of his two pure strategies. The expected payoff to each is then zero. A refinement of the information sets in this game is given if we assume that player 2 is informed of player 1's move before player 2 is called upon to move; but the refined game has only a pure-strategy equilibrium.

It can be shown that as information is refined in a game represented by a finite game tree, the number of nondegenerate mixed-strategy equilibria remains the same or decreases, until with perfect information all nondegenerate mixed-strategy equilibria disappear (Dubey and Shubik, 1981).

If there are chance moves in the game, and if we vary information of the traders regarding each other's moves only, while their information about chance moves is held fixed, then the proposition continues to hold. However, the example in figure 9.13 shows that the result breaks down outside of this case. Here 0 represents a chance move that selects L and R with probabilities 9/10 and 1/10, respectively. The first (second) component of the payoff vector is the payoff

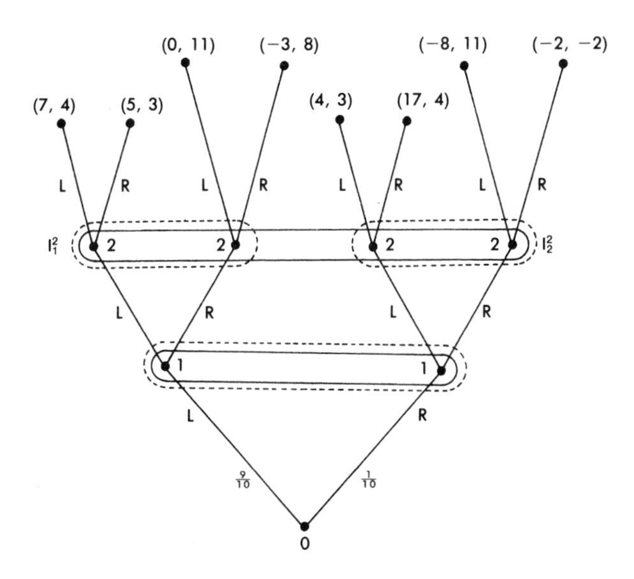

Figure 9.13
Information and exogenous uncertainty.

to 1 (2). In the game with unbroken information sets the unique noncooperative equilibrium is: 1 chooses L; 2 chooses L. In the game with broken information sets the unique noncooperative equilibrium is: 1 chooses L; 2 chooses L on I_1^2 and R on I_2^2.

9.2.5 When more information hurts

Jack and Jill have chipped in to buy a ticket in a raffle, agreeing to share equally anything they win. Since the prizes offered are mostly indivisible objects (a ticket can win at most one prize), they plan to decide ownership by the toss of a coin. When (for simplicity) only two prizes remain, let us suppose that the raffler draws their ticket from the bowl and proclaims it a winner; he then prepares to draw from another bowl the marker that determines which prize has been won.

At this precise point in time Jack's expected utility is $(u^1(A) + u^1(B))/4$ and Jill's is $(u^2(A) + u^2(B))/4$, where A and B are the remaining prizes and we have normalized $u^1(\varnothing) = u^2(\varnothing) = 0$. Now Jill suddenly realizes that their tastes are different:

$$u^1(A) > u^1(B) > 0, \qquad u^2(B) > u^2(A) > 0$$

(figure 9.14a) and drags Jack into the next room, where they cannot hear the raffler's next announcement. She then proposes a new division scheme: if the prize turns out to have been B, she takes it; if it turns out to have been A, Jack gets it. Since this increases their expectations to $u^1(A)/2$ and $u^2(B)/2$, respectively, Jack quickly agrees, and they return and claim their prize.

The point of this story is that purposeful disregard of freely available information can sometimes be beneficial to all concerned.

EXERCISE

9.12. *Show that the set of possible division schemes is represented by the shading in figure 9.14c. In what sense was the scheme chosen "best"? "fairest"?*

9.3 Games with Incomplete Information

The requirements of formal game theory are strict regarding the rules of the game and the portrayal of exogenous uncertainty. All rules are given, and exogenous uncertainty is portrayed by given objective probabilities describing Nature's acts. Where these probabilities come from, how individuals form subjective probabilities, and

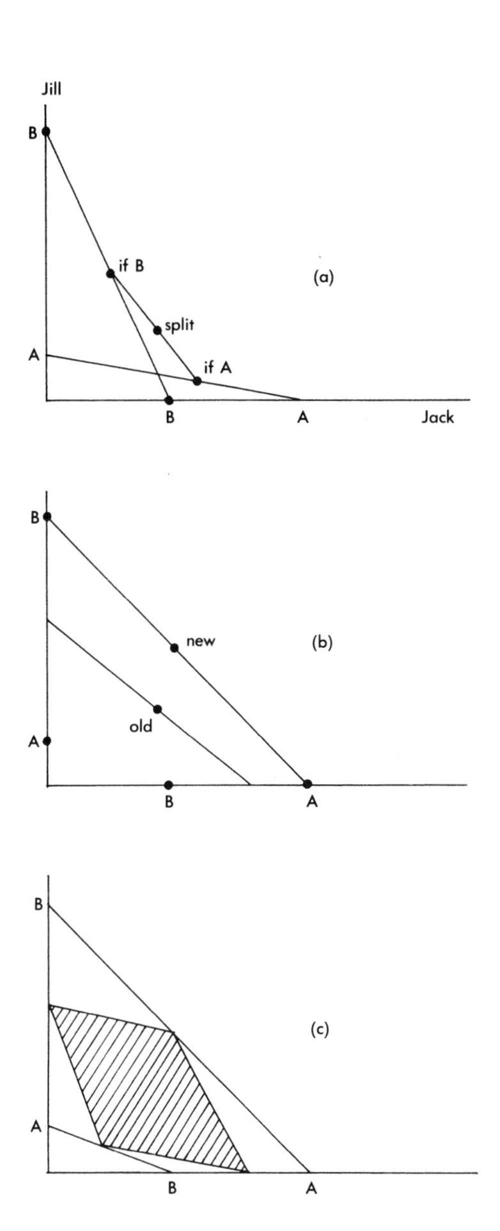

Figure 9.14
An example in which disregard of information proves beneficial.

how they update their probabilities pose deep problems in modeling, psychology, and decision theory. We shall not attempt to cover these topics here, our major concern being the n-person aspects of competition and cooperation; a summary is given in appendix C.

9.3.1 *n*-Person games with incomplete information

Many strategic situations cannot initially be modeled as well-defined games because players lack information about (1) available strategies, (2) payoff functions, or (3) outcomes resulting from various strategies. One way to convert such situations so that they can be modeled as well-defined games is to introduce a move by Nature. The lack of definition in the rules of the game is compensated by modeling each player's uncertainties concerning the rules as subjective probabilities.

The first model in which an ill-defined strategic situation was enlarged and defined as a Bayesian game was, to the best of my knowledge, Vickrey's (1961) study of bidding in which the players were not informed of each other's costs. Instead each player was given a probability distribution over the others' costs. The same method was utilized by Griesmer, Shubik, and Levitan (1967). A simple version of the bidding model is given in the following exercise.

EXERCISE

9.13. *A single object is to be sold at auction. There are two bidders. Each is required to submit his bid secretly to the seller. Each has $100, and they both know this. Each knows the value of the object to himself and assigns a rectangular distribution over the range 0–$100 for his subjective estimate of the worth of the object to the other. Each bidder is aware of the other's state of information and estimation. Suppose that the actual values to bidders 1 and 2 are* a_1 *and* a_2*. Calculate the equilibrium point. Is it unique?*

A systematic development of games with incomplete information played by Bayesian players has been given by Harsanyi (1967, 1968a,b). The players may lack information about the outcomes, the utility functions, or parts of the strategy spaces. The uncertainty is treated by means of a joint subjective probability distribution that is shared by all players at the start of the game. Rather than develop a considerable amount of special notation, we illustrate this approach to games with incomplete information by example.

Suppose that two players P_1 and P_2 are about to play a simple 2×2 matrix game G. They are uncertain about the payoffs of the game, though, and in actuality there are four possible games they could be playing. Denote these by G_{11}, G_{12}, G_{21}, and G_{22}.

One way to approach this situation is to assume that each of the players consists of a two-person team. Depending upon which agent from each team is selected to play, the game played will be different. The uncertainty in the original game is now represented by the uncertainty of each player concerning the selection of agents.

The agents are denoted by P_1^1, P_1^2 and P_2^1, P_2^2, respectively. Figure 9.15 shows the extensive form in which Nature randomizes to select the agents but only the information concerning the selection of one's own agent is realeased to the players. Thus, for example, when player 2 is called upon to play, even if he knows that he is represented by his first agent, he does not know whether he is playing against P_1^1 or P_1^2, that is, in game G_{11} or G_{21}. For simplicity in the example we assume that each of the four games is zero-sum:

G_{11}

	1	2
1	−1	1
2	1	−1

G_{12}

	1	2
1	3	4
2	−4	−4

G_{21}

	1	2
1	2	0
2	1	1

G_{22}

	1	2
1	0	0
2	0	3

We may now construct the normal form of the game and also consider the different games as viewed by the agents. The strategic form of the game has four pure strategies for each player. The strategy for player 1 is:

If I am called to send agent P_1^1, I shall play i, where $i = 1, 2$.

If I am called to send agent P_1^2, I shall play j, where $j = 1, 2$.

Call the strategies for player 1 x_{ij} and for player 2 y_{ij}. Then the strategic form is:

	y_{11}	y_{12}	y_{21}	y_{22}	min
x_{11}	1	5/4	5/4	5/4	1 max
x_{12}	3/4	7/4	5/4	9/4	3/4
x_{21}	−1/4	−1/4	−5/4	−5/4	−5/4
x_{22}	−1/2	1/4	−1	−1/4	−1
max	1 min	7/4	5/4	9/4	

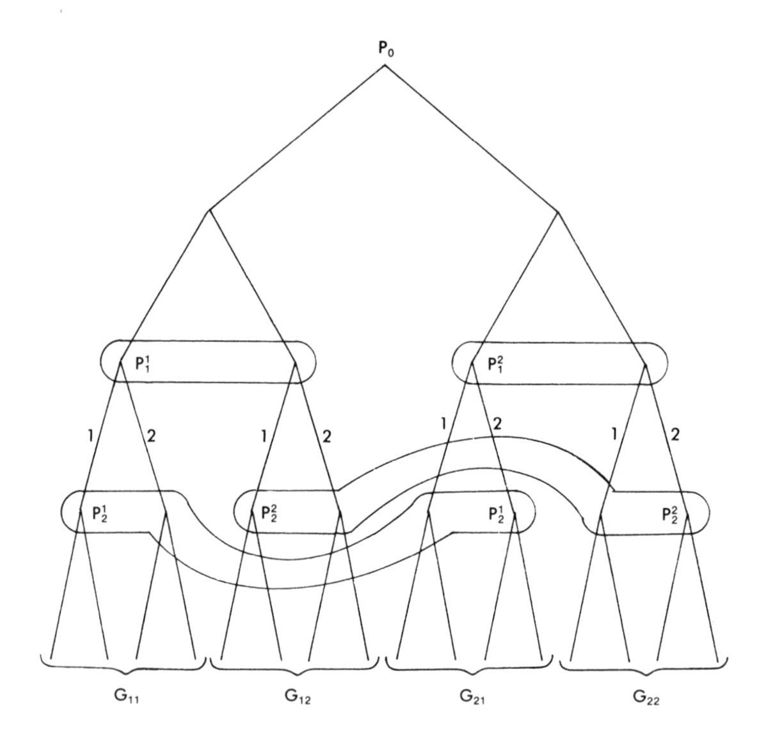

Figure 9.15
A game with incomplete information.

The entries in this matrix are calculated as follows. Suppose that the a priori probabilities for the occurrence of the various games are 1/4 each. Before knowing which of their agents have been selected, if player 1 uses x_{11} and player 2 uses y_{11}, the payoff will be

$$\frac{1}{4}(-1) + \frac{1}{4}(3) + \frac{1}{4}(2) + \frac{1}{4}(0) = 1.$$

A similar calculation can be made for the other entries.

In normal or strategic form (which we denote by G^*) this is a zero-sum matrix game, and it can be shown to have a saddlepoint at (x_{11}, y_{11}), which is also a (Bayesian) noncooperative equilibrium point.

Associated with the solution to a zero-sum game in strategic form is the concept of a safety level. No matter what the other player does, player 1 can guarantee himself the payoff given by the maxmin strategy. Here, however, we encounter a conceptual difficulty that

Harsanyi terms "delayed commitment": Why should a player use a fully mixed strategy rather than a behavior strategy that makes full use of his knowledge of which of his agents is selected?

Consider the game G from the point of view of the two agents of P_1. If P_1^1 is selected, then he faces the following expected outcomes against P_2's pure strategies:

	y_{11}	y_{12}	y_{21}	y_{22}
$i = 1$	$\frac{1}{2}(-1) + \frac{1}{2}(3)$	$\frac{1}{2}(-1) + \frac{1}{2}(4)$	$\frac{1}{2}(1) + \frac{1}{2}(3)$	$\frac{1}{2}(1) + \frac{1}{4}(4)$
2	$\frac{1}{2}(1) + \frac{1}{2}(-4)$	$\frac{1}{2}(1) + \frac{1}{2}(-4)$	$\frac{1}{2}(-1) + \frac{1}{2}(-4)$	$\frac{1}{2}(-1) + \frac{1}{2}(-4)$

or

1	3/2	2	5/2
−3/2	−3/2	−5/2	−5/2

If P_1^2 is selected, he faces

	y_{11}	y_{12}	y_{21}	y_{22}
$j = 1$	1	1	0	0
2	1/2	2	1/2	2

The safety levels in these two situations are 1 and 1/2, respectively, and P_1^2's second choice is his optimal strategy.

In this example the probability for any of the four games G_{11}, G_{12}, G_{21}, or G_{22} is 1/4. Thus the selection of agents for P_1 and P_2 is noncorrelated. We can easily include the possibility of correlation. For example, there might be a high probability that if P_1's first agent is chosen, then P_2's first agent will also be chosen.

Let r_{ij} be the probability for the occurrence of the various games. Assume

$$r_{11} = 0.45, \quad r_{12} = 0.05, \quad r_{21} = 0.05, \quad r_{22} = 0.45.$$

Player 1 faces the following contingent probabilities:

$$\Pr(G_{11}|P_1^1) = 0.9, \quad \Pr(G_{12}|P_1^1) = 0.1,$$
$$\Pr(G_{21}|P_1^2) = 0.1, \quad \Pr(G_{22}|P_1^2) = 0.9,$$

where $\Pr(G_{11}|P_1^1)$, for example, is the probability that the game is G_{11} given P_1^1. A similar calculation gives the conditional probabilities for player 2 given that he knows which of his own agents has been selected.

The concept of noncooperative games with misconceived payoffs was formulated by Milnor (1951), who extended the definition of noncooperative equilibrium to this case. Aumann (1974) noted that the Harsanyi analysis can be extended to games in which the players do not have consistent subjective probabilities, so that they might have different probabilities attached to the same event. He suggested that the presence of an outside event D, such as the election of a particular candidate, for which P_1 has a subjective probability $\alpha_1 > 1/2$ and P_2 has a subjective probability $\alpha_2 < 1/2$, could turn a zero-sum game into one in which both sides can gain. Suppose the game is Matching Pennies:

1	−1
−1	1

P_1 and P_2 could agree to play as follows: P_1 always plays 1; P_2 plays 1 if D occurs and 2 if D does not occur. The expectation for P_1 is

$$\alpha_1 - (1 - \alpha_1) = -1 + 2\alpha_1 > 0$$

and that for P_2 is

$$-\alpha_2 + (1 - \alpha_2) = 1 - 2\alpha_2 > 0.$$

Although initially two individuals may in fact have different subjective probabilities for the same event, a fully satisfactory game-theoretic model must allow for a reconciliation of these probabilities via the mechanism of the game. The normal form of the game obscures the difficulties involved in the description of the dynamics of information exchange and the updating of subjective probabilities.

9.3.2 The tracing procedure

Harsanyi (1975b) has suggested a means for selecting a unique equilibrium point from any game based upon an extension of his ideas on Bayesian players. The *tracing procedure* is essentially a computational method whose outcome depends upon the initial conditions. The initial conditions are the players' initial subjective probabilities concerning how each of the other players will select their pure strategies.

Harsanyi assumed at first that the priors are given as a datum. In a later paper (Harsanyi, 1976) he proposed a procedure for calculating the initial priors based upon the basic parameters of the game. It should be noted that Bayesian decision theorists fall into two groups: those who regard the decision maker's priors as data, and those who feel that the priors themselves can be determined from the initial data provided by the problem. Harsanyi identifies himself with the latter position and admits to looking for *the* solution to an n-person game, that is, a single perfect equilibrium point that a "rational advisor" can prescribe.

Because the strategic or normal form of the game does not provide enough structure to differentiate perfect equilibria from others, and because the full extensive form is cumbersome to work with, Harsanyi and Selten have suggested an *agent-normal form* of the game. In this form each player is represented by as many agents as the number of information sets he has in the extensive form of the game, each agent has a payoff function identical with that of his main player, and the pure strategies of each agent are the choices available to the main player at that information set.

Using the agent-normal form, we may search for perfect equilibria by imposing an error term on the selection of pure strategies by an agent i. If the agent wishes to use his pure strategy a^i_k, he can do so only with a probability $1 - \epsilon$. Here ϵ represents a small probability that by error he will select one of his other pure strategies; the probability of mistakenly choosing a specific pure strategy is $\epsilon/(K_i - 1)$, where K_i is the number of i's pure strategies. Given these errors we have a game in *disturbed agent-normal form* in which all information sets will be reached with positive probabilities. The perfect equilibria of the original game are the limit equilibria of the disturbed agent-normal form as $\epsilon \to 0$.

The tracing procedure is basically an extension of one-person Bayesian decision theory to an n-person noncooperative game. Let r

be an n-person game in disturbed agent-normal form with the players being agents (one player per information set). Each player i has K_i pure strategies. Let a_k^i be the kth pure strategy of the ith agent. A mixed strategy of player i, s_i, has the form $s_i = (s_i^1, s_i^2, \ldots, s_i^{K_i})$, where s_i^k is the probability assigned by i to his pure strategy a_i^k. A mixed-strategy n-tuple is written as $s = (s_1, s_2, \ldots, s_n)$, and $\bar{s}_i = (s_1, s_2, \ldots, s_{i-1}, s_{i+1}, \ldots, s_n)$ denotes the $(n - 1)$-tuple with the ith component left out from s. The payoff function to i is denoted by H_i.

A *prior probability distribution* $p_i = (p_i^1, p_i^2, \ldots, p_i^{K_i})$ is defined for each player i as the initial expectation of i's action, shared in common by all the other players. Mathematically the p_i all look like mixed strategies, but their interpretation is clearly different. Here the p_i express subjective uncertainty about the strategic behavior of others. Let p denote the n-tuple $(p_1, p_2, \ldots, p_n)$, and $\bar{p}_i = (p_1, p_2, \ldots, p_{i-1}, p_{i+1}, \ldots, p_n)$. A tracing procedure $T(r, p)$ selects a unique equilibrium.

A fictitious game r^0 is defined in which each player i has the same set of strategies as in r but with a payoff $H_i(s_i, \bar{p}_i)$. This is essentially a set of one-person games and (setting aside degeneracies) should have a single equilibrium point. Given r^0, a one-parameter family of games r^t is defined as follows. Each player has strategies as before but a payoff of the form

$$r^t = tH_i(s_i, \bar{s}_i) + (1 - t)H_i(s_i, \bar{p}_i), \qquad 0 \leq t \leq 1;$$

clearly $r^0 = r^0$ and $r^1 = r$. All r^t have equilibrium points. Let E^t be the set of all equilibrium points of r^t, and let Ω be the graph of the correspondence $t \rightarrow E^t$, $0 \leq t \leq 1$. Each point x of Ω will have the form $x = (t, s(t))$, where $s(t)$ is an equilibrium point of r^t. In general Ω will be a piecewise algebraic curve. Harsanyi has shown that Ω will always contain a path connecting a point $x^0 = (0, s(0))$ with a point $x^1 = (1, s(1))$. The *linear tracing procedure* follows this path.

A second version of this procedure is the *logarithmic tracing procedure,* obtained by selecting a small positive number ϵ^* and constructing another one-parameter family of games r_*^t with a payoff function for each player having the form

$$tH_i(s_i, \bar{s}_i) + (1 - t)H_i(s_i, \bar{p}_i) + \epsilon^* \sum_{k=1}^{K_i} \log s_i^k.$$

The graph Ω^* of the correspondence will be an algebraic curve. Intuitively the log function keeps individuals away from boundaries, since $\log 0 = -\infty$. It has been proved that the graph Ω^* will contain one and only one path connecting a point $x^0 = (0, s^*(0))$ with a point $x^1 = (1, s^*(1))$, where $s^*(0)$ is the unique equilibrium point of r^0_* and $s^*(1)$ is an equilibrium point of $G^1_* = G$.

The interpretation of the tracing procedure is that it represents an intellectual process by which rational players manage to coordinate their plans and their expectations about each other's plans and to make these converge to a specific equilibrium point as *the* solution of the game.

As a simple example of the linear tracing procedure with three branches, consider the following game:

	a^2_1	a^2_2
a^1_1	4, 1	0, 0
a^1_2	0, 0	1, 4

There are three equilibrium points: (1, 0; 1, 0), (0, 1; 0, 1), and (4/5, 1/5; 1/5, 4/5), where the first (second) two numbers in the parentheses are the probabilities utilized by P_1 (P_2) on his pure strategies a^1_1, a^1_2 (a^2_1, a^2_2).

The pairs of mixed strategies (s^1_1, s^1_2; s^2_1, s^2_2) can be represented by points in the square as shown in figure 9.16. The point A represents (1, 0; 1, 0); C, (0, 1; 0, 1); and E, (4/5, 1/5; 1/5, 4/5).

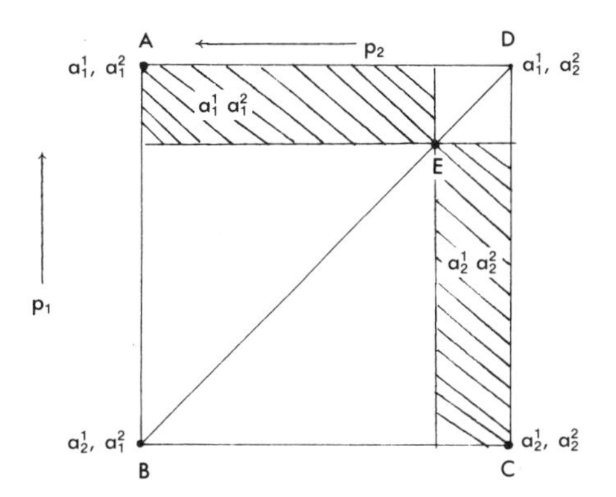

Figure 9.16
Representation of the mixed strategies in a two-person game in which each player has two possible strategies.

The shaded rectangle a_1^1, a_1^2 is the domain of stability of equilibrium point A. If player 1 expects that a_2^1 is played with probability $p_2 > 1/5$, his best reply is a_1^1; similarly for player 2, if $p_1 > 4/5$, his best reply is a_2^1. The shaded rectangle a_2^1, a_2^2 has the same interpretation for C. The mixed-strategy equilibrium denoted E has only itself as a domain of stability.

Suppose that originally the priors of the players are given by ((1/2, 1/2), (1/2, 1/2)). The pure Bayesian game r^0 can be contrasted with the actual game r^1 as shown in figure 9.17a. The generic game r^t is shown in figure 9.17b. The arrows around the matrices provide an indication of best replies to various strategies. Thus, for example, if player 1 uses a_1^1 in r^0, player 2 should choose a_2^2, but in r^1 he should choose a_1^2.

The generic game r^t has payoffs derived from a convex combination of r^0 and r^1 with weights $1 - t$ and t. When a_1^1 is used for r^t, the best reply by player 2 will reverse from a_2^2 for $t < 3/5$ to a_1^2 for $t > 3/5$. An equilibrium point is indicated by vertical and horizontal

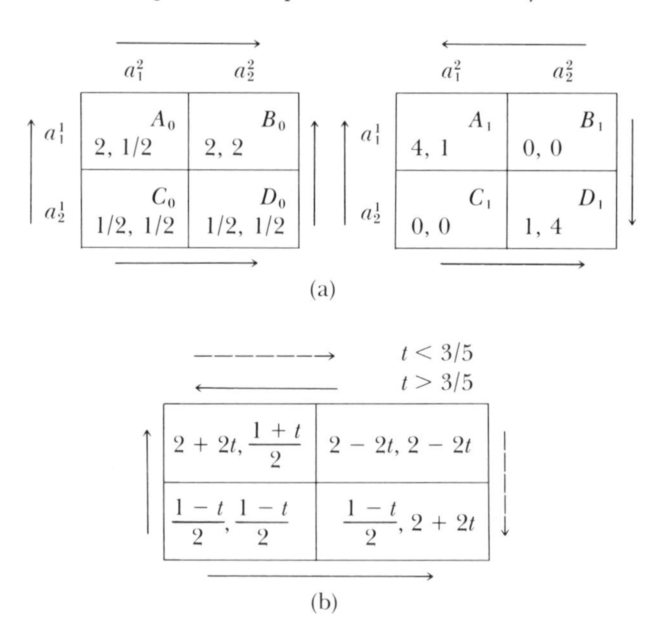

Figure 9.17
(a) A fictitious Bayesian version (left) of an actual game (right). (b) The generic game.

arrows pointing toward each other. Game r^0 has a unique equilibrium point at (a_1^1, a_2^2); r^1 has two pure-strategy equilibria at (a_1^1, a_1^2) and (a_2^1, a_2^2). For $t > 3/5$ there are two pure-strategy equilibria and a mixed-strategy equilibrium with $p_1 = (3 + 5t)/10t$ and $p_2 = (5t - 3)/10t$, where p_i is the probability associated with a_1^i.

Figure 9.18 shows the trace of the equilibrium points for this simple 2×2 matrix game. The four points A_0, B_0, C_0, D_0 represent the four cells in the pure Bayesian game r^0; A_1, B_1, C_1, D_1 the cells in the game r^1; and $A_{0.6}$, $B_{0.6}$, $C_{0.6}$, $D_{0.6}$ the cells in the game r^t, where $t = 0.6$.

For $0 \leq t \leq 0.6$ the trace has one branch from D_0 to $D_{0.6}$. At $t = 0.6$ there are three branches. In branch I player 2 switches from $D_{0.6}$ to $A_{0.6}$, then for $0.6 \leq t \leq 1$ continues to A^1. In branch II player 1 switches from $D_{0.6}$ to $C_{0.6}$, then for $0.6 \leq t \leq 1$ continues to C_1. In branch III the mixed strategy for player 1, $p_1 = 5/12$, appears, with $p_2 = 0$; then for $0.6 \leq t \leq 1$ this is modified until at $t = 1$ it reaches the mixed-strategy pair at E_1 with $p_1 = 4/5$ and $p_2 = 1/5$.

In this example the linear tracing procedure does not select a unique path. If we had started with initial expectations at any point off the diagonal $B_1E_1D_1$, a unique equilibrium point would have been selected. If the logarithmic procedure is used here, then E_1 is selected.[15]

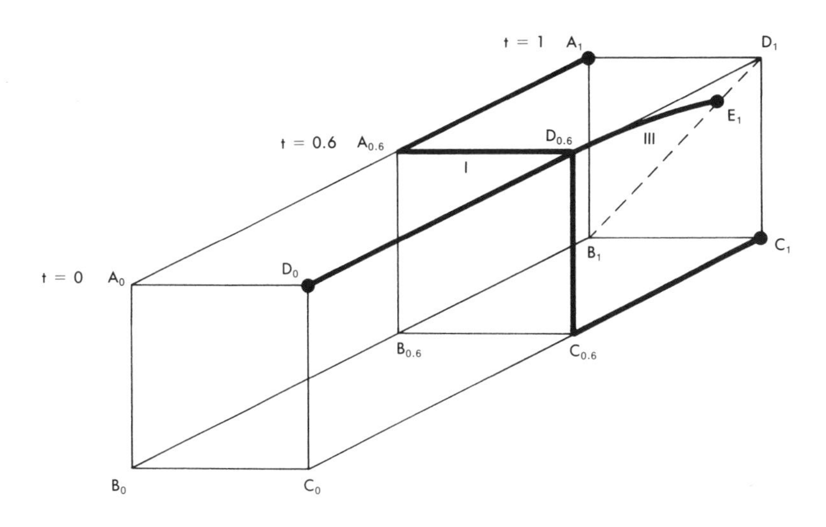

Figure 9.18
Equilibrium points for a 2×2 matrix game.

9.4 Games with an Indefinite Horizon

The difficulties that are encountered with solutions for games with finite moves in extensive form are compounded when games of indefinite length are considered.

Cooperative solution concepts such as the core, value, or stable sets do not adapt with ease even to the finite extensive form, let alone infinite-horizon games. This is not necessarily a fault of these solutions. They address general questions involving possible cooperation without going into details concerning the dynamics of communication and bargaining or the form of cooperative agreements. In contrast, any attempt to model in extensive form is tantamount to specifying process.

There is no generally satisfactory dynamic theory of games. Given that individuals do not live forever, so that we may wish to consider games with a sequence of individuals, even the elemental definitions of optimality require careful definition. Learning and adaptation are of considerable importance to much of the dynamics of human behavior. Yet it is difficult to reconcile much of the theory of games with learning (Simon, 1956).

Perhaps the most fruitful approach is to concentrate on hand-tailored solutions designed ad hoc to fit special classes of games, leaving the development of a general theory of dynamic games to come later. There are natural simplifications which make stochastic games or differential games fit specific problems even if their domain of application is limited. Many mass economic processes have sufficient structure to make it worthwhile to consider special solutions that do not necessarily generalize well to broader classes of games.

9.4.1 The Prisoner's Dilemma revisited

In section 9.2 we considered the Prisoner's Dilemma and finite iterations of this game. By means of a backward induction it is possible to show that the highly nonoptimal unique equilibrium in the one-shot game has its nonoptimal counterpart in the multistage game. Consider the following Prisoner's Dilemma matrix:

5, 5	−5, 10
10, −5	0, 0

We note the equilibrium point at (2, 2) with payoffs of (0, 0). For any finite version the perfect EP gives (0, 0) as a payoff.

When the game can last for an indefinite time, there are four natural ways of modeling it so that individuals can try to optimize a well-defined bounded payoff function (Shubik, 1970b):

1. Introduce a positive termination probability at every stage.
2. Introduce a positive discount factor.
3. Replace the game by a game with a finite horizon and an end-value function.
4. Optimize the average return per period.

All of these methods enable us to consider both indefinite play of matrix games and natural extensions of equilibrium points in terms of both stationary and historical strategies.

In section 9.4.3 we give a formal definition of general n-person stochastic games. Here both for ease of exposition and to save on notation we illustrate the different methods of modeling by means of simple examples involving extensions of the Prisoner's Dilemma.

The termination probability Let p_t be the probability that the game will terminate after period t, and let $\hat{p}$ be the smallest of the probabilities p_t. Let H_{it} be the payoff to P_i for period t, and let H_i be the payoff to P_i for the whole game. Figure 9.19 shows the extensive form for the first stage of the iterated Prisoner's Dilemma.

We consider two types of strategy. We look first at a highly simplified strategy in which each individual considers only the game with which he is presently confronted. Here there is only one state of the system that is of strategic consequence: the game is either over or the players must play the Prisoner's Dilemma game. A stationary strategy prescribes that whenever you find yourself in a given state, take the same action you have previously taken in that state. It is easy to show that the strategy pair (2, 2) forms the only stationary equilibrium and that this is an infinite-game analog of a perfect equilibrium.

We could also define the state of the game in a more complicated manner that makes use of a historical summary of the payoff experience of the players. We do this by defining a state for P_i at time T as consisting of the game he is about to play and his average payoff up to that point.

Consider the strategy that gives as a policy: Play 1 to start, and

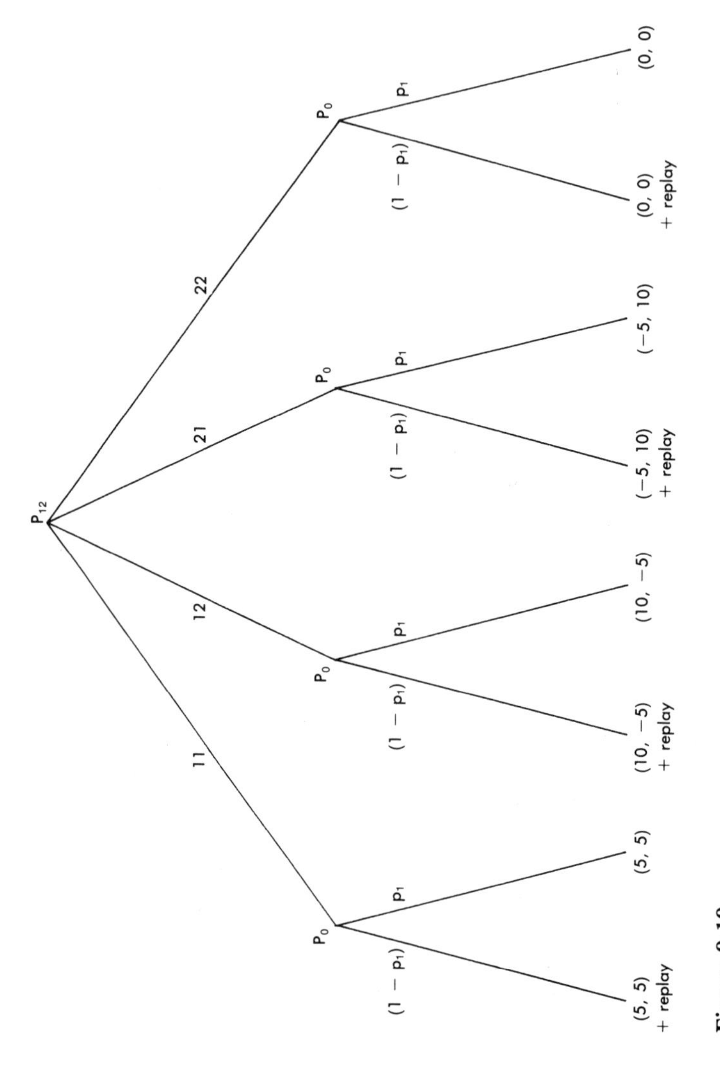

Figure 9.19
A game with a termination probability.

continue to do so as long as your average payoff is 5 or more; whenever your average payoff is less than 5, switch to 2; whenever it reaches 5 or more, switch back to 1. For a sufficiently low $\hat{p}$, if both players adopt this strategy, they will reach an equilibrium with an expected payoff to each of 5 per period. If $\hat{p} > 0.5$, this is not an equilibrium, since a constant playing of 2 against this strategy yields 10 to the individual playing 2 from the start; and the high probability of terminating reduces the worth of cooperation to an expected value less than 5 per period.

The discount factor The *discount factor,* denoted by ρ, where $0 \leq \rho < 1$, plays the same mathematical role in bounding the payoff functions as does the termination probability. The two strategies suggested above as leading to different equilibria when the game has a probability of terminating, lead to analogous equilibria in the game with a discount factor.

An end-value function We may wish to "model away" the infinite horizon by considering a game that ends at time T, where each player i receives an end-of-the-game bonus Q_i. The Q_i may be any general function of the play up until T. This method of treating the infinite horizon is related to methods in dynamic programming in which a salvage or terminal value is assigned to a finite ending position (Denardo, 1981); the interpretation is that the value represents the worth of the program from that time to infinity. Ideally we expect that the solutions to the finite games behave smoothly as T increases.

The technique of defining an end value appears to be particularly reasonable in modeling games with sequences of players, where the end value for one player may be interpreted as the value attached by one individual to leaving resources to his successor. Thus we may model games of indeterminate length with sequences of players each of whom lives for a finite period. The end value for each may be determined by introspective factors such as altruism or desire to leave inheritances, or by outside factors such as taxes and subsidies, law or custom (Shubik, 1980, 1981).

Returning to our particular example, suppose that the bonus is 5 each. If at time T (1, 1) is selected, and -15 each otherwise, then both of the strategies noted before lead to equilibria, except that now the repeated play of (2, 2) is even less plausible than before. The bonus or penalty at the end of the finite game has provided a self-policing mechanism that enables individuals to achieve optimal outcomes without innate trust. Escrow arrangements, policing systems, laws, and social and political pressures are frequently designed to do precisely this.

Optimizing the average payoff per period If the payoffs in every period are bounded by some constant C, then clearly the average payoff per period is also bounded by C. Hence if we define the goals of the players to be the optimization of their average payoffs per period, then we have redefined the original payoffs to the infinite-horizon game in a manner that keeps them bounded. Given this transformation of the game, both of the strategies considered before lead to equilibrium points in the new game. This approach has been considered by Aumann (1959), and Aumann and Shapley (in an unpublished work) have shown that every point in the payoff space can be established as the payoff from an equilibrium point involving only behavior strategies. The example in section 9.2.2 shows the proliferation of noncooperative equilibria.

The criterion that calls for the optimization of the average per-period payoff is reasonable as a heuristic rule, especially when there is no interest rate involved.

On limited memory An interesting way to approach dynamics is by imposing limited memory on the participants. Aumann and Kurz (private communication) have offered an example of a Prisoner's Dilemma game played an indefinite number of times by individuals who only remember one period in the past and whose strategies therefore take into account just one set of contingencies. We can represent this by the 8×8 matrix associated with the 2×2 matrix in figure 9.20 (the payoffs in the 8×8 matrix are averages over the two periods).

In the 8×8 matrix the eight strategies are denoted by HHH, HHG, . . . , as noted on the left of the matrix. They are numbered 1–8 on the right of the matrix. We may read the strategy HHG as follows: "I play H to begin with; if he plays H, I play H; if he plays G, I play G." This is essentially a cooperative "tit-for-tat" policy.

The 8×8 matrix has several equilibrium points, denoted by asterisks. There is, however, a unique perfect equilibrium point. This can be seen by noting that in a perfect equilibrium weakly dominated

3, 3	0, 4
4, 0	1, 1

	HHH	HHG	HGH	HGG	GHH	GHG	GGH	GGG	
HHH	3, 3	3, 3	0, 4	0, 4	3, 3	3, 3	0, 4	0, 4	1
HHG	3, 3	3, 3*	2, 2	1, 1	3, 3	2, 2	2, 2	1, 1	2
HGH	4, 0	2, 2	2, 2	0, 4	4, 0	2, 2	0, 4	0, 4	3
HGG	4, 0	1, 1	4, 0	1, 1*	4, 0	1, 1	4, 0	1, 1*	4
GHH	3, 3	3, 3	0, 4	0, 4	3, 3	3, 3	0, 4	0, 4	5
GHG	3, 3	2, 2	2, 2	1, 1	3, 3	1, 1	2, 2	1, 1	6
GGH	4, 0	2, 2	4, 0	0, 4	4, 0	2, 2	2, 2	0, 4	7
GGG	4, 0	1, 1	4, 0	1, 1*	4, 0	1, 1	4, 0	1, 1*	8

Figure 9.20
An iterated Prisoner's Dilemma game.

strategies cannot be active. By inspection, for P_1 strategy 6 is dominated by 2, and 3 by 7. Similarly for P_2. We continue to look for weak dominance on the remaining 6×6 matrix (figure 9.21a). Here strategy 1 is weakly dominated by 2, as is 5 by 2, leading to a 4×4 matrix (figure 9.21b). But here strategy 7 is weakly dominated by 2, so that we are left with the 3×3 matrix in figure 9.21c. Hence strategies 4 and 8 are weakly dominated by 2, and the unique perfect equilibrium is the tit-for-tat policy.

An example of escalation Suppose that an auctioneer offers to auction a dollar bill under the following conditions. The dollar will be sold to the highest bidder, but both the highest bidder and the second highest bidder must pay their bids to the auctioneer. Bidding cannot be in intervals of less than five cents. If no one bids within five minutes after a bid, the game is over. If no one bids at all, the auctioneer declares the auction over after five minutes and keeps the dollar (Shubik, 1971b).

3, 3	3, 3	0, 4	3, 3	0, 4	0, 4	1
3, 3	3, 3	1, 1	3, 3	2, 2	1, 1	2
4, 0	1, 1	1, 1	4, 0	4, 0	1, 1	4
3, 3	3, 3	0, 4	3, 3	0, 4	0, 4	5
4, 0	2, 2	0, 4	4, 0	2, 2	0, 4	7
4, 0	1, 1	1, 1	4, 0	4, 0	1, 1	8

(a)

3, 3	1, 1	2, 2	1, 1	2
1, 1	1, 1	4, 0	1, 1	4
2, 2	0, 4	2, 2	0, 4	7
1, 1	1, 1	4, 0	1, 1	8

(b)

3, 3	1, 1	1, 1	2
1, 1	1, 1	1, 1	4
1, 1	1, 1	1, 1	8

(c)

Figure 9.21
Deduction of the unique perfect equilibrium for the game in figure 9.20.

The cooperative form of this game is easy to analyze. Suppose that there are only two bidders 1 and 2 and that the auctioneer is 3. The characteristic function is

$$V(1) = 0, \quad V(2) = 0, \quad V(3) = -0.95,$$
$$V(12) = 0.95, \quad V(13) = 0, \quad V(23) = 0,$$
$$V(123) = 0.$$

The auctioneer appears to be at a clear disadvantage in that he may be forced to sell a dollar for five cents.

When we consider the game in extensive form, however, it is potentially of indeterminate length. It has two pure-strategy equilibrium points in which either 1 or 2 bids a dollar and the other does not bid. However, if both bid positive amounts, an interesting possibility for escalation occurs. Suppose, for example, that 1 has bid 90 cents and 2 has bid 95 cents. Player 1 is motivated to bid a dollar, thus having a chance of breaking even rather than losing 90 cents. But 2 is then motivated to bid $1.05 for a dollar, cutting his losses from 95 cents to 5 cents. From here on there is no natural bound to the escalation!

9.4.2 Threats, talk, theory, and experiment

The key element in enforcing equilibria, other than ones that make no direct use of history, is the use and communication of threat. In much of actual bargaining and negotiation, communication about contingent behavior is in words or gestures, sometimes with and sometimes without contracts or binding agreements.

A major difficulty in applying game theory to the study of bargaining or negotiation is that the theory is not designed to deal with words and gestures—especially when they are deliberately ambiguous—as moves. Verbal sallies pose two unresolved problems in game-theoretic modeling: (1) how to code words, and (2) how to describe the degree of commitment.

Skilled negotiators and lawyers thrive on verbal ambiguity. Formally the ambiguity of language can be handled by treating sentences as moves with ill-defined meanings. For example, the statement "one step closer and I shoot" is a threat which may or may not contain an element of bluff. Neave and Wiginton (1977), using fuzzy sets (Zadeh, 1965), have formalized the relationships among verbal statements that involve threats (or promises), degrees of belief, and action.

There are now hundreds if not thousands of articles and experiments on iterated two-person non-constant-sum games in general and the Prisoner's Dilemma in particular. The experimental evidence heavily supports the proposition that actual players of even finite-length Prisoner's Dilemma games do not play the nonoptimal perfect equilibrium. Personality, limited memory, limited ability to plan or process data, and nuances in experimental control all play their roles. There has even been a Prisoner's Dilemma strategy competition (Axelrod, 1980) to which many experimenters and game theorists submitted their pet strategies for how one should behave. The strategies were then matched against each other. The "best" in the sense of yielding the highest score to the player was the simple tit-for-tat strategy submitted by Anatol Rapoport with the first move being cooperative (see the example of limited-memory play in section 9.4.1).

Papers such as those of Sanghvi and Sobel (1976) and Smale (1980) have suggested models with players of limited rationality, aggregating historical information or behaving according to relatively simple decision rules. However, the problem does not appear to be primarily mathematical, but involves the selection and justification of a solution concept on normative or behavioristic grounds. In the latter instance this calls for modeling based on empirical evidence.

9.4.3 Stochastic games

There is a natural extension of the concept of a two-person zero-sum stochastic game to n-person non-constant-sum games (Sobel, 1971). Let N be a set of n players, S a set of states, and A_s^i a set of moves available to P_i, where $i \in N$ when the system is in state $s \in S$. All sets are assumed to be nonempty. The actions of all players when the system is in state s can be described as an element $C_s = \Pi_{i \in N} A_s^i$. Let $a = (a^1, a^2, \ldots, a^n)$. An outcome of a stochastic game is a sequence $s_1, a_1, s_2, a_2, \ldots$, where $a_t \in C_{s_t}$ for all t.

The dynamics are determined by the decision rules used by the players and by a set of probability measures $\{p(\cdot|s, a) : (s, a) \in W\}$, where $W = \{(s, a) : a \in C_s, s \in S\}$.

A two-person zero-sum matrix game played repeatedly is a special case of a stochastic game in which S has only one element.

If we consider nonanticipative decision rules (rules that at most make use of history, but contain no threats), then for some problems *stationary* strategies or policies may be considered, in which the basis at time t for the selection of a_t depends at most upon $s_1, a_1, s_2, a_2, \ldots, a_{t-1}, s_t$.

Existence results have been obtained for equilibria that use stationary policies in n-person non-constant-sum stochastic games (Rogers, 1969; Sobel, 1971; Parthasarathy and Stern, 1977).

There have been some applications of stochastic game models to problems in oligopolistic competition, but at the price of a reduction in the complexity of the models. One way to cut down on complexity is to impose a leader-follower behavior on the players from the start of the game (Lippman, 1977); with perfect information the problem becomes equivalent to a one-person Markov decision process.

Another type of simplification can be made if the original dynamic game can be reduced to a static game. Sobel (1978) has defined a *myopic equilibrium* for a dynamic game as an equilibrium that can be regarded as an infinite repetition of an equilibrium point in an associated static game. He has considered the sufficient conditions for the existence of myopic equilibria. Several oligopoly models satisfy the criteria (Kirman and Sobel, 1974).

9.4.4 Economic dynamics

Games of economic or social survival A specific form of stochastic game designed to model the twin features of survival and profit maximization or "living well" is the game of economic or social survival. Games of survival (Milnor and Shapley, 1957) have only two outcomes for any player: he is ruined or he survives. In many ecological, social, political, and economic processes, however, the goals reflect an intermix of the desire to survive and the quality of existence while surviving.

We can model the twin goals of survival and "living well" as follows. An n-person game of economic survival is described by

$(a(t); W_t, B_t, S_t, L_t, \rho_t, V)$,

where

$a(t) \equiv (a^1(t), a^2(t), \ldots, a^n(t))$ are the single-period payoffs faced by the players at time t (we assume these are bounded);

$W_t \equiv (W_t^1, W_t^2, \ldots, W_t^n)$ are the wealth holdings of the players at time t;

$B_t \equiv (B_t^1, B_t^2, \ldots, B_t^n)$ are the ruin conditions or bankruptcy levels (if the assets of P_i at time t drop below B_t^i, he is out of the game);

$S_t \equiv (S_t^1, S_t^2, \ldots, S_t^n)$ are the survival values (if P_i is the sole survivor at time t, then S_t^i is the present value of the remaining one-person game);

$L_t \equiv (L_t^1, L_t^2, \ldots, L_t^n)$ are the liquidation values (if P_i is ruined, he may still have residual assets at the time of ruin);

$\rho_t \equiv (\rho_t^1, \rho_t^2, \ldots, \rho_t^n)$ are discount factors;

$V \equiv (V^1, V^2, \ldots, V^n)$ are the payoff functions.

The payoff functions can be specified in several different ways that lay different stresses on the tradeoffs between survival and optimization. Pure survival as a goal is represented by

$$V^i = \begin{cases} 1 & \text{if } W_t^i \geq B_t^i \text{ for all } t, \\ 0 & \text{otherwise.} \end{cases}$$

Pure optimization involves a maximization of expected payoffs regardless of the probability of being ruined. Several simple games of economic survival have been investigated (Miyasawa, 1962; Shubik and Thompson, 1959; Shubik and Sobel, 1980; Shubik, 1959b; Gerber, 1972).

A simple example of a one-person game against Nature illustrates the operational difference in goals. We assume that Nature always plays (1/2, 1/2):

2	−1
E	E
−1	2

Here $W_1 = 1, B = 0, L = 0, \rho < 1, 0 < E < 1/2$. If the goal is survival, then

$$V = \begin{cases} 1 & \text{if not ruined,} \\ 0 & \text{if ruined,} \end{cases}$$

and the strategy for P_1 is 2. If the goal is to maximize expected return, P_1's policy will depend on E and ρ and may involve the use of all three strategies. Let C_t be the consumption of P_1 during t. Then

$$V = \sum_{t=0}^{\infty} \rho^{t-1} C_t,$$

$$W_{t+1} = W_t - C_t + a_{ij},$$

and P_1 is ruined if $W_t \leq 0$. His strategy in each period involves both playing the matrix game and selecting a level of consumption.

Supergames and reaction functions Aumann (1959) introduced the term "supergame" to describe a sequence of static games that are not contingent on the past actions of the players. Investigations have centered mostly upon the repeated play of the same game. The iterated Prisoner's Dilemma provides an example. This can be viewed as a nonterminating stochastic game with a single state. J. W. Friedman (1977) has analyzed this case for oligopoly models. He considers reaction-function strategies in which each player's decision is a function of the other players' previous actions and, in general, not of his own actions.

There is a large pre-game-theory literature in economics leading up to the formal analysis of reaction functions. In particular, Chamberlin (1962) and Fellner (1949) provide examples of "if, then" chains of reasoning to describe the adjustments of firms in oligopolistic competition. This literature may be regarded as somewhat related to the method of fictitious play for finding an equilibrium point, noted in chapter 8.

A simple example of a Cournot duopoly market serves as an illustration. Let Π_1 and Π_2 be the profits of the duopolists and q_1 and q_2 the prices they charge their customers. Now suppose that

$$\Pi_1 = [30 - (q_1 + q_2)]q_1, \quad \Pi_2 = [30 - (q_1 + q_2)]q_2.$$

The unique noncooperative equilibrium is (10, 10), yielding a payoff of (100, 100). At time t let the duopolists play $q_{1,t}$ and $q_{2,t}$. Then two simple optimal reaction functions are

$$q_{1,t+1} = (30 - q_{2,t})/2, \quad q_{2,t+1} = (30 - q_{1,t})/2.$$

As can be seen from figure 9.22, this process converges to the equilibrium point: $(0, 15) \rightarrow (7.5, 15) \rightarrow (7.5, 11.25) \rightarrow (9.875, 11.25) \rightarrow \ldots \rightarrow (10, 10)$. But it is well known that in general one can expect neither a unique equilibrium nor convergence.

Differential games The study of differential games grew out of investigations of pursuit and evasion games by Isaacs (1952, 1965). The original work was formulated in terms of two-person zero-sum games, and a parallel literature has developed in optimal control theory. The concept of a Nash equilibrium has also been extended to non-zero-sum differential games (Starr and Ho, 1969; A. Fried-

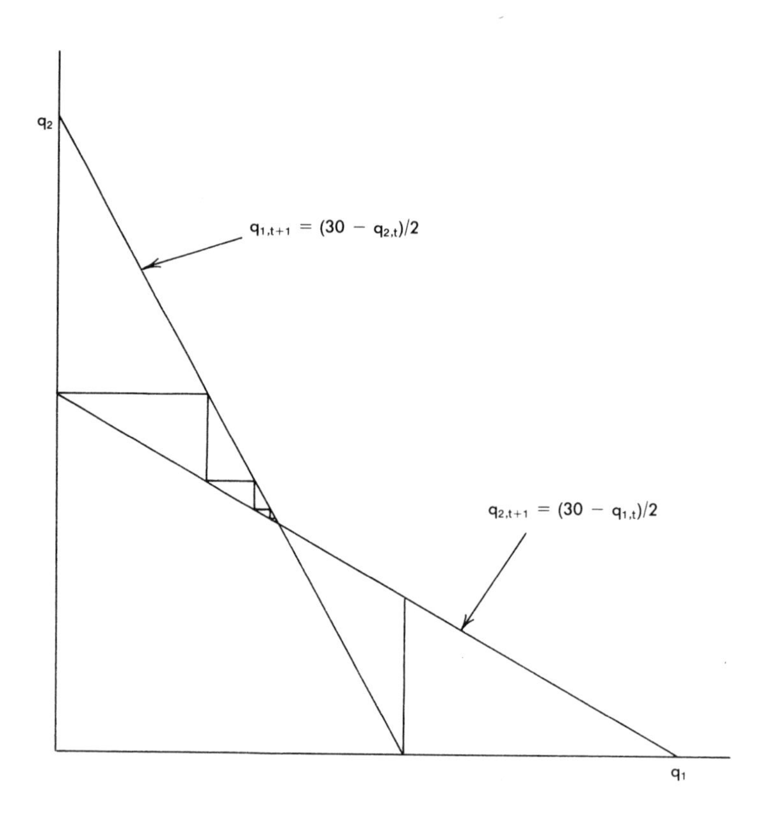

Figure 9.22
The Cournot duopoly model.

man, 1971), and this has led to applications of differential games to economics (Kuhn and Szegö, 1971; Case, 1979). Some economic problems may in fact be more naturally cast in continuous time than in discrete time; but none of the multitude of conceptual difficulties in interpreting noncooperative equilibrium points are removed by the introduction of continuous time.

9.5 Modeling and Games in Strategic Form

The strategic form of a game is highly useful in many applications. In particular, it is appropriate to the study of many tactical problems and a large range of problems in economics. In most applications it is more natural to consider the game in strategic form as basic rather than as having been derived from a complex extensive form.

This chapter has dealt primarily with single simultaneous-move games in which, by the very definition of the structure of the game, there can be neither implicit nor explicit communication. This is not true for multimove games. This observation appears to have formed the basis of Schelling's (1960a) criticism that game theory fails to take communication into account. He apparently misunderstood how the strategic form of a game can be derived from an extensive form. Nevertheless, the questions he raised were and still are critical in forging the link between games played in the extensive form and games played in the strategic form.

Game theory has a static normative model for how Chess *should* be played. Sooner or later computational methods may become sufficiently good that an optimal strategy for Chess will be found. Until then the game as actually played will continue to involve implicit communication, "ploy," bluff, and many of the features of psychological warfare. The normative theory for non-constant-sum games is not universally persuasive, even leaving aside modeling difficulties.

The game in strategic form as described here assumes that all individuals know all rules and that all relevant information is given by the payoff matrices. In many possible applications individuals may not know certain outcomes or their values, or even how many strategies their opponents have. These features pose modeling difficulties more than mathematical difficulties. One can postulate subjective probability distributions to patch up all holes in knowledge. The basic problem is how well this can be done. How are the initial probabilities obtained? Do we obtain a theory that is useful either as a normative or as a behavioral guide to multiperson interaction?

Modeling any aspect of an economy, polity, or society as a game in strategic form is tantamount to specifying the institutions of the society implicitly as rules of the game. The way a vote takes place, how a market works, the rules of the auction appear within the rules of the game.

9.5.1 One, two, few, and many

The n-person problem begins in full force with the two-person non-constant-sum game. Several distinctions are fruitful, however, in both applications and the development of theory. In particular, a natural way to categorize games is by the number of players, as follows:

1. One-player games are problems involving maximization.

2. In two-player games the emphasis is on negotiation, bargaining, fair division, duopoly, and social psychology. In almost all of these games communication is of considerable importance.

3. Few-player games are ones in which the players number anywhere from three to around twenty, with the specific cutoff point for "fewness" determined in a more or less ad hoc manner. Oligopoly theory, the formation of political coalitions, international negotiations, ecological struggles, and cartel problems all fall into this category.

4. "Many" may start at twenty to thirty. The investigator is interested primarily in mass properties of the game. Individuals are treated as anonymous. The study of mass political, economic, or social behavior calls for special methods entailing the use of limiting properties of large games.

It is important to stress that, in each of the distinctions made above, modeling considerations call for different emphases as numbers change. Thus face-to-face communication may be an important factor in few-person games, where individuals are in a position to personalize threats or promises; this can hardly be so for games with thousands of players. Noncooperative game theory appears to be particularly useful for the study of mass phenomena in which communication between individuals must be relatively low and individuals interact with a more or less anonymous and faceless economy, polity, or society.

9.5.2 Further examples

Fallacies of composition Consider n individuals who have joint access to a commons used for grazing cattle (Hardin, 1968). The output per animal is a function of the total number of cattle grazing. An individual is concerned with how many other cattle are grazing. Suppose that there are four individuals with two cattle each. A strategy is to send zero, one, or two to graze. Consider the payoff to P_i to be

$$q_i\left(10 - \sum_{j=1}^{4} q_j\right),$$

where q_i is the number of cattle sent by P_i. The payoffs are shown below in matrix form for P_i, where we can add the cattle (that is, the effect of the others' strategies) together:

	0	1	2	3	4	5	6
0	0	0	0	0	0	0	0
1	9	8	7	6	5	4	3
2	16	14	12	10	8	6	4

This example is similar to that in section 9.1.7, and the unique noncooperative equilibrium point is (2, 2, 2, 2). This is a glorified Prisoner's Dilemma game.

Another example is the block party for which everyone promises to bring a gallon of wine and pour it into a common keg. Suppose that the strategies for each individual are to pour either wine or water![16]

In bank panics or market crashes it may appear to each individual that it is "rational" to get out because he is so small with respect to the market as a whole that his actions can only benefit himself and not harm others. If all reason in this manner, the independence is illusory. The freedom apparently available to the individual does not exist for the system as a whole. In financial systems the concept of liquidity provides a key example of the fallacy of composition.

Implicit communication, self-fulfilling prophecy, and social custom In their development of the stable-set solutions for n-person games von Neumann and Morgenstern were highly concerned with capturing the idea of social stability in the context of games in cooperative form. The same goal holds for n-person noncooperative games. A simple example will illustrate the interaction of social custom, self-fulfilling prophecy, and implicit communication.

Some years ago it was forbidden to walk on the lawns in many of the parks in Vienna. A stranger who walked on the lawns had a high probability of being chased off by an irate local citizen. This in general was less pleasant than not walking on the lawn. The situation might be viewed as an n-person noncooperative game with many anonymous players: it was as though all individuals knew that there was a high probability that a threat would be carried out against them if they violated the equilibrium of not walking on the lawn. It is unlikely that such inhibitions would exist in New York. A more detailed understanding of this phenomenon calls for a consideration of the game in extensive form.

9.5.3 Behavior and norms

In exploring the properties of the noncooperative equilibria of games in strategic or extensive form there are many criteria that can be applied to examine the plausibility and robustness of the equilibria and the model. A brief listing of some of the considerations of the models is given below:

1. Aesthetic properties: uniqueness; symmetry; value.

2. Attractive norms: Pareto optimality; independence from irrelevant alternatives.

3. Sensitivity: sensitivity of the equilibria to perturbation of payoffs, the "trembling hand" or error in strategy, information leaks, and sequencing of moves.

4. Limit properties: the influence on equilibria of large numbers of players, an infinite horizon for the game, changes in information conditions, and aggregation of players.

5. Types of goals and preferences: effect of players with preference orderings, utility functions defined up to a linear transformation, or comparable utilities; fiduciary players playing for others; bureaucratic players trying to minimize corporate blame for errors, subject to bureaucratic survival.

6. Limited abilities of players: effects of limitations on the complexity of strategies, the memory limits of players, and the data-processing abilities of players.

7. Levels of ignorance: lack of knowledge of the number of players, the payoffs, and the available sets of moves.

8. Communication conditions: costly communication; coding problems.

9. Availability of enforcement mechanisms: escrow arrangements; exchange of hostages or other assets.

It is suggested here that much of conscious optimizing behavior takes place in a relatively short time span as compared with the construction of the games that are played. The games are models of the institutions of society and the environment of the decision maker. Even if we regard the behavior of individuals as reasonably modeled by some form of noncooperative behavior appropriately modified for limited ability, levels of ignorance, and a variety of goals, then the social engineer or applied game theorist may have the task of designing the game to satisfy normative considerations.

The construction of a game requires a specification of its rules, which describe how the mechanisms of society and its institutions function. Societal norms may be achieved by designing *self-policing* systems. The contrast between cooperative and noncooperative solution theories is not one between normative and behavioral approaches; but there is a change in emphasis. For example, cooperative solution theories have Pareto optimality as an axiom. The norm of Pareto optimality may still be present when we consider noncooperative solutions, but it may be manifested in the game design and not as a condition for the solution.

Notes to chapter 9

1. The term "noncooperative game" is something of a misnomer; more precise would be "game viewed in the light of a noncooperative solution theory." It is at the solution level, rather than the description level, that the question of degree of cooperation properly arises. To be sure, some feature of the situation that we are modeling may invite the application of a noncooperative solution concept, but the formal rules of the game are neither cooperative nor noncooperative in themselves. (Even in the drafting of legislation, where formal mathematical rigor is seldom sought, it is notoriously difficult to make rules either requiring cooperation or forbidding it.) A similar criticism applies, of course, to the term "cooperative game." Nevertheless we shall sometimes use these terms.

2. One sometimes encounters the terms "Nash point" or "Nash solution," but these risk confusion with the Nash bargaining point (see Nash, 1950a) and the Nash two-person cooperative solution (see Nash, 1953).

3. One selects a square submatrix, finds the equalizing solution for that submatrix if it exists, then tests the solution against the full matrix (cf. Shapley and Snow, 1950). It is not practical to work through all square submatrices, however, unless the game is very small or very symmetric. Bimatrix games, unlike matrix games, are not directly equivalent to linear programming problems. Using a method akin to the "complementary bases" method of mathematical programming, though, Lemke and Howson (1964) have devised a finite algorithm that will find at least one EP of any bimatrix game. With more than two players, it is possible to have several EPs using the same sets of pure strategies (see exercise 9.2).

4. See theorem 3 of Bohnenblust, Karlin, and Shapley (1950). This result generalizes in a curious way: it can be shown that almost all bimatrix games have an odd number of EPs (Harsanyi, 1973).

5. In a "random" 2×2 bimatrix game it can be calculated that with probability 3/4 there is a unique pure-strategy EP, with probability 1/8 there is a unique mixed EP, and with probability 1/8 there are exactly three EPs, two pure and one mixed (Shapley, unpublished notes). Harsanyi (1964) has proposed a systematic method for eliminating nonuniqueness in the noncooperative solution, by invoking additional conditions based on an idea of "risk dominance."

An interesting mathematical curiosity is that there is a unique equilibrium point in the three-player $2 \times 2 \times 2$ matrix game if all the equilibria have totally mixed strategies (see Parthasarathy and Raghavan, 1974).

6. The example does not depend on coincidental relationships among the numbers used; the three EPs described are not significantly affected by small perturbations.

7. The pair (M, M), though no longer equalizing, remains as an EP, giving us a total of four EPs. This is an exceptional occurrence—the number is usually odd.

8. Writers on differential games often use the term "optimal" in reference to EP strategies in non-constant-sum situations, in an unfortunate and uncritical analogy with control theory usage (see Blaquiere, Gerard, and Leitman, 1969; Starr and Ho, 1969; and Kuhn and Szegö, 1971).

9. Rapoport and Guyer (1966) have calculated that if one considers optimal payoffs $a > b > c > d$ there are 78 strategically different 2×2 bimatrix games.

10. For example, in Matching Pennies, if you know that your opponent is randomizing fifty-fifty between Heads and Tails, then you might as well always play Heads, or always Tails.

11. Even a small reaction in the right direction—increasing even slightly the probability of the preferred alternative—is sufficient in these examples. Thus, in the example in the text, if P_2 merely shifts to (0.51, 0.49), instead of putting all his weight on the first column ("keeping his guard up" as it were), the new payoff vector is (0.0062, 0.0009).

12. Rather than use words such as "polymatrix game" or "n-dimensional matrix game," we use the term *matrix game* to stand for a game with any number of players each of whom has a finite set of pure strategies. The dimension of the matrix will be clear from the context. *Bimatrix game* is accepted usage for a two-person non-constant-sum game with finite sets of pure strategies, and we occasionally use this expression.

13. There are some other, asymmetrical EPs wherein one farmer ships 90 and the others 91. If fractional bushels were allowed, all these EPs would coalesce into a single EP wherein everyone would ship $90\frac{10}{11}$ bushels and the price would be $90\frac{10}{11}$ cents.

14. An interesting article by Forst and Lucianovic (1977) considers the actual evidence of what happens in single defendant and codefendant trials.

15. The tracing procedure, combined with a method for selecting the starting point or the initial subjective probabilities and an interpretation of any cooperative game as a noncooperative game, is the subject of a solution concept for all games proposed by Harsanyi and Selten. The reader is referred to the recent work of Harsanyi (1979) and a book projected by Harsanyi and Selten.

16. See Schelling (1978, essays 2 and 3) for many mass behavior models.

10 Other Noncooperative Solutions and Their Applications

10.1 Fair-Division Games

10.1.1 Three approaches

In chapter 7 various schemes for the cooperative division of joint gains were discussed. The solutions suggested were variants of the value, and the models presented were games in characteristic-function form or in a related coalitional form.

It is also possible to suggest fair-division schemes which make no pretense to connect individuals' rewards or shares to their strategic or coalitional roles or to their contribution to the joint welfare. The injunction "to each according to his needs, from each according to his abilities" provides an example in which the claims to product are not directly related to individuals' contributions to production.

Fair-division procedures can be approached in three ways: as a cooperative game; by means of a rule or set of axioms that select a division but do not depend upon game-theoretic considerations; or by specifying a game that is played in a noncooperative manner to determine the outcome.

The second approach has been considered by several economists in reference to models that implicitly or explicitly contain a simple economic structure. These include Lerner (1944, 1970), Guilbaud (1952a,b), R. Frisch (1971), Sen (1973), and Arrow (1973) in an interpretation of the work of Rawls.

Lerner and subsequently Frisch considered a model that amounts to assuming that the judge, or arbiter for a society, is in a position to divide all (monetary) income among its members, who may then buy all goods from a set of markets. It is assumed that utility functions for money are concave and that the sum of the utilities is to be maximized, but that the judge does not know the utility functions of the individuals. It is then argued that the probable total satisfaction can be maximized by an equal distribution of money. Sen (1973) dispenses with the assumption of joint maximization of utility and obtains this result. This completely separates the production and distribution aspects of the society.

Knaster suggested a fair-division scheme applicable when (1) the objects to be distributed are indivisible, (2) a money value can be attached to each object by each participant, (3) these values are additive, and (4) monetary side payments can be made among the players. The rule is simple. Suppose there are M objects and n players and the value attached by player i to object j is v_{ij}. A fair share for i should be $(\Sigma_j v_{ij})/n$. According to Knaster's rule, each object will be assigned to the player who values it most (randomize for ties). Then side payments summing to zero are made to give each an equal surplus over his self-estimated fair share. This amounts to

$$\frac{1}{n}\left(\sum_j \left(\max_i v_{ij}\right) - \frac{1}{n}\sum_i \sum_j v_{ij}\right). \tag{10.1}$$

The simple example with two players and two items in table 10.1 illustrates the procedure.

The Knaster procedure is not a fair-division game if the actual valuations of the players are known. It becomes a game that can be played noncooperatively if each has as his strategy a statement of his values to the referee in which each is at liberty to lie. This problem arises in practice when competing departments or research groups apply to a central body for funds (Shubik, 1970a).

10.1.2 Fair division as a noncooperative game

Is it possible to design a procedure that can be played noncooperatively as a game in extensive form and that will meet certain criteria

Table 10.1
A simple example illustrating Knaster's fair-division scheme

	Item A	Item B
Player 1's valuation	10,000	6,000
Player 2's valuation	2,000	8,000
Total value	12,000	14,000
Fair share	6,000	7,000
Item A to 1	10,000	
Item B to 2		8,000
Surplus	4,000	1,000
Side payment	−1,500	+1,500
Final surplus	2,500	2,500

or axioms of fairness? This is the central problem in the design of fair-division games.

The simplest example of a fair-division game between two individuals is one in which one party divides the goods to be distributed into two piles and the other chooses. This has frequently been referred to as the cake-cutting problem. There are many variants to this game, hence it is important to be explicit in specifying all assumptions. We informally present the assumptions below.

ASSUMPTION 1. *The cake is divisible, and no matter how it is divided into two pieces, at least one piece will be acceptable to each player in the sense that it is deemed to yield him as much as or more than half of the value he places on the whole cake.*

ASSUMPTION 2. *The divider can cut the cake in such a way that both pieces are acceptable to him.*

ASSUMPTION 3. *The preference assumptions are: (a) only a preference ordering is given; or (b) utility scales are specified up to a linear transformation; or (c) utility scales are given and interpersonal comparisons are made.*

ASSUMPTION 4. *The state of knowledge that the players have about each other's preferences is: (a) they know them; or (b) they do not know them.*

The weakest set of assumptions consists of 1, 2, 3a, and 4b. Paradoxically the cut-and-choose fair division may be intuitively more appealing if the players do not know each other's preferences than if they do. If they do not know each other's preferences, then we may formulate the problem as a well-defined game by considering that each player has a prior distribution over what he thinks are the preferences of the other (Harsanyi, 1967; Steinhaus, 1949), and each is aware of the priors of the other.

Suppose they know each other's preferences. Then the one who divides will have an opportunity to exploit the other. Consider a cake made of 9 parts sponge and 1 part marzipan. Player A values sponge and marzipan equally. He has a value of 10 for the whole cake. Player B values the marzipan alone at 10 times his value for sponge. His total for the cake is 19. If they both know this, and if A cuts, he will offer two pieces, one consisting of all of the sponge plus a small amount of marzipan, and the other with slightly more than 19/20 of the marzipan. He can expect to gain approximately 90 percent of his value for the whole cake while B obtains 50 percent. If B cuts, he offers two pieces, one containing a little more than 5/9 of the sponge and the other with all the marzipan and almost 4/9 of the sponge. A will obtain approximately 50 percent of his value for the cake, and B around 74 percent.

Banach and Knaster (reported in Steinhaus, 1949) extended the two-person fair-division game to an n-person game as follows. The players randomize to determine who cuts first. The first cutter, A, cuts a piece which is then offered in sequence to the other players, who each have an opportunity to diminish its size. If no one does so, then A obtains the slice and the process is repeated on the rest of the cake with A excluded. If, however, some player diminishes the slice and passes it on, the last diminisher must take the slice he has diminished. Dubins and Spanier (1961) prove the fairness of this procedure, and Kuhn (1967) suggests a somewhat different procedure.

Summarizing, we stress the following points:

1. Fair-division procedures differ from fair-division games.
2. A fair-division game is a game usually played noncooperatively in extensive form.
3. Assumptions concerning the players' knowledge of each other's payoffs are critical.
4. A division procedure whereby players can lie to the referee becomes a game of strategy.

10.1.3 The design of cheat-proof systems

The theory of social choice has passed through two major phases in recent years. The first phase was initiated by the well-known impossibility result of Arrow in the early 1950s. The objects of study were functions that mapped profiles of individual preferences into a single choice or a choice set, or more generally into a preference ranking for the society as a whole. Typical results concerned the existence or nonexistence of social choice functions that satisfy given sets of axioms and the cataloging of properties held by particular social choice functions.

The problem of determining the preferences of the society's members was not addressed in this early work. This problem must clearly be answered, though, if one wishes to apply a social choice function in making a real decision. One can think of a usable social choice *mechanism* as a noncooperative game: the players (e.g., voters) choose strategies (according to their preferences), and an outcome (a social choice, or the determination of a societal preference ranking) results.

It is well known that most common voting mechanisms are strategically manipulable, in the sense that it is occasionally in a voter's best interest to "misrepresent" his preferences. Are there any reasonable voting mechanisms which are not manipulable? Gibbard (1973) and Satterthwaite (1975) have independently shown that the answer is negative. Specifically there is no nonmanipulable voting mechanism for choosing among three or more alternatives in which no voter has dictatorial powers. That is, no such mechanism has an equilibrium point in dominant strategies: for some voter, with some particular set of preferences, it will happen that his individually best action depends on the actions of the other voters. The proof of this result is quite close to the proof of Arrow's impossibility theorem.

Suppose that each of a number of individuals has knowledge of his own attributes and preferences, but not of the attributes and preferences of the others. A social designer might have a preconceived notion of what outcome he would wish to associate with each possible profile of attributes and preferences. The *implementation question* is whether there is a mechanism (also called a *game form*) with a distinguishable equilibrium point (dominant strategy, or strong, or merely Nash) such that each social profile is associated (when the players follow their equilibrium strategies) with the desired outcome. In a variety of contexts the work of such authors as Shubik (1970a), Hurwicz (1977), Maskin (1977), Schmeidler (1973), Peleg (1978), and Postlewaite (1979) relates to the implementation question.

10.1.4 Auctions and fair division

Related to the work of Lerner and Knaster discussed in section 10.1.1, but independent of interpersonal comparisons of utility, is the use of the auction as a fair-division scheme.

Suppose that a group of individuals with different and unknown preferences are left equal shares in an estate. One form of fair division is for the referee to hold an auction. After the auction is over, each individual is credited with $1/n$ of the proceeds. He then pays or receives the balance between his income and what he has spent.

The formulation of this procedure as a well-defined game is left to the reader. This fair-division auction has a close relative in actual trading when a group of art dealers forms a "ring" to buy collusively at an auction, after which they hold a private auction to divide the proceeds.

There is a somewhat different version of the Knaster cake-cutting problem in which the cutter moves the knife over the cake, increasing the portion until some player cries "cut." That player obtains the piece cut, and the division proceeds in the same manner for the rest of the cake. This process has as its direct analog the Dutch auction, in which all individuals watch the price on an item gradually drop on a clock until someone indicates that he will buy, at which time the clock stops and the person buys at the last price indicated.

Auction and fair-division procedures and games are particular mechanisms of an economy devoted primarily to distribution problems. For this reason, before any attempt is made to generalize from simple problems like dividing an estate or selling tulip bulbs to problems of general welfare, care must be taken in specifying the ownership claims and the role of production as well as distribution.

There is a large literature on auctions and noncooperative bidding games; Stark and Rothkopf (1979) offer a bibliography.

10.2 Special Classes of Games in Strategic Form

10.2.1 A nonstrategic solution and an incompletely formulated game

There is one important solution concept in economic theory that is not a game-theoretic solution per se but is related to many game-theoretic solutions when the game to be analyzed can be interpreted as a model of an economic market. This is the *price system,* also referred to as the *competitive equilibrium.*

Many of the applications of game theory to economics have involved the relationship between the price system and different game-theoretic solutions to markets with many participants. This relationship is so basic that it is manifested in connection with both cooperative and noncooperative solutions. The cooperative solutions are based upon the market modeled in coalitional form, whereas the noncooperative solutions call for a description in strategic form.

In the development of economic theory the price system has been presented in three different formulations associated with the names of Cournot (1838), Edgeworth (1881), and Walras (1874). The first two are essentially game-theoretic in nature and serve as clear precursors of the solution concepts of the noncooperative equilibrium and the core. The third explicitly avoids the game-theoretic formulation by employing an assumption that the individual economic agent has no strategic power whatsoever (there is a game-theoretic interpretation of the Walrasian formulation for games with a continuum of players, but not for economies with a finite number of traders). Here we shall specify the Walrasian model which, though not strictly game-theoretic, has much in common with the noncooperative viewpoint.

Assume a market with n traders in m commodities. Assume that all traders have preferences that can be represented by quasiconcave utility functions. Assume that trader i has an initial endowment of goods $a^i = (a^i_1, a^i_2, \ldots, a^i_m)$, where all $a^i_j \geq 0$ and

$$a_j = \sum_{i=1}^{n} a^i_j > 0 \quad (j = 1, 2, \ldots, m) \tag{10.2}$$

is the total supply of commodity j. *Walras's law* then asserts that there exists at least one set of fixed prices $p = (p_1, p_2, \ldots, p_m)$ such that if each trader buys and sells to satisfy his individual desires, subject only to his budgetary restrictions, then the result of these n independent maximization processes is a final distribution $x^i = (x^i_1, x^i_2, \ldots, x^i_m) \geq 0$ for each trader i that represents a balance of supply and demand,

$$\sum_{i=1}^{n} (x^i_j - a^i_j) = 0 \quad (j = 1, 2, \ldots, m), \tag{10.3}$$

and is, moreover, Pareto-optimal.

This so-called competitive equilibrium can be regarded as a device for decentralizing an economy by using a price system. The various proofs of the existence of a competitive price system (Debreu, 1952; Arrow and Debreu, 1954) are generally nonconstructive; they do not supply any insight into the dynamic evolution of the prices, and they are totally independent of the number of traders in the economy. In contrast, the various game-theoretic approaches to the price system depend explicitly and often delicately upon the numbers of traders in the market.

In their search for the existence proof of a price system in a competitive economy, Arrow and Debreu (1954) made use of a central theorem concerning noncooperative games proved by Nash (1950c). This is a (fixed-point) proof of the existence of a noncooperative equilibrium in an n-person game. Subsequently Debreu (1959a) dispensed with this artifact of using a result from noncooperative game theory and provided an elegant direct proof which obviated the need for any discussion at all of the strategic role of the players.

It is instructive nevertheless to consider the model proposed by Arrow and Debreu. The abstract economy they considered has $m + n + 1$ players divided into three classes: m consumers, n producers, and 1 market governor. Each consumer i strives to maximize his utility $u_i(x_i)$, where x_i is a consumption vector chosen by i that satisfies his budget constraint. Each producer j strives to maximize his profits $p_j y_j$.

The market governor attempts to maximize a quantity pz, where $z = x - y - \xi$ can be interpreted as having as its components the excess of demand over supply for the various commodities. The symbols x, y, and ξ can be interpreted as the vectors representing demand, supply from production, and supply from the initial holdings of the consumers, respectively. The strategy of the market governor is to name a vector of prices p which has as its components a price for each commodity in the economy. All prices are constrained to be nonnegative.

The budget constraint of any consumer includes not only the market value of his initial bundle of resources but also a share in the profits made by the production units. Consumer i obtains a percentage α_{ij} of the profits of producer j; that is, α_{ij} represents i's percentage of shares in j.

Unfortunately a consumer cannot know if a potential strategy is feasible unless he knows his budget constraint, which depends upon the choices of the other players. The game is thus not completely defined unless we specify the information conditions and the order of moves. We must also specify what the payoffs are when the system is not in equilibrium. Who gets paid what if a consumer names a nonfeasible consumption? Is there a rationing scheme or an iterative process that corrects his error?

Arrow and Debreu ignored the difficulties associated with the details noted above. They were interested in establishing the existence of the competitive equilibrium as a noncooperative equilibrium and were not concerned with the strategic features of their model.

Paradoxically a model that does fit the Arrow–Debreu description and can be well defined involves not a competitive economy but a completely price-controlled economy (Shubik, 1977). An example of the extensive form is given in figure 10.1 drawn for the case where $m = n = 1$. This is a noncooperative game in which the market governor moves first and announces his prices, the producer moves second and announces his production, and then the consumer buys. This can be interpreted as a price-controlled economy in which a central planning board broadcasts prices to producers and consumers who are constrained to take these prices as given. For this game it is necessary to specify what happens if players make blunders or otherwise do not play optimally. If it is strategically possible to do the wrong thing, the consequences must be specified in a well-defined game.

Cournot (1838, 1897) sketched the first limiting argument linking the competitive to the noncooperative equilibrium. The relationship can be seen intuitively in the idea that as the number of competitors increases, each individual progressively becomes strategically weaker, until he has virtually no strategic power to influence the welfare of others independently and ends up by accepting the market price as given.

10.2.2 Strategic market games

Market games are a class of games with a special structure designed to reflect basic aspects of trade among owner-traders. Solutions such as the core and the value show economic properties, and as market games are replicated, a relationship between these solutions and the

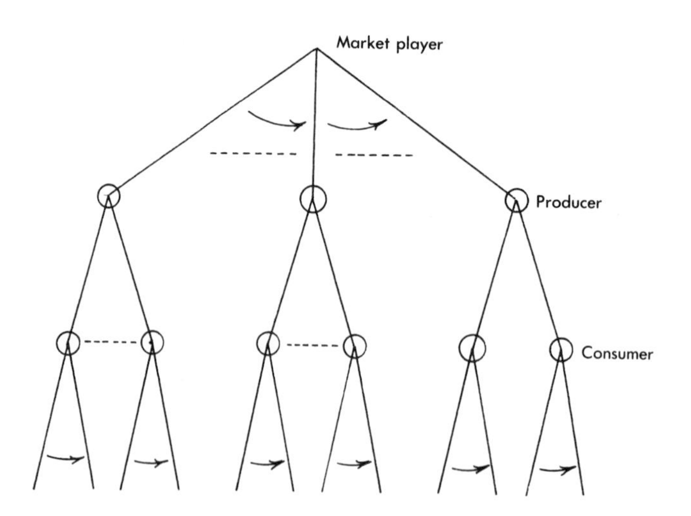

Figure 10.1
A centralized economy.

competitive equilibrium emerges. As noted in section 6.2.2, market games are c-games, that is, they are well represented by the characteristic function. Once a set of traders S has decided to trade together, what it can do is independent of the actions of its complement $\bar{S}$.

Market games represent an exchange economy in cooperative form. No exchange technology is postulated, and no description of individual strategy sets is supplied. We can take the same basic economic description that characterized the class of market games and produce a class of *strategic market games,* that is, games in strategic form which provide a model for economic exchange and whose properties can be studied by means of the noncooperative equilibrium solution. In this process an important new feature appears and a very special property of market games is lost. In particular, it becomes natural to distinguish one commodity for strategic purposes and to have it play the role of a means of exchange or a money. By introducing a market structure and a trading technology with a price-formation mechanism, we are able to define a strategic market game but we lose the c-game "property." If we begin with a strategic market game and then construct the characteristic function of this game, it will not be the same as the characteristic function of a market game as defined in chapter 6. In a market game, once trading groups have been formed, the traders are independent of all other individuals. In a strategic market game the traders are interlinked by the price-formation mechanism. This difference will be illustrated after strategic market games have been discussed.

A strategic market game is essentially a game in strategic form in which players have strategies that can be interpreted as bids or offers, direct or contingent statements to buy or sell goods. There is a market mechanism that processes all of the messages and produces final trades. More formally (see Dubey and Shubik, 1978b), we consider n individuals trading in m commodities:

1. Each individual i has an initial endowment of the m resources given by $a^i = (a^i_1, a^i_2, \ldots, a^i_m)$.

2. A *market* is a mechanism that processes messages and produces trades. We call the set of all markets in an economy e the *market structure* of that economy. This set is denoted by M_e, and $m_e = |M_e|$ is the number of markets in e.

3. A *trade* is a mapping of a market input by i into a market output to i.

4. To specify a *strategic market game* for the n traders with initial endowments a^i and utility functions u^i, we must specify the market structure M_e and identify the messages with the strategy sets S^i for each i.

Under relatively general conditions it can be proved that pure-strategy noncooperative equilibria exist for strategic market games (Dubey and Shubik, 1978a; Dubey, Mas-Colell, and Shubik, 1980). Furthermore, under the appropriate circumstances, as a strategic market game is replicated, or if it is considered to have a continuum of nonatomic traders, a relationship emerges between the competitive equilibria of the underlying exchange economy and the noncooperative equilibria of the strategic market game. The outcomes resulting from the competitive equilibria are approached by, or become outcomes associated with, noncooperative equilibria of a strategic market game (Shapley and Shubik, 1977; Dubey and Shubik, 1977b; Shapley, 1976).

For illustrative purposes we restrict our further remarks to a simple specific instance of a strategic market game, which has the advantage that we are able to employ the Edgeworth box diagram to illustrate much of the mechanism and the noncooperative equilibrium.

The Edgeworth box was introduced in section 6.3.1. It has proved to be a versatile diagram for illustrating basic properties of no-side-payment market games; and in a non-game-theoretic context it can be used to illustrate the competitive equilibria of an exchange economy. Here its use is extended to the portrayal of a simple strategic market game.

We conceive of a market structure in which there are $m - 1$ markets or trading posts where the traders are required to deposit, for sale, all of their initial holdings of the first $m - 1$ goods. Each uses the mth good to bid for the other goods. Thus a strategy by player i is an $(m - 1)$-dimensional vector of bids, $b^i = (b^i_1, b^i_2, \ldots, b^i_{m-1})$, where, if no credit is available,

$$\sum_{j=1}^{m-1} b^i_j \leq a^i_m \quad \text{and} \quad b^i_j \geq 0 \quad (j = 1, 2, \ldots, m - 1). \tag{10.4}$$

If we denote the total supply of commodity j by $\bar{a}_j = \Sigma^n_{i=1} a^i_j$ and the total amount bid on that commodity by $\bar{b}_j = \Sigma^n_{i=1} b^i_j$, then price is defined as

$$p_j = \bar{b}_j/\bar{a}_j \quad (j = 1, 2, \ldots, m - 1), \qquad p_m = 1. \tag{10.5}$$

The amount of the jth good that trader i obtains for his bid b_j^i is given by

$$x_j^i = \begin{cases} b_j^i/p_j & \text{if } p_j > 0 \quad (j = 1, 2, \ldots, m-1), \\ 0 & \text{if } p_j = 0 \quad (j = 1, 2, \ldots, m-1). \end{cases} \tag{10.6}$$

His final amount of the mth good, taking into account his sales as well as his purchases, is given by

$$x_m^i = a_m^i - \sum_{j=1}^{m-1} b_j^i + \sum_{j=1}^{m-1} a_j^i p_j. \tag{10.7}$$

Thus player i's payoff, expressed in terms of strategies, is

$$H_i(b^1, b^2, \ldots, b^n) = u^i(x_1^i, x_2^i, \ldots, x_m^i). \tag{10.8}$$

The Edgeworth Box The case $m = n = 2$ lends itself to a simple two-dimensional descriptive analysis based on the familiar Edgeworth Box. Much of this geometry will apply also to the general case, with many goods and traders, because of the way in which the operation of the market decouples most of the interactions among traders and among goods.

In figure 10.2 the first trader's holdings are measured up and to the right of the point O^1; the second trader's holdings are measured down and to the left of O^2. The dimensions of the box are $\bar{a}_2$ high by $\bar{a}_1$ wide. The point R represents a typical *initial allocation*, hence $a_1^1 = |M^1R|$, $a_2^1 = |G^1R|$, etc. (Here $|M^1R|$ indicates the length of the segment M^1R.) The points S^1 and S^2 represent typical *strategy choices*, hence $b_1^1 = |M^1S^1|$ and $b_1^2 = |M^2S^2|$. Bids are restricted to lie along the edges M^1O^1 and M^2O^2, respectively, since we are requiring all of good 1 to be sent to market. (If we did not, the strategies would be arbitrary points in the rectangles $RM^1O^1G^1$ and $RM^2O^2G^2$, respectively.)

Now consider the line joining S^1 and S^2. Its slope is $(b_1^1 + b_1^2)/(a_1^1 + a_1^2)$, which is just the price p_1. Moreover, it divides the line M^1M^2 into two segments, equal in length to x_1^1 and x_1^2, the amounts of good 1 purchased by traders 1 and 2, respectively. Similarly it divides the line G^1G^2 into segments equal to their final holdings x_2^1 and x_2^2 of the monetary commodity or "cash." The *final allocation* is thus represented by the point F. The vector RF represents the actual transaction that takes place; its slope, naturally, is equal (in absolute value) to the price p_1.

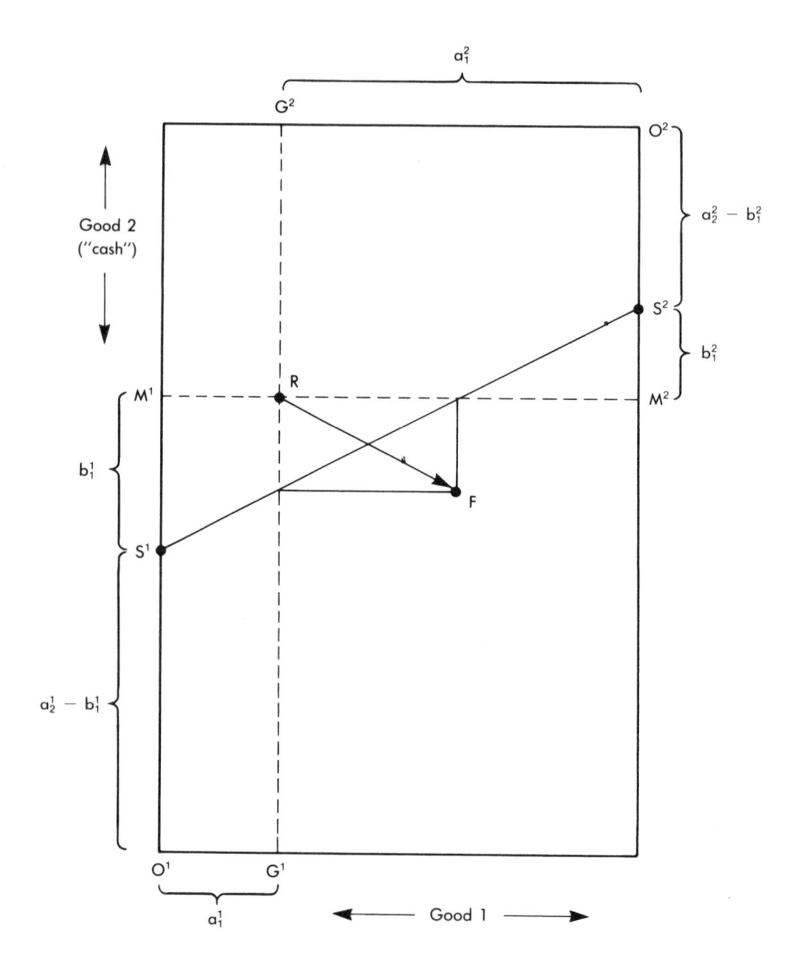

Figure 10.2
Price formation in an Edgeworth Box.

Figure 10.3 shows the effect of holding S^2 fixed while varying S^1. As S^1 moves over the interval from M^1 to O^1, the point F traces out the curve A^1RB^1. This curve is a portion of a hyperbola whose asymptotes are the horizontal and vertical axes through S^2. If we reverse the process and move S^2, holding S^1 fixed, we trace out the curve A^2B^2, which is part of a similar hyperbola centered at S^1. The endpoints A^1 and A^2 correspond to zero bids; the endpoints B^1 and B^2 reflect the upper limit on the amount a trader can bid. It happens in this case that we did not allow trader 2 enough cash to buy back his original holdings when trader 1 plays S^1, so the curve A^2B^2 stops short of the point R.

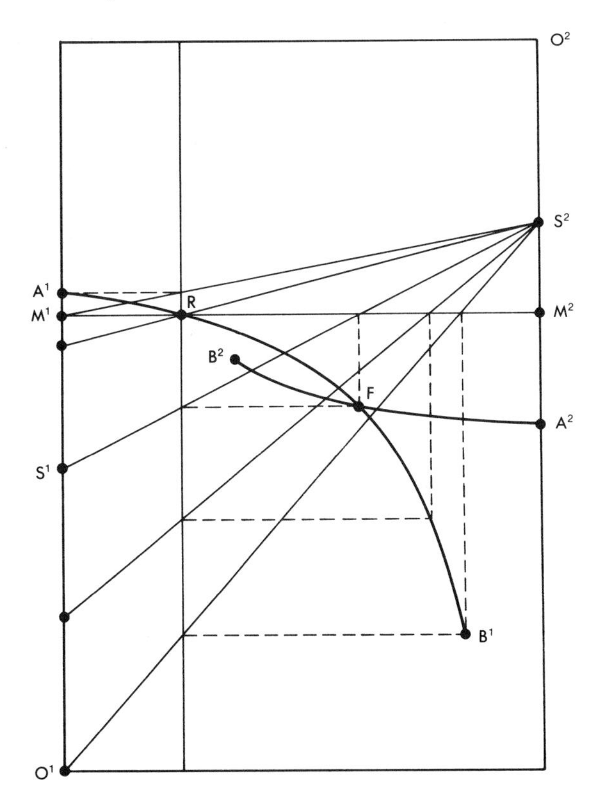

Figure 10.3
Budget sets.

These traces are comparable to the price rays or "budget sets" that confront a trader in the classical Walrasian model with its fixed prices. The difference is that in the present case the price is not constant but reacts to variations in a trader's own decisions; we therefore get a curve instead of a line. The curve is concave, as one would expect, so that if a trader increases his purchase he drives the price up, and if he bids less the price falls. The connection between the two approaches can be illustrated in figure 10.3 by supposing the second trader's holdings and bid to be very large, pushing the points O^2 and S^2 far off the page. The first trader's hyperbolic "budget set" would then approximate a straight line through R and F, reflecting the fact that he now has little influence on the price.

Two kinds of equilibrium So far we have discussed only the mechanics of the rules of exchange. We are now ready to add the traders' preferences to the picture by superimposing the contours of the utility functions u^1 and u^2. As shown in figure 10.4, S^1 happens to be the *best response* to S^2, since A^1B^1 is tangent at F to the contour of u^1. (Note that the curvature of A^1B^1 is such that there is always a unique point of tangency.) Similarly S^2 is the best response to S^1, since A^2B^2 is tangent at F to one of the contours of u^2. Thus figure 10.4 illustrates a noncooperative or "Nash" equilibrium (NE) for the market. Neither trader, knowing the strategy of the other, would wish to change.

A striking feature of this kind of equilibrium is its nonoptimality. Since the curves A^1B^1 and A^2B^2 are not generally tangent to each other, the point F cannot be expected to be Pareto-optimal or "efficient." (Pareto optimality in an NE can occur only at corners of the indifference curves or at special points such as $F = R$ or $F = B^2$.) In effect, the traders are working with unequal marginal prices, represented by the unequal slopes of A^1B^1 and A^2B^2 at F. Any outcome in the shaded region would be preferred by both traders to the NE allocation at F. In particular, they would both benefit from increased trade at the average price p_1, represented by the slope of RF.

A reader familiar with the Edgeworth Box will recognize the *contract curve* $C^1EP^2C^2$ in figure 10.4, which is a subset of the more extensive *Pareto set* $O^1P^1C^1EP^2C^2O^2$. The competitive equilibrium (CE) is represented by the competitive price ray RE, which is tangent to both indifference curves at the competitive allocation E. The situation illustrated seems to be typical: there is less volume of trade at the NE than at the CE. But the reverse is also possible.

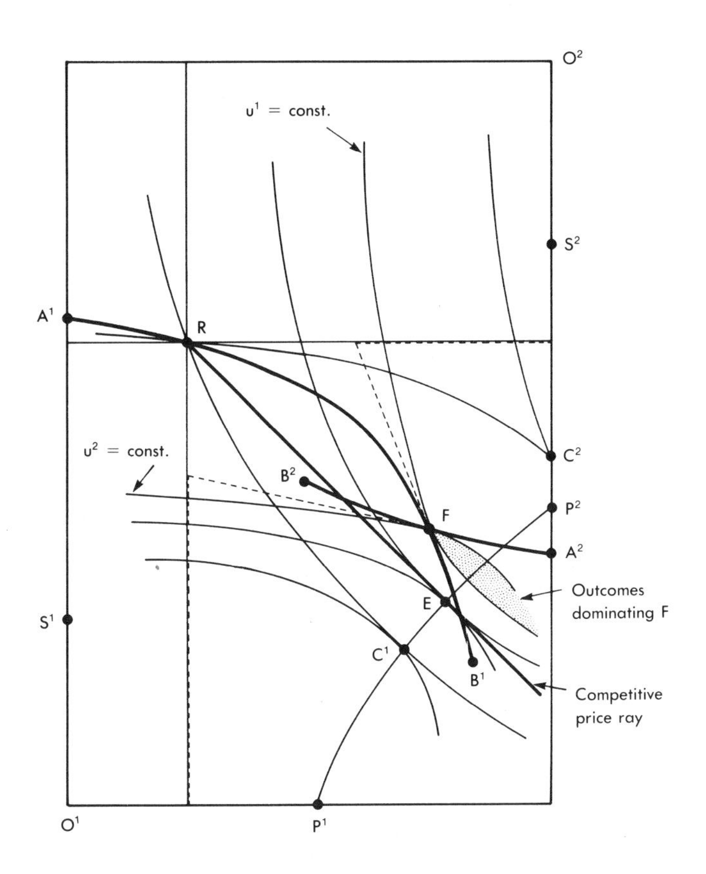

Figure 10.4
Noncooperative (F) and competitive (E) equilibria.

The properties of strategic market games will be investigated in more detail in the next volume (see also Dubey and Shubik, 1977a,b); these games have proved useful in investigations of problems in the development of the theory of money and financial institutions. Here the point to be stressed is that, despite the many difficulties encountered with the noncooperative equilibrium in general, as a solution concept applied to strategic market games it yields valuable insights which are different from but clearly related to an existing body of economic theory.

10.2.3 Extended Colonel Blotto games and lobbying games

In chapter 7 we considered ways of assigning values to players in n-person games played essentially cooperatively. Here we consider two models that convert an n-person game into a two-person constant-sum game or into a somewhat special type of n-person game. Our purpose is twofold. Several interesting applications are suggested, and a link between noncooperative and cooperative solutions is established (Shubik and Young, 1978; Young, 1978a; Shubik and Weber, 1981a,b).

Consider an n-person game in cooperative form with a characteristic function v. Imagine two superplayers I and II with resources a and b, respectively, which they can use to try to control the players in the n-person game. Suppose that a strategy by I is to allocate his resources to the n players as $a = (a_1, a_2, \ldots, a_n)$. Superplayer II's allocation is $b = (b_1, b_2, \ldots, b_n)$. Let the probability that player k supports superplayer I be $p_k(a_k, b_k)$. The payoff to superplayer I is the value $v(S)$ of the coalition S of players who actually support him. The payoff to II is $v(N) - v(S)$, where N is the set of all players.

We can immediately give two interpretations for application. The n players may be considered voters and the two superplayers can be regarded as lobbyists out to buy them. The "support functions" p_k model the chances for buying any individual k. A different interpretation is in terms of network defense. We may regard the n players as nodes in a complex network with $v(S)$ measuring the overall performance of the system. If, for example, the original game is a simple game (see chapter 6), then the characteristic function essentially measures the redundancy in a network, that is, it tells us what units can be knocked out and still leave the system functioning. In a military context $p_k(a_k, b_k)$ can be interpreted as the battle condition at target

k; it measures the probability that the target is captured or destroyed as a function of the resources a_k expended on defense and the resources b_k utilized in attack.

The expected payoff to superplayer I is

$$P_{\text{I}}(a, b) = \sum_{S \subset N} \left[\prod_{i \in S} p_i(a_i, b_i) \prod_{j \in S} (1 - p_j(a_j, b_j)) \right] v(S), \tag{10.9}$$

and the payoff to superplayer II is

$$P_{\text{II}}(a, b) = v(N) - P_{\text{I}}(a, b). \tag{10.10}$$

In section 7.7.3 we defined a t-value for any player i of an n-person game as

$$\phi_i^{(t)}(v) = \sum_{S \subset N \setminus i} t^s (1 - t)^{n-s-1} [v(S \cup i) - v(S)], \tag{10.11}$$

where N is the set of all players and $s = |S|$.

Shubik and Weber (1981a) have shown that for an n-person game without dummies, if the two-person competitive supergame based upon v has pure-strategy equilibria, the resources at equilibrium will be distributed in proportion to some t-value. This will hold if $p_i(a_i, b_i) = p(a_i, b_i)$ for all players i, where p is continuously differentiable, strictly monotonic in each variable, and homogeneous of degree zero. In other words, the class of t-values that form a cooperative solution can be obtained as the strategies from a noncooperative equilibrium. This offers an additional economic justification for the value allocation.

A different noncooperative approach has been suggested by Young (1978b). Suppose there is one rich lobbyist who wishes to purchase a winning coalition at least cost. If a voter can be bought, we may assume he will try to maximize his price.

If $p_i \geq 0$ is the price of player i, then $p(S) = \Sigma_{i \in S} p_i$ is the cost of buying the set S. We may imagine an n-person game in which all voters simultaneously announce their prices and then the lobbyist buys a minimum-cost winning set automatically. We can ignore the lobbyist as a strategic player in this formulation and replace him by a mechanism.

Let $L(p)$ be the lobbyist's purchase rule, and suppose that there is a positive minimum price $p_i^0 > 0$ for each voter i. Then it can be shown that for a simple game without veto players there will always be at least one strong noncooperative equilibrium (see section 9.1.8).

An example provided by Young (1978b) illustrates this result. Consider an eight-player voting game with the players called 1, 2, . . . , 8 and the following minimal winning sets (see chapter 6):

$\{1, 2, 3\}$, $\{1, 4, 5, 6\}$, $\{2, 4, 5, 6, 7\}$, $\{3, 4, 5, 6, 7\}$.

Suppose that the floor prices are $p_i^0 = 1$ ($i = 1, 2, \ldots, 8$). As a tie-breaking rule suppose that $\{1, 2, 3\}$ is chosen whenever it is one of several minimum-cost sets.

EXERCISES

10.1. *Show that (2, 1, 2, 1, 1, 1, 1, 1) and (3, 2, 1, 1, 1, 1, 1, 1) are strong equilibria.*

10.2. *Why is the result described above not true if there is a veto player?*

10.2.4 Some open questions

In the last two sections we have considered two classes of games in strategic form with special structure. In both instances the noncooperative-equilibrium solution yielded results of some applied interest. Are there other classes of games in strategic form which have both an interpretable special structure of interest and equilibrium points that appear to give insight?

One such class of games has already been noted but not analyzed, namely *games of status.* Another natural class consists of *strategic simple games.* Suppose that there are n individuals and m "bills" or items on which to vote. Any individual i will vote on some or all of the items. Thus his strategy set will contain 2^r ($0 \leq r \leq m$) strategies (yes or no for each item he votes on), or 3^r if abstention is permitted. There are 2^m different outcomes to be evaluated by each voter; and voting rules must be specified to determine the conditions for the passage or defeat of each item.

For $m = 1$ the analysis of these games is trivial for both equilibria and strong equilibria. This is not true for $m \geq 2$.

A key element in the description of a strategic market game is the individual's ability to aggregate the strategies of all other players. This is, however, a property of political and social as well as economic processes. The individual is frequently involved in situations in which he is concerned with his actions and what "they" have done. How do we aggregate the actions of others? In the case of Cournot's model of economic behavior this is easy. Strategies are identified with quantities offered to a market. Quantities of a commodity can be treated additively. Thus each individual views "his game" as a payoff function of two variables:

$$H_i(q_1, q_2, \ldots, q_n) = \tilde{H}_i\left(q_i, \sum_{j \neq i} q_j\right).$$

More generally, if s_i is a strategy of player i, we may consider games with the property that

$$H_i(s_1, s_2, \ldots, s_n) = \tilde{H}_i(s_i, f_i(s \backslash s_i)),$$

where $s \backslash s_i$ stands for the vector $(s_1, s_2, \ldots, s_n)$ with the ith component removed.

Games with the above property clearly have much more structure than a game selected at random. How this structure influences the equilibrium points has not yet been explored in depth.

10.3 Other Noncooperative Solutions and Applications

The three solutions discussed here are not new solutions but the noncooperative-equilibrium solution applied to special mappings from outcomes to utility functions. The mappings are selected for their relevance to specific problems.

10.3.1 Maxmin-the-difference and beat-the-average

The entries in the payoff functions of the two-person non-constant-sum games that are studied in experiments or otherwise are usually expressed in terms of points or money or some other physical entity that has value to the individual. Sometimes an assumption is made that over the small range covered by the game an individual's utility for money is approximately linear, hence the "money" numbers in the matrix can be regarded as also reflecting the individual's utility for money.

Especially when the stakes are low, the assumption that the entries in the matrices can be interpreted in this way may be inadequate. In particular, the competitiveness of the players may be such that their goals are to maximize the difference in their point score or monetary gain. Thus the bimatrix or non-constant-sum two-person game characterized by (a_{ij}) and (b_{ij}) is replaced by a constant-sum game characterized by $(a_{ij} - b_{ij})$, where we can regard the new entries as reflecting the utilities attached by the players to the outcomes.

In experimental work a safe procedure is to see how "maxmin the difference" compares with the noncooperative equilibrium as a predictor of behavior. Furthermore, when experimenting with small matrices such as the 2 × 2, one should check for the possibility that different solutions predict the same outcome. For example, the Prisoner's Dilemma matrix transforms into the constant-sum game shown below. This has the same pair of strategies for the saddlepoint as the equilibrium pair in the nontransformed game:

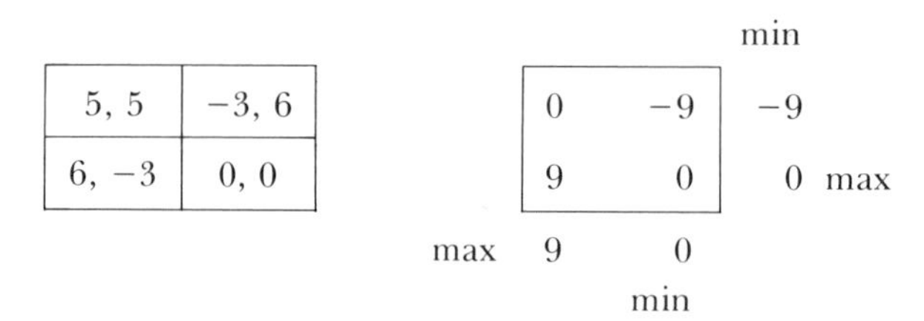

The maxmin-the-difference solution has an economic interpretation in terms of "illfare economics," where the parties are concerned with measuring the damage exchange rate between them if they carry out threats. In the value calculations (see chapter 7) the maxmin-the-difference game serves as a means for establishing the initial point from which the gains from cooperation are measured. This point lies on a Pareto-minimal surface (Shubik, 1959b).

The maxmin-the-difference approach literally transforms a two-person non-constant-sum game into one of pure opposition. The damage exchange rate feature is natural to the economics of tactical military combat. But even in military application care must be taken in the correct specification of the measure of opposition. Convoy defense provides an example. One measure might be the number of ships sunk versus the number of submarines sunk. But this attrition index may not be as appropriate a measure as a specific function of the type and amount of cargo destroyed.

The transformation of a two-person non-constant-sum game treats the payoffs in the original matrices as though they were points, with utility or worth being defined on the difference in point score. When there are more than two players, we may consider an analogous generalization by defining a "beat-the-average" payoff or weighted beat-the-average payoffs. The beat-the-average payoff to i is defined as follows. Let $H_i(s_1, s_2, \ldots, s_n)$ be the original payoff matrix to each player i, where the payoff is in points or money or some other score.

The subjective payoff to i is then defined as

$$\pi_i = H_i(s_1, s_2, \ldots, s_n) - \frac{1}{n-1} \sum_{j=1}^{n} H_j(s_1, s_2, \ldots, s_n). \tag{10.12}$$

10.3.2 The cooperation parameter

The maxmin-the-difference solution is defined for a two-person game, and the related beat-the-average solution is defined for games with more than two players. These both belong to a class of solutions which are all really nothing more than the noncooperative equilibrium applied to a transformed set of outcome matrices that are treated as containing the players' valuations of the outcomes.

Consider an n-person game in strategic form, with the outcome matrix of each player i given by $H_i(s_1, s_2, \ldots, s_n)$. We introduce a set of n^2 parameters θ_{ij}, which may be interpreted as the concern of player i for the outcome to j. (Edgeworth called this the "coefficient of sympathy.") Without loss of generality we may set $\theta_{kk} = 1$ ($k = 1, 2, \ldots, n$). Payoff matrices (in utility terms) are defined by n new functions of the form

$$\pi_i(s_1, s_2, \ldots, s_n) = \sum_{j=1}^{n} \theta_{ij} H_j(s_1, s_2, \ldots, s_n) \quad (i = 1, 2, \ldots, n). \tag{10.13}$$

Suppose that $\theta_{ij} = -1/(n-1)$ for $i \neq j$. We apply the noncooperative-equilibrium solution to the π_i. For $n = 2$ this immediately yields the maxmin-the-difference solution. For $n > 2$ we have a beat-the-average situation. When the players are, in some appropriate sense, of commensurate size and strength and when the outcomes can be reasonably compared, as in the case of the monetary earnings of corporations in the same industry, this utility transformation of the outcomes may be reasonable. In other instances it is easy to construct examples in which this type of utility for the outcomes appears to be pathological.

If the θ_{ij} are set so that

$$\theta_{ij} = \begin{cases} 1 & \text{for } i = j, \\ 0 & \text{for } i \neq j, \end{cases}$$

and if each individual maximizes π_i, we have an "inner-directed" noncooperative-equilibrium solution in the sense that the outcomes to other players have no influence on the preferences of each individual.

If the θ_{ij} are selected so that $\theta_{ij} = 1$ for all i and j, the joint maximum has been defined. In this instance each individual values the outcome to others as to himself.

10.3.3 Games of status

The new solutions suggested above have in fact not been new solutions but specific transformations linking outcomes with payoffs. The resultant games have utility scales for each player defined up to a linear transformation.

It is possible and not unreasonable in some instances to consider a transformation of the outcomes that gives rise to a preference ordering without a utility scale. This might be the case if the only concern of the players is rank as measured by an ordering of their point score.

For a two-person game it is clear that the appropriate solution concept must be the noncooperative equilibrium, and that the values of the outcomes will consist only of win, lose, or draw.

With three or more players the possibility for coalition becomes greater, and other solution concepts must be considered. Cooperative solutions to games of status have been considered elsewhere (Shubik, 1971c).

Leaving aside ties, an n-person game of status has n payoffs for each player corresponding to being first, second, . . . , nth. If we denote the payoffs by 1, 2, . . . , n, then the entries in any payoff cell of a game of status in strategic form with a finite set of strategies is a permutation of these numbers.

10.3.4 Metagames

Nigel Howard (1966, 1971) has argued in his theory of metagames that a noncooperative equilibrium should include the possibility of higher-order expectations of the players about the dependency of the choice of one on the behavior of others. He thereby constructs a series of *minorant games* and examines their equilibrium points. To illustrate, consider the following 2×2 bimatrix game:

	1	2
1	3, 3	−1, 4
2	4, −1	0, 0*

It has two first-order metagames, namely player 1's metagame, which is a 2 × 4 matrix arising from his belief that player 2 is committed to a policy dependent on player 1's selection,

	(1, 1; 2, 1)	(1, 1; 2, 2)	(1, 2; 2, 1)	(1, 2; 2, 2)
1	3, 3	3, 3	−1, 4	−1, 4
2	4, −1	0, 0	4, −1	0, 0*

and player 2's metagame, which is similarly a 4 × 2 matrix:

	1	2
(1, 1; 2, 1)	3, 3	−1, 4
(1, 1; 2, 2)	3, 3	0, 0
(1, 2; 2, 1)	4, −1	−1, 4
(1, 2; 2, 2)	4, −1	0, 0*

Note that no new outcomes are produced; we merely enlarge one or the other player's strategy space, in effect giving him foresight of the other player's choice.

The equilibrium points are indicated by an asterisk in each game. It is evident that a pure-strategy equilibrium point must exist in these first-order metagames, even if there is none in the original. In this example, which is of the Prisoner's Dilemma type, no new equilibria have appeared. It is now possible, however, to form higher orders. The next two are the 1,(1, 2) game in which player 1 believes that player 2's policy is based upon what he does, and he selects his policy based upon player 2's policy; and similarly the 2,(2, 1) game. These can be represented as a 16 × 4 and a 4 × 16 matrix, respectively.

The 1,(1, 2) game is illustrated in figure 10.5. This game has three pure-equilibrium points, two of which are Pareto-optimal. It can be shown that the corresponding 2,(2, 1) game would have equilibrium points producing the same outcomes as the equilibrium points in the 1,(1, 2) game, though in general this is not the case. Howard has proved that no further enlargement of the set of outcomes arising from equilibria will come from a consideration of higher orders of "metagames."

	1	2	3	4
1	3, 3	−1, 4	3, 3	−1, 4
2	4, −1	0, 0*	0, 0	4, −1
3	4, −1	0, 0	0, 0	−1, 4
4	4, −1	0, 0	3, 3*	4, −1
5	4, −1	0, 0	3, 3	−1, 4
6	4, −1	−1, 4	0, 0	4, −1
7	4, −1	−1, 4	0, 0	−1, 4
8	4, −1	−1, 4	3, 3	4, −1
9	3, 3	−1, 4	3, 3	−1, 4
10	3, 3	0, 0	0, 0	4, −1
11	3, 3	0, 0	0, 0	−1, 4
12	3, 3	0, 0	3, 3*	4, −1
13	3, 3	0, 0	3, 3	−1, 4
14	3, 3	−1, 4	0, 0	4, −1
15	3, 3	−1, 4	0, 0	−1, 4
16	3, 3	−1, 4	3, 3	4, −1

Figure 10.5
The 1,(1, 2) metagame.

This generalizes to n players. One can take a finite n-person game in strategic form and construct its metagames by "expanding" the strategy space of each player in turn. There are $n!$ ways of completing this process, corresponding to the various orders in which the players can be allowed, in effect, to anticipate the other players' decisions. Howard's result is that expanding more than once in the process on the same player adds nothing new to the set of equilibrium outcomes that are generated.

Although this analysis does bring in player expectations in an interesting manner, and in the example of the Prisoner's Dilemma enlarges the equilibria to encompass a Pareto-optimal outcome, nevertheless most of the difficulties with the noncooperative equilibrium still remain. Indeed, the lack-of-uniqueness problem is exacerbated.

10.3.5 $\{k_i\} - \{r_j\}$ stability

Until recently most of the work on the noncooperative equilibrium has involved static models, even though much of the discussion has concerned games in an implicitly dynamic context. It would of course be desirable to catch the dynamics of various forms of competition more explicitly, but it would also be of interest to devise stronger stability properties than those possessed by the simple noncooperative-equilibrium solution. In particular, stability has been considered so far in terms of individual behavior. In some applications, particularly applications to economic competition, it is desirable to consider disturbances of the equilibrium that might be caused by the efforts of more than one individual acting in concert. This might be termed partially cooperative behavior in an otherwise noncooperative context.

Shubik (1959b) has suggested a somewhat ad hoc noncooperative solution specifically for application to problems in oligopolistic competition. This is a modification of the Nash noncooperative equilibrium, which has stability properties specified only in terms of individual maximization. Group stability properties are considered for "$\{k_i\} - \{r_j\}$ stability." Here $\{k_i\}$ and $\{r_j\}$ stand for specific sets of players to be defined below. Two versions are noted. The first and simplest to define is for a game in strategic form; the second calls for a dynamic formulation.

Consider an n-person game in strategic form. We arbitrarily impose a coalition structure on the players and assume that the new game is played noncooperatively among the coalitions, with the further condition that no side payments are made. (A somewhat different solution can be defined if side payments are permitted within coalitions.) If we consider all partitions, an astronomical number of noncooperative games are formed: for a symmetric game a split into two competing groups gives rise to n games for n players; if we consider partitions, then for $n = 10$, 50, and 100 there are 42, 173,525, and 104,569,292 different games, respectively.

Coalition structure may be specified exogenously given institutional or other information. This approach is adopted by Luce (1954, 1955) in his definition of Ψ-stability for a game played cooperatively in characteristic-function form.

A simple coalition structure might involve k of the players acting alone and the remaining $n - k$ players acting in concert. It might be hoped that for interesting classes of games we could find equilibrium points with higher levels of stability than stability against an individual. Furthermore, it would be desirable if a higher level of stability implied all lower levels. Unfortunately this is highly unlikelv in general.

The three-person game shown below, where A picks the rcw, B the column, and C the matrix has two equilibria arising from strategies (1, 1, 1) and (2, 2, 2). The first is stable against defections by 1, 2, or 3 players; the second is stable only against defections by one player:

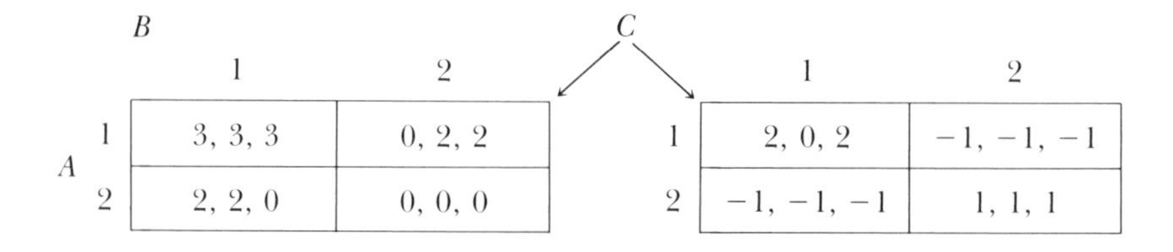

$C = 1$:

A \ B	1	2
1	3, 3, 3	0, 2, 2
2	2, 2, 0	0, 0, 0

$C = 2$:

A \ B	1	2
1	2, 0, 2	−1, −1, −1
2	−1, −1, −1	1, 1, 1

The motivation for the $\{k_i\} - \{r_j\}$ stability solution arose in the study of oligopolistic markets considered as multiperiod games of economic survival (Shubik, 1959b; Shubik and Thompson, 1959). Consider a market through time such that in each period the firms (and potential entrants) face the same market and have the same costs.

We divide the players into three sets: the policing set $\{k_i\}$, the violating set $\{r_j\}$, and the neutrals $N - \{k_i\} - \{r_j\}$. Our concern is limited to stationary equilibria. The neutrals are presumed to have announced a stationary strategy regardless of what the others do.

The policing set announce a "threat strategy" which consists of a plan to support the status quo as long as no one else violates it. If the status quo is violated, then after having been informed of the violation they will punish the violators. The violators will disturb the equilibrium if the threat is not strong enough.

The Aumann strong equilibrium (see section 9.1.8) is essentially the same as a $\emptyset - \{r_j\}$ stable solution for all sets $\{r_j\}$. (Here $\emptyset$ is the empty set, the coalition of no one.) In a test for a strong equilibrium we consider all possible violating sets of players acting against the remainder who are neutral, that is, who stick to a stationary strategy.

The "unanimity game" provides an example of a strong equilibrium point, or equivalently a $\emptyset - \{r_j\}$ stable solution. Each of n players has two strategies, "yes" or "no." If all agree, the payoff is (1, 1, . . . , 1); in all other circumstances it is (0, 0, . . . , 0). The n-tuple of strategies where all say "yes" is a strong equilibrium.

The stability of the $\{k_i\} - \{r_j\}$ equilibrium depends upon the size of the policing group, the size of the violator group, the future discount, and the punitive strength of the threat strategy. This last item raises problems concerning how a threat strategy is communicated and, in situations where there is no contract or guarantee that it will be carried out, the extent of belief that it will indeed be utilized.

The problem with this solution is not demonstrating the existence of an equilibrium but the possibility that too many equilibria exist. Although it may offer an intuitively appealing description of the policing of a market, it has a low resolution power.

Closely related to this group-stability approach to games in extensive or repeated strategic form is the work on strategic voting by Farquharson (1969).

10.4 On Models and Solutions

In closing this last chapter on noncooperative solutions, let me reiterate my position that the search for a single unifying solution theory to apply to all games is akin to the search for the philosopher's stone. Not only is it intuitively unlikely that one solution concept will be reasonable in all instances; even if it were, it would still be unlikely that the solution theory would yield a unique solution point that would satisfy all properties deemed desirable for a solution.

Even in chapter 7, where value solutions designed to reflect properties of justice and fairness were discussed, it was not really feasible to obtain a one-point solution without assuming the existence of

Table 10.2
Summary of solution properties

	Games in general	Strategic market games	Two-person constant-sum games
Existence	always for finite games	yes	yes
Pure- or mixed-strategy EPs	a high probability for the existence of pure EPs when at least two players have many strategies	pure EPs	a high probability for mixed EPs
Uniqueness	rare	rare	yes
Value	rare	rare	yes
Interchangeability	rare	rare	yes
Pareto optimality	rare	for finite numbers usually no, but yes for some games; in the limit yes for many games	yes, by definition, for all outcomes
Strong	extremely rare	no for most games, but yes for some	yes by definition
Perfect	always exist for finite games and form a subset of the EPs	yes	yes

EP = equilibrium point.

linear transferable utility. In these chapters it has been shown that in general, for virtually any desirable property, such as optimality, uniqueness, or stability, a game can be constructed with noncooperative equilibria that lack the property. The existence of counterexamples does not, however, vitiate the usefulness of the noncooperative-equilibrium solution or its modifications.

The fact that some games have no pure-strategy equilibria while others have a surfeit of equilibria does not mean that there exist no substantial classes of games, some of applied interest, for which the noncooperative-equilibrium solution is reasonable. There is even experimental evidence in favor of the noncooperative equilibrium in some games (see section 12.4). Even so, from our discussion of the many facets of the noncooperative-equilibrium solution, we must conclude that it offers a first step toward the construction of a dynamic solution theory. We have still not progressed far in this direction, as indicated by the difficulties in modeling communication and in reconciling solutions to games played in extensive form with solutions to games played in strategic form.

Table 10.2 offers a summary of some of the properties that have been considered and examines two special classes of games for these properties.

11 Further Cooperative Solution Concepts

11.1 Other Cooperative Solutions Based on the Characteristic Function

In this chapter we return to the cooperative branch of game theory and survey some of the additional solution concepts that have been proposed. While we cannot go into as much detail as many of them deserve, we shall take pains to stress the logical and conceptual relationships among the various kinds of solution, since "intersolutional analysis" seems destined to become increasingly important in applications.

11.1.1 Reasonable outcomes

Before turning to the next major group of cooperative solution concepts—the "bargaining-set" family—we shall introduce one more kind of presolution for characteristic-function games. It is roughly at the same level as the concepts of individual rationality and Pareto optimality (see section 6.3), though its rationale is slightly less direct.

The principle of individual rationality expressed by equation (6.3) holds that a player should not accept less from the game than he can take by his own unaided efforts. On the other hand, one might argue that he should not take more from the game than the amount he can contribute to the worth of at least one coalition. A payoff that exceeds all his marginals—that is, all the numbers $v(S) - v(S - \{i\})$ for $S \ni i$—might well be considered unreasonable for player i in any cooperative solution. Accordingly a payoff vector $\alpha = (\alpha_i : i \in N)$ is defined to be *reasonable* if and only if it satisfies

$$\alpha_i \leq \max_{S \ni i} [v(S) - v(S - \{i\})] \quad \text{for all } i \in N. \tag{11.1}$$

The set of all feasible Pareto-optimal payoff vectors that are reasonable in this sense will be denoted by R.

This concept was introduced by John Milnor in 1952. That he considered it a presolution is shown by his remark that "it is better to have the set too large rather than too small . . . it is not asserted that all points within one of our sets are plausible as outcomes; but only that points outside these sets are implausible." It can be shown, by relatively simple arguments, that the value, the core, and all the stable sets must be contained within R (Milnor, 1952; Luce and Raiffa, 1957; Gillies, 1953c, 1959). The same will hold for most of the other cooperative solutions to be discussed below.

Geometrically the set R is a simplex that is inverted with respect to the simplex of imputations since it sets upper rather than lower bounds on what the individual players receive (see figure 11.1). The effect is to "chip off the corners," eliminating some of the more extreme imputations from consideration. For some games (e.g., simple games without dummies) R is so large that there is no reduction: every imputation is "reasonable." For other games the reduction may be considerable. But R is never so small that its vertices lie strictly within the imputation space; nor do the latter's vertices ever lie strictly within R.

Milnor also suggested two other criteria of "reasonableness," setting upper and lower limits on what any coalition might reasonably claim in a cooperative solution. Their properties have not yet been sufficiently developed to warrant discussion here (see Luce and Raiffa, 1957; Shapley, 1971b). We shall only mention a simple generalization of (11.1), dubbed the *anticore,* defined as the set of all imputations that are "coalitionally reasonable" in the sense that for every S,

$$\sum_{S} \alpha_i \leq \max_{T \supseteq S} [v(T) - v(T - S)]; \qquad (11.2)$$

that is, no coalition gets more than it contributes to the worth of at least one coalition. The anticore relates to the simplex R in much the same way as the core itself relates to the simplex of imputations. It can be shown that the anticore is larger than the core and that it exists for some but not all games that do not have cores (Shapley, 1971b).

EXERCISE

11.1. *Show that for a three-person game the anticore is merely the intersection of R with the imputation space.*

11.1.2 The nucleolus

In chapter 6 some extensions of the core concept were discussed—in particular, the notion of an ϵ-*core.* The number ϵ was, one might say, an amount charged against the effective worth of a coalition. In

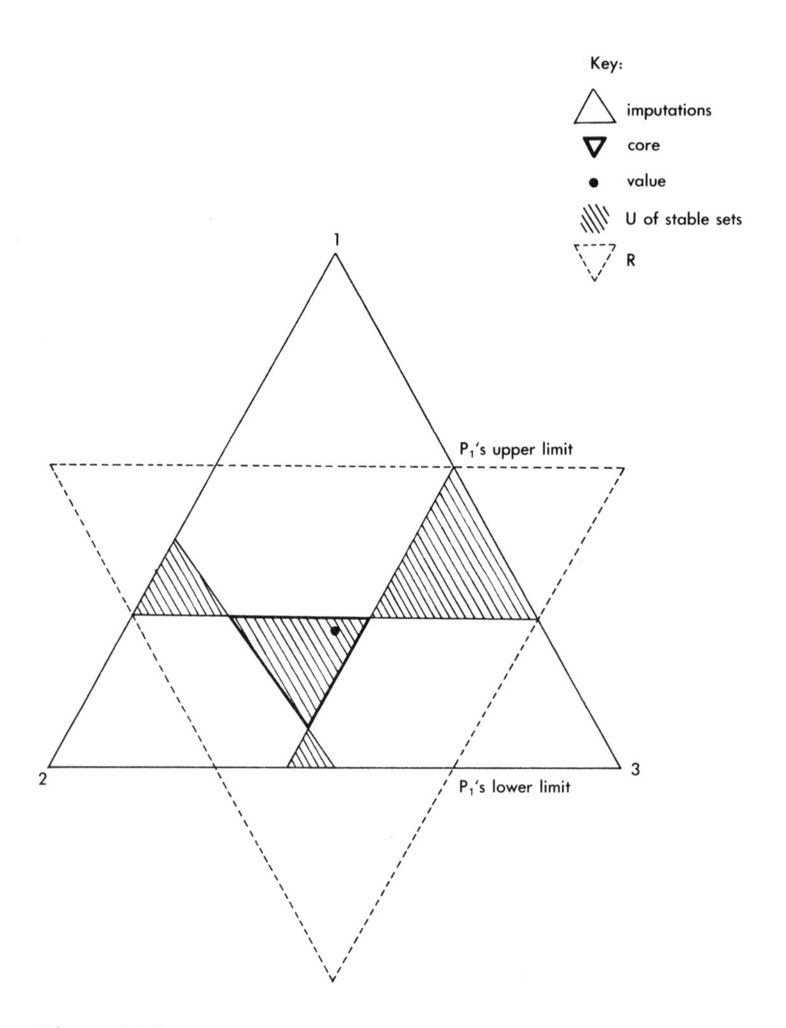

Figure 11.1
"Reasonable" outcomes in a three-person game.

practical applications it might represent expenses of communication or organization (if $\epsilon > 0$) or an agreed level of concession, by the most difficult-to-satisfy coalitions, in the interest of a compromise corelike solution, when no true core exists. There are several solution concepts that can be related to this idea of systematically modifying the powers or demands of the effective coalitions.

A technical term will be helpful here. By the *excess* of a coalition at a given payoff vector we shall mean the amount (positive or negative) by which the worth of the coalition exceeds its payoff; formally,

$$e(S, \alpha) = v(S) - \sum_{i \in S} \alpha_i. \tag{11.3}$$

Thus the core is the set of imputations at which no excess is greater than zero. The ϵ-core is the set of imputations at which no excess is greater than ϵ (except possibly for $\varnothing$ or N, both of which have excess automatically 0, while ϵ may be negative). The *near-core* has been defined as the smallest nonempty ϵ-core; this is precisely the set of imputations at which the maximum excess (again ignoring $\varnothing$ and N) is as small as possible. The solution we are about to describe is a refinement of the near-core.

The near-core is of course a convex set—possibly a single point. The maximum excess at each point has a constant value throughout the set. But if the set does not consist of a single point, there will be a coalition, in fact many coalitions, whose excess is not constant throughout the set. Considering just these coalitions of variable excess, we can proceed to minimize their maximum excess, always staying within the near-core. This will give us a smaller convex set, in fact a set of lower dimension than the near-core. If this is not yet a single point, we can minimize maximum excess a third time, among those coalitions whose excess still varies even within the reduced set. Since the dimension always drops by at least one, this repeated procedure must devolve in the end to a single point, called the *nucleolus*.

As a rough geometrical description, the nucleolus is found by pushing the walls of the core inward, stopping the movement of each wall just in time to avoid obliterating the set. If the core does not exist, any (strong) ϵ-core that does exist—for example, the near-core—may be used instead (Maschler, Peleg, and Shapley, 1972).

The above construction can be summarized loosely by saying that the nucleolus minimizes dissatisfaction, with priority to the coalitions that are most dissatisfied. Like the near-core, the nucleolus is contained in the "reasonable" set R and is in the core whenever the core exists. Like the value, the nucleolus is a continuous function of the game, that is, of the characteristic values $v(S)$ (Schmeidler, 1969b; Kohlberg, 1971); but where the value is a linear function, as in (7.1), the nucleolus is piecewise linear. In this respect it is analogous to the median of a set of numbers, whereas the value resembles a mean or average, being sensitive to variation in every $v(S)$. Needless to say, the value and the nucleolus rarely coincide.

Intuitively the nucleolus represents, as nearly as any single imputation can, the *location* of the core of the game. That is, if the core exists, the nucleolus is its effective "center"; and if the core does not exist, then the nucleolus represents its "latent" position, the center of the spot where it would first appear if the characteristic values were adjusted (diminished) at a constant rate, for all coalitions except $\varnothing$ and N.

In the Three-Cornered Market, first considered in section 6.4.1, the near-core coincides with the core. This can be seen from the fact that the core is of lower dimension than the imputation space. Thus, as shown in figure 11.2, the strength of the coalition $\overline{FS}$ is such that it forces the core to the right side of the simplex. Any increase in $v(\overline{FS})$ would wipe out the core, as would any increase in $v(\overline{M})$. Both $\overline{FS}$ and $\overline{M}$ have excess zero, constant throughout this set. In order to determine the nucleolus, we must consider the remaining four coalitions (other than $\varnothing$ and N) and their excesses.

At the upper endpoint of the near-core (α in the figure), the payoffs are 2, 0, and 1 to F, M, and S, respectively. Thus we calculate from (11.2):

$$
\begin{aligned}
e(\overline{F}, \alpha) &= 1 - 2 = -1, \\
e(\overline{S}, \alpha) &= 0 - 1 = -1, \\
e(\overline{FM}, \alpha) &= 2 - 2 = 0, \\
e(\overline{MS}, \alpha) &= 0 - 1 = -1.
\end{aligned}
$$

Note that since we are in the core, no excesses are positive. Indeed the excess for $\overline{FM}$ would become positive if we moved above α, showing that this point is the upper limit of the core. Making the same calculation for the lower endpoint (β in the figure), and for the point ν midway between α and β, we obtain the following table:

	α	ν	β
$\overline{F}$	−1	−1.5	−2
$\overline{S}$	−1	−0.5	0
$\overline{FM}$	0	−0.5	−1
$\overline{MS}$	−1	−0.5	0

Since the variation is linear, it is clear that the maximum excess, at any point other than ν, wll be greater than −0.5. Hence the maximum excess is minimized only at the point ν, and this point is therefore the nucleolus.

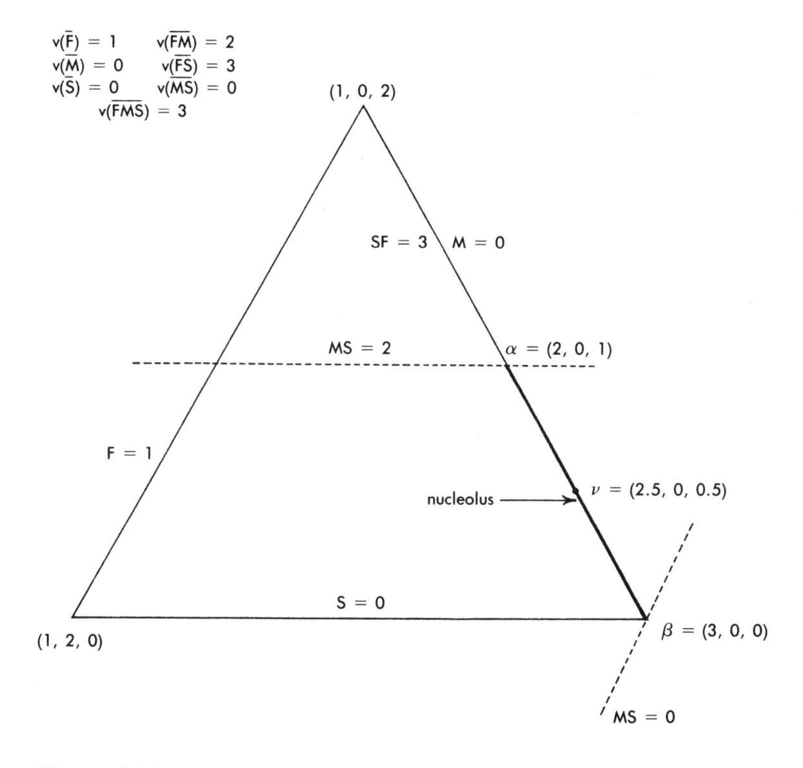

Figure 11.2
Nucleolus of the Three-Cornered Market.

Sobolev (1975) has suggested an approach to interpreting the nucleolus by means of a functional equation in which a coalition S may decide which of the following choices is more profitable: to act by itself without player i or to include i and receive the payoff for the larger coalition minus a stipulated payoff to i.

Intuitively the justification for the nucleolus appears to depend heavily upon the assumption of interpersonal utility comparisons. Given this assumption we might regard the nucleolus as the location of the taxation or subsidy point at which the maximum complaint against favored treatment of any group has been minimized. Thus, when money can be considered a good approximation to a transferable utility with constant marginal worth, the nucleolus has a satisfactory interpretation in terms of taxation or subsidy.

When there are no side payments, the definition of the nucleolus is not clear. We can, however, use the λ-transfer as in chapters 6 and 7 to define an *inner core* and *no-side-payment value*. Given the Pareto-optimal surface of the no-side-payment game, at each point we construct an associated side-payment game and calculate the nucleolus for each of these games. We then define the no-side-payment nucleolus to be the set of nucleoli of the side-payment games that lie at points of contact between the no-side-payment and the side-payment games. In general there may be more than one point.

11.1.3 The kernel

Having defined the excess of a coalition, we may look for a solution that depends on its use as a bargaining weapon among players jockeying for shares in the final outcome.

The *surplus* of one player against another (with respect to a given imputation) is defined as the largest excess of any coalition that contains the one player and not the other. This surplus, if positive, measures a kind of bargaining pressure: "If you do not give me more, I shall try to organize a coalition that can upset the present arrangement and increase its take by the amount of my surplus against you." If negative, it represents a defensive position: "Do not push me too far, or I will organize a coalition. . . ." But the second player will also have a surplus against the first. If these two surpluses happen to be equal (whether positive, negative, or zero), a certain balance can be said to exist—a stand-off or cancellation of opposing arguments regarding possible adjustment of the payoffs.

This balance, applied simultaneously to all pairs of players, characterizes the kernel solution. Formally the *kernel* (Davis and Maschler, 1965) can be defined as the set K of all imputations α such that for every two players i and j,

$$\max_{\substack{S \ni i \\ S \not\ni j}} e(S, \alpha) = \max_{\substack{T \ni j \\ T \not\ni i}} e(T, \alpha). \tag{11.4}$$

The expressions on either side of (11.4) are the respective surpluses of the two players against each other. This definition is a simplification of the one usually found in the literature, though it is equivalent to the other for superadditive games [compare the "prekernel" of Maschler, Peleg, and Shapley (1972)].

Note that (11.4) imposes $(n^2 - n)/2$ conditions on the variables α_1, $\alpha_2, \ldots, \alpha_n$, if there are n players. It is therefore rather remarkable that a solution to (11.4) is always possible. In fact Peleg (1965) has shown that *the nucleolus is always in the kernel.*

Let us consider the kernel of a game that has a core. For a point inside the core the various surpluses are all negative, since all excesses are negative except those for $\varnothing$ and N, neither of which "separates" i from j or j from i. Through such an interior point we can draw a "bargaining line" for any two players along which payoff is transferred from one to the other, keeping the remaining player(s) fixed (figure 11.3). When the boundary of the core is reached, one player will have given exactly his surplus to the other. The two boundary limits therefore represent the "defensive positions" mentioned earlier, and equating the two surpluses means taking the geometrical midpoint of the interval in which the bargaining line meets the core. Thus, in figure 11.3 at the point α only one pair of players (the pair $\overline{12}$) is "in balance," while at the point ν (the nucleolus) all three pairs are in balance.

For any core one can define the "bisecting surface in the i-j direction" as the locus of midpoints of all the parallel lines intersecting the core in the direction corresponding to transfers between players i and j. This is illustrated in figure 11.4. It is a remarkable geometrical fact that for any possible shape of the core these surfaces intersect in at least one common point, even though the number of surfaces is in general considerably more than the number of "degrees of freedom" (dimensions) in the core. (See Maschler, Peleg, and Shapley, 1970.)

For three-person games the kernel is always a single point, namely

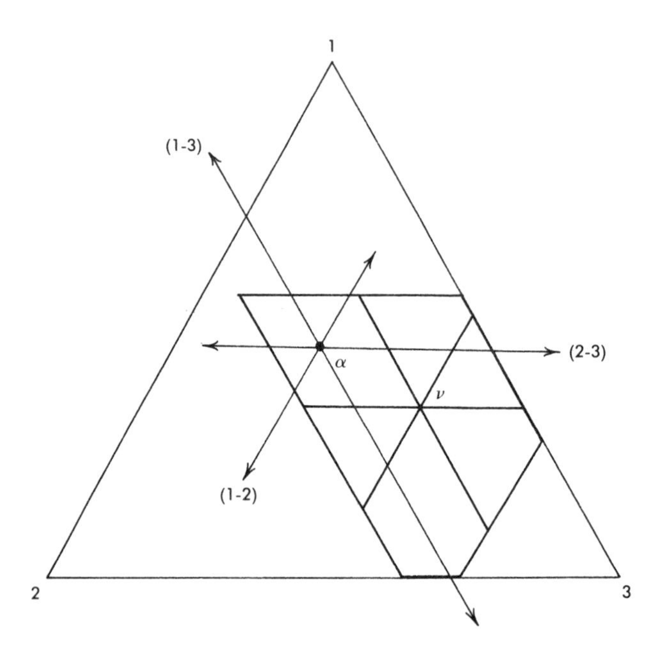

Figure 11.3
Bargaining lines in a three-person game.

the nucleolus. For larger games, however, there may be other points in the kernel. For example, in a four-player version of our land-utilization example, with two farmers offering to sell, the core is again a linear segment with the nucleolus at its center, but the kernel now fills the whole core. In other examples the kernel may extend beyond the core, and it need not even be a connected set (see Kopelowitz, 1967). However, the kernel never extends beyond the "reasonable" zone (11.1) (Wesley, 1971; Maschler, Peleg, and Shapley, 1970).

EXERCISE

11.2. *Extend the Three-Cornered Market by adding another farmer F', symmetric with F. Then the characteristic function is given by the data in figure 11.2, the same data with F' replacing F, and the values $v(\overline{FF'}) = 0$, $v(\overline{FF'M}) = 2$, $v(\overline{FF'S}) = 3$, and $v(\overline{FF'MS}) = 5$. Show that the core consists of the line segment joining the imputations (2, 2, 0, 2) and (1, 1, 1, 2), corresponding to a price range from \$100,000 to \$200,000 (Shapley and Shubik, 1972a; Böhm-Bawerk, 1891), and that every point in the core has the kernel property (11.4).*

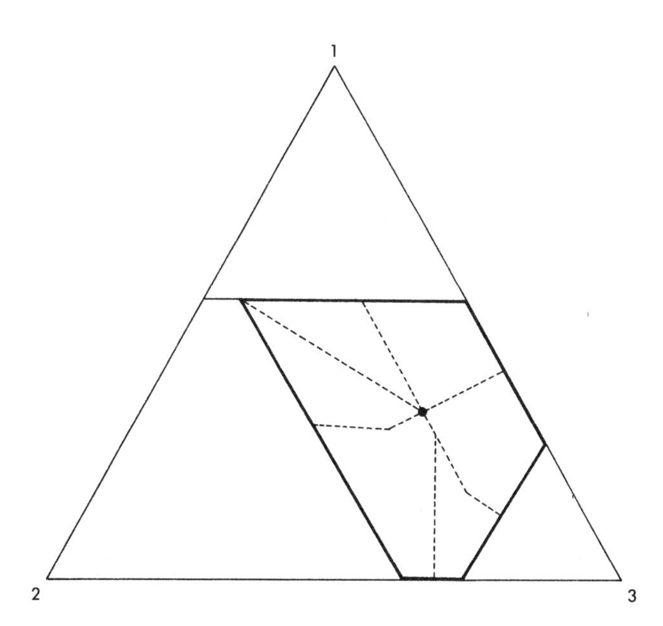

Figure 11.4
Bisecting surfaces in the core of figure 11.3.

An example of a four-person game in which the kernel does not coincide with the nucleolus is given by:

$v(i) = 0,$

$v(1, 2) = v(3, 4) = 0, \quad \text{otherwise } v(i, j) = 2,$

$v(i, j, k) = 2 + \epsilon, \text{ where } 0 \leq \epsilon \leq 1,$

$v(1, 2, 3, 4) = 4.$

This is illustrated in figure 11.5, where *CD* shows the core whose range is $(2 - \epsilon, 2 - \epsilon, \epsilon, \epsilon)$ to $(\epsilon, \epsilon, 2 - \epsilon, 2 - \epsilon)$. The kernel and core coincide. The nucleolus is at *K* with coordinates (1, 1, 1, 1). The diagram is drawn for $\epsilon = 0$; if $\epsilon = 1$, then the core, kernel, and nucleolus coincide.

Maschler's Aunt: A five-player game Davis and Maschler (1965) have posed the following five-person example of the kernel: *A* (Maschler's aunt), *I* (Maschler), and three other players *P*, *Q*, and *R* play a game in which $v(AI) = 100$, $v(AP) = v(AQ) = v(AR) = 100$, and $v(IPQR) =$

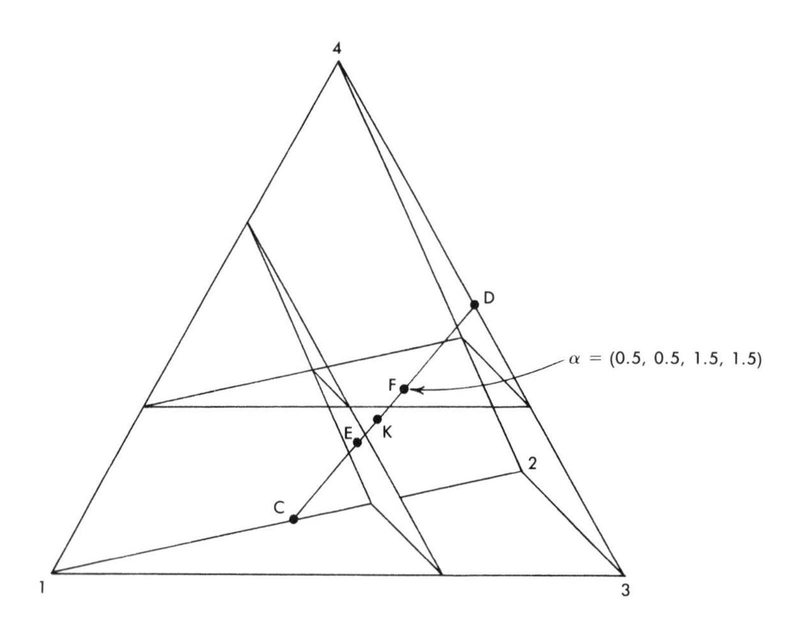

Figure 11.5
The kernel of a four-person game.

100. The kernel yields (50, 50, 0, 0, 0; *AI*, *P*, *Q*, *R*), where the five numbers are the payoffs to the players and the letters are the coalition structures. Two other points in the kernel are (25, 25, 25, 25, 0; *IPQR*, *A*) and (300/7, 100/7, 100/7,100/7, 100/7; *AIPQR*). Intuitively we may feel that *A* is considerably stronger than *I*, yet the first point noted in the kernel does not reflect this.

The phenomenon of transfer Intuitively we might guess that if we put together two games that are totally independent strategically, then their solutions should combine in a straightforward manner. This is not so for the stable set (von Neumann and Morgenstern, 1944) and is also not so for the kernel. Peleg (1965d) provides the following example. Consider two three-person games:

$$v_1(12) = v_1(13) = v_1(23) = v_1(123) = 1;$$

$$v_2(ab) = v_2(ac) = v_2(bc) = 3/4, \quad v_2(abc) = 1.$$

The kernel of each of these games consists of the symmetric points (1/3, 1/3, 1/3). The six-person game, however, has a kernel of (1 −

$t/3$, $1 - t/3$, $1 - t/3$, $1 + t/3$, $1 + t/3$, $1 + t/3$), where $-1/4 \leq t \leq 1/8$. Thus we have a transfer of resources from one three-person set of players to the other.

Kernels of convex games We noted in chapter 6 a special class of convex games in which

$$v(\emptyset) = 0, \tag{11.5}$$

$$v(S) + v(T) \leq v(S \cup T) + v(S \cap T) \quad \text{for } S, T \subset N. \tag{11.6}$$

In a convex game there is considerable incentive to form large coalitions. It has been shown that the kernel is a single point and that it coincides with the nucleolus (Maschler, Peleg, and Shapley, 1972).

A no-side-payment kernel By establishing a relationship between no-side-payment games and side-payment games, Billera (1972a) has suggested a generalization of the kernel for side-payment games. Little is known of its properties, however, beyond its existence.

11.1.4 The bargaining set

A *bargaining point* of (N, v) is an imputation with the property that for each pair $i, j \in N$, any "objection" that might be raised by i against j can be met by a "counterobjection" by j against i. Here an *objection* consists of a coalition S, containing i but not j, and an imputation feasible for S that is preferred to the given imputation by every member of S. The *counterobjection* consists of another coalition T, containing j but not i, and an imputation feasible for T that is (weakly) preferred to the objection of every member of $T \cap S$ and is (weakly) preferred to the original imputation by every member of $T - S$.

Formally, if α is the original imputation, (S, β) is the objection, and (T, γ) is the counterobjection, we have

$$\sum_{k \in S} \beta_k \leq v(S) \quad \text{and } \beta_k > \alpha_k \quad \text{for all } k \in S \tag{11.7}$$

and

$$\sum_{k \in T} \gamma_k \leq v(T) \quad \text{and} \begin{cases} \gamma_k \geq \beta_k & \text{for } k \in T \cap S, \\ \gamma_k \geq \alpha_k & \text{for } k \in T - S. \end{cases} \tag{11.8}$$

This solution concept was inspired by observing players in an experimental situation. It exists in the literature in several variants. The set M of bargaining points includes both the core and the kernel (and hence the nucleolus). In the core there are no objections to be raised, since (11.7) entails a positive excess for S at α. As for the kernel, it is easy to show that an objection of i against j can always be countered if j has as large a surplus against i as i has against j, and so the kernel conditions (11.4) are sufficient (but not necessary) to ensure that an imputation is a bargaining point.

It will be noted that (11.7) is exactly the condition for *domination,* as used in the definition of stable set in chapter 6, and that (11.8) is closely related. We should stress, however, that the present concept concerns the stability of a single imputation, whereas the von Neumann–Morgenstern concept concerns the stability of a set of imputations. In a certain sense, then, the bargaining set M (like the core and the kernel) is not a solution but a set of solutions—the collectivity of all possible outcomes using the particular solution concept. In contrast, each stable set in toto is a single solution, and the collectivity of all outcomes using this concept (the union of all stable sets) is generally not a stable set.

Our definition of the bargaining set corresponds to the Aumann–Maschler definition of the bargaining set $\mathcal{M}_1^{(i)}$, where the objections and counterobjections involve single individuals. We may modify these definitions to create a bargaining set $\mathcal{M}^{(i)}$ in which, instead of single players i and j, we consider a set of players I objecting against the set J, where I and J are disjoint sets. In this definition I is not permitted to use members of J. In the counterobjection I is allowed to use part but not all of J.

Peleg (1967) has proved that for all side-payment games the set $\mathcal{M}_1^{(i)}$ will always have at least one element. Furthermore Aumann and Maschler (1964) have shown that $\mathcal{M}^{(i)} \subset \mathcal{M}_1^{(i)}$. Unless we state otherwise, we refer to $\mathcal{M}_1^{(i)}$.

For $n = 2$ the bargaining set will be

$$(0, 0; 1, 2),$$

$$(\alpha_1, \alpha_2; 12), \quad \alpha_1 + \alpha_2 = 1, \quad \alpha_1 \geq 0, \quad \alpha_2 \geq 0.$$

For $n = 3$, for the constant-sum game

$$v(1) = v(2) = v(3) = 0,$$

$$v(12) = v(13) = v(23) = v(123) = 1,$$

the bargaining set is

$$0: \quad (0, 0, 0; 1, 2, 3),$$

A: (1/2, 1/2, 0; 12, 3),
B: (1/2, 0, 1/2; 13, 2),
C: (0, 1/2, 1/2; 1, 23),
D: (1/3, 1/3, 1/3; 123).

This game has no core, and the von Neumann–Morgenstern stable-set symmetric solution consists of the points A, B, and C, but not D.

The idea of an objection is closely related to Vickrey's (1959) "heretical imputation," which he uses to suggest a modification to the von Neumann–Morgenstern stable set.

Bargaining sets of convex games Maschler, Peleg, and Shapley (1972) have proved that the bargaining set coincides with the core for a convex game.

A no-side-payment bargaining set The bargaining set $\mathcal{M}_1^{(i)}$ for cooperative games with side payments has been generalized to no-side-payment games (Peleg, 1963c, 1969), with some modification to the notions of objection and counterobjection. Several variations have been suggested, including ordinal and cardinal definitions of bargaining sets for no-side-payment games (Asscher, 1976, 1977; Billera, 1970c). Asscher (1975) has explored the three-person no-side-payment game, but beyond that little is known.

Dynamic solutions for the kernel and bargaining set Cooperative solutions neglect process. Bargaining, haggling, and communication are not dealt with directly. The construction of the bargaining set and the kernel was, however, to some extent motivated by observations on simple experiments (Aumann and Maschler, 1964; Davis and Maschler, 1965), and they are therefore more closely tied to process.

Stearns (1968) showed that one can define a discrete transfer procedure which, starting from any distribution, converges to a stable point, that is, a point in the kernel or in the bargaining set.

The excess has been defined in (11.3). We may define *surplus* as the maximum excess of i over j in S:

$$s_{ij}(\alpha) = \max[e(S, \alpha) \mid \{i\} \subseteq S \subseteq N - \{j\}]. \qquad (11.9)$$

It represents the largest potential payoff that P_i can offer members of some coalition S to act for their own advantage against the interests of P_j.

If α is an imputation, i and j are distinct players of the game, and $x \geq 0$, then an imputation β results from α by a *transfer* of x from j to i if

$$\beta_i = \alpha_i + x, \tag{11.10}$$

$$\beta_j = \alpha_j - x, \tag{11.11}$$

$$\beta_k = \alpha_k \quad \text{for } k \neq i, j. \tag{11.12}$$

Here β is said to be the result of a *K-transfer* of size x if equations (11.10)–(11.12) hold together with

$$s_{ij}(\beta) \geq s_{ji}(\beta). \tag{11.13}$$

Essentially a K-transfer reduces the differences in excess of j over i. A sequence of K-transfers leads to a point in the bargaining set.

In order to obtain convergence to the kernel, the K-transfer paid by j to i amounts to a "split-the-difference" measure of relative excesses.

Intuitively at each payoff α we measure a number $s_{ij}(\alpha)$ which may be regarded as the demand of P_i against P_j. A payoff point is stable if, for all pairs of players, the demand is zero. Stearns describes a finite process based upon this idea, whereas Billera (1972b) considers a continuous process described by the differential equations

$$\frac{d\alpha_i}{dt} = \sum_j s_{ij}(\alpha) - \sum_j s_{ji}(\alpha),$$

which leads to the same outcome. [For stability properties of the nucleolus see Kalai, Maschler, and Owen (1975).]

11.1.5 Market games

The nucleolus, kernel, and bargaining-set concepts provide solutions of interest when applied to market games. In particular, since all market games have a core and the nucleolus is within the core, it follows that if the core "shrinks" under replication, the nucleolus and the core approach the same limit, which is the imputation given by the competitive equilibrium of an exchange economy related to the market game.

The kernel always intersects the core but may extend beyond it even for market games. An example of such behavior is given below (Shapley, 1973b).

A nonconverging kernel in a replicated market game We consider a market with two commodities and u-money, and with two types of traders having utility functions $u^1(x_1, x_2, m) \equiv u^2(x_1, x_2, m) \equiv \min(x_1, 2x_2) + m$ and initial endowments (1, 0) and (0, 1), respectively. (Here x_1, x_2, and m are the amounts of goods 1 and 2 and of u-money, respectively.) The characteristic function is

$$v(S) = \phi(\sigma^1, \sigma^2) = \min(\sigma^1, 2\sigma^2),$$

where (σ^1, σ^2) is the profile of the coalition S (see figure 11.6). We assume k traders of each type.

For all $k > 1$ the core consists of the single imputation $(1, 0)^k$. [This notation means the vector (1, 0) replicated k times.] Since the kernel always intersects the core if the latter exists, $(1, 0)^k$ must also be a kernel point.

Let α be a typical kernel payoff. Since players who are substitutes get equal amounts at any kernel point, α must be of the form $(a, b)^k$,

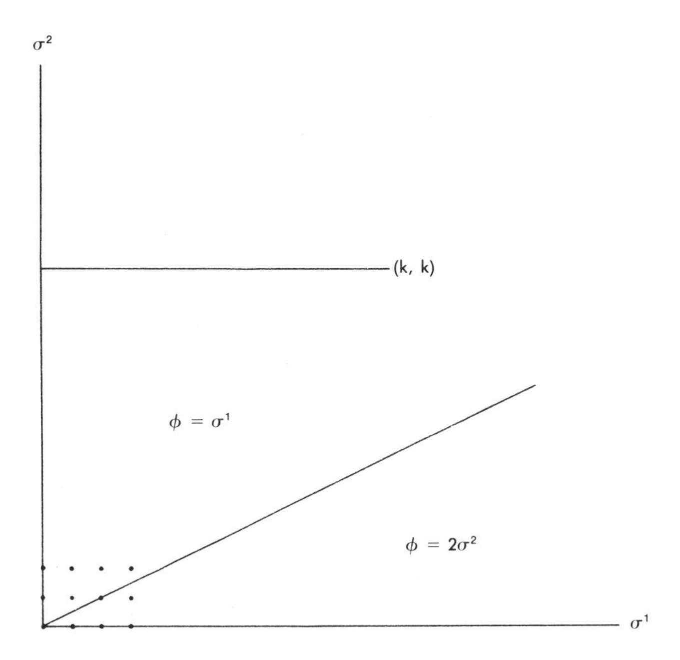

Figure 11.6
The profile of a coalition.

where $b = 1 - a$. Since the kernel is individually rational, $a \geq 0$ and $b \geq 0$. Let i and j be traders of type 1 and 2, respectively. Equating their surplus over each other will give us the remaining conditions for the kernel. The surplus of i over j is

$$s_{ij} = \max_{\substack{S:i\in S \\ j\notin S}} [v(S) - \alpha(S)] = \max_{\substack{1\leq\sigma^1\leq k \\ 0\leq\sigma^2\leq k-1}} [\phi(\sigma^1, \sigma^2) - \sigma^1 a - \sigma^2 b].$$

If k is even, the maximum is attained at $\sigma^1 = k$, $\sigma^2 = k/2$, and we have

$$s_{ij} = k(1 - a - b/2) = kb/2 \quad (k \text{ even}).$$

If k is odd, the maximum is attained at $\sigma^1 = k - 1$, $\sigma^2 = (k - 1)/2$—and also at $\sigma^1 = k$, $\sigma^2 = (k + 1)/2$—and we have

$$s_{ij} = (k - 1)(1 - a - b/2) = (k - 1)b/2 \quad (k \text{ odd}).$$

Similarly the surplus of j over i is

$$s_{ji} = \max_{\substack{T:j\in T \\ i\notin T}} [v(T) - \alpha(T)] = \max_{\substack{0\leq\tau^1\leq k-1 \\ 1<\tau<k}} [\phi(\tau^1, \tau^2) - \tau^1 a - \tau^2 b].$$

For k even we set $\tau^1 = k - 2$, $\tau^2 = (k - 2)/2$ and obtain

$$s_{ji} = (k - 2)b/2 \quad (k \text{ even}),$$

and for k odd we set $\tau^1 = k - 1$, $\tau^2 = (k - 1)/2$ and obtain

$$s_{ji} = (k - 1)b/2 \quad (k \text{ odd}).$$

Equating s_{ij} to s_{ji}, we find that $b = 0$ when k is even and that b is unrestricted when k is odd. We conclude as follows:

1. When k is even, the kernel is exactly the core:
$\mathcal{K} = \{(1, 0)^k\} = \mathcal{C}$.

2. When k is odd, the kernel is the set of all symmetric imputations:
$\mathcal{K} = \{(a, 1 - a)^k : 0 \leq a \leq 1\} \supset \mathcal{C}$.

The nondifferentiability of ϕ is essential to this behavior. If ϕ were differentiable, the kernel would converge to a single point along with the bargaining set, as indicated below.

Convergence of the bargaining set It has been shown that a modified bargaining set that contains the set $\mathcal{M}_1^{(i)}$ is within the ϵ-core for certain sequences of side-payment market games in which the trader types are fixed while the number of traders of all types increases without limit (Shapley and Shubik, 1972b). An essential condition for this to hold is the differentiability of the characteristic function in profile form. This condition also ensures that the core shrinks to a single point, which is the payoff generated by the competitive price vector.

Geanakoplos (1978) has investigated the interrelationship between the bargaining set and the core for nonstandard infinite market games.

The bargaining set catches the essence of politicoeconomic bargaining and, to some extent, the adversary process in the context of a larger society. The active proponent of change (the leader of the objection) lines up his forces. This challenge to the status quo invokes a response, and others form an adversary group. The conditions under which a price system will emerge from this adversary process are reasonably general.

The bargaining set and the core In chapter 7 we considered the following game:

$$v(1) = v(2) = v(3) = 0,$$

$$v(12) = 0, \quad v(13) = v(23) = 100,$$

$$v(123) = 100.$$

We observed that the core has a unique point (0, 0, 100), which in an economic context does not intuitively appear to be always reasonable. The value, in contrast, yielded (100/6, 100/6, 400/6). The bargaining set includes the possibilities (α, α, $100 - 2\alpha$; 12, 3), where $0 \leq \alpha \leq 50$.

The proposition that in this game P_1 and P_2 should each be able to obtain something is reflected in the value, the stable set, and the bargaining-set solutions but not in the core or in a competitive-market representation of the game (see also Maschler, 1976).

11.2 Conditions on the Coalition Structure

The original approach of von Neumann and Morgenstern (1944) to n-person cooperative games was to consider a constant-sum game in which the most that a coalition S can obtain is naturally described by calculating the maxmin for S in a two-person game of S against its complement $\bar{S}$. They then showed that an n-person non-constant-sum game could be regarded as strategically equivalent to an (n +

1)-person constant-sum game with an extra dummy player "Nature" who absorbs residual gains or losses in such a way that the game is constant-sum. This provides a method for calculating the characteristic function but, as we noted in chapter 6, does not supply a sufficient justification for the characteristic function unless the underlying game is a c-game.

The economist may recognize that the acceptability of the characteristic function (or its no-side-payment equivalent) is intimately related to the presence of externalities to members of a group. The class of c-games has been defined as having the property that once a coalition has formed, either the excluded members have no directly strategic influence on it whatsoever or as a group they are directly opposed to the coalition and are strategically interlinked. Market games provide an example of the first type, and constant-sum games an example of the second type. An economy with organized markets in which individuals are required to trade will not, in general, model as a c-game.

If the von Neumann–Morgenstern characteristic-function representation is not always adequate, what alternatives do we have? In section 7.3.3 we discussed one way to modify the characteristic function to take care of threats. There are several other ways to modify the coalitional description of a game either by considering partitions or by limiting coalition formation.

11.2.1 Games in partition-function form

Thrall (1962), Lucas (1963), and Thrall and Lucas (1963) have considered games in partition-function form. An n-person game in partition-function form is defined as follows: Let $N = \{1, 2, \ldots, n\}$ be the set of n players. Let $P = \{P_1, P_2, \ldots, P_r\}$ be an arbitrary partition of N into r coalitions denoted by $P_1, P_2, \ldots, P_r$. The set of all partitions of N is denoted by $\Pi = \{P\}$. Following the notation of Thrall and Lucas, denote the real numbers by R^1. Then for each partition P we specify an outcome function

$$F_P: P \rightarrow R^1,$$

which assigns the (real-number) outcome $F_P(P_i)$ to coalition P_i given the partition P. The function $F: \Pi \rightarrow \{F_P\}$ that assigns to each partition its outcome function is called the *payoff function* or the *partition function* of the game. The ordered pair $\Gamma = (N, F)$ is then called an n-person game in partition-function form.

For each nonempty subset M of N the value of M is defined as

$$v(M) = \min_{\{P|M \in P\}} F_P(M).$$

In words, the value of the coalition is the minimum it can achieve as a coset in the partitions to which it belongs. The value for the empty set is defined as zero, and we assign numbers v_i to the one-person sets, so that $v(\{i\}) = v_i$ for each $i \in N$.

The concepts of imputation, domination, and stable-set solution are introduced in a manner similar to that of von Neumann and Morgenstern. However, the function v here is not necessarily superadditive: it may actually cost some coalitions a loss in payoffs if they form.

Thrall and Lucas (1963) prove that if (N, v') is a von Neumann–Morgenstern game in which N is the set of players and v' is a superadditive characteristic function, then there exists a game (N, F) in partition-function form that has the same stable-set solutions. Thus the characteristic-function games can be regarded as a special case of the partition-function games.

It is trivial to observe that for $n = 2$ the characteristic function and partition function coincide. The first case of interest is $n = 3$. The values of the outcome functions can be denoted as

$$F_{P^0}(N) = c,$$

$$F_{P^i}(\{i\}) = d_i, \quad F_{P^i}(\{j, k\}) = e_i \quad (i = 1, 2, 3),$$

$$F_{P^4}(\{i\}) = g_i \quad (i = 1, 2, 3),$$

giving coalition values of

$$v(\varnothing) = 0,$$

$$v_i = \min\{d_i, g_i\} \quad (i = 1, 2, 3),$$

$$v(\{j, k\}) = e_i \quad (i = 1, 2, 3),$$

$$v(N) = c.$$

All stable sets for three-person games have been determined. In general, there are fewer members of stable sets for games in partition-function form than for games in characteristic-function form. Lucas (1968a) has even demonstrated the existence of an n-person game in partition-function form without a stable set; and with a slight change to this example there is a game without either a stable set or a core.

The core and value have been considered for games in partition-function form (Myerson, 1977); and a model of an n-person game in partition-function form without side payments has been proposed (Lucas and Maceli, 1977).

A critical reason for considering games in this form is that the partition function offers the investigator a more flexible representation of coalition structure than does the characteristic function. The partition function has many more degrees of freedom (possibly too many!), allowing the investigator to do far more ad hoc modeling.

There is a clear way to reduce a game of finite length in extensive form to a game in normal or strategic form. The way to make the reduction to a game in cooperative form is not well-defined in general. The partition function structure offers an alternative to the characteristic function.

The worth of the partition-function form of a game depends to a great extent upon the importance of being able to utilize this structure as a simpler alternative to constructing process models of coalition formation (see Maschler, 1963). For some special classes of games the problem may not arise. In particular, it should be noted that the standard economic model for trade among n individuals, which may be portrayed as a market game, will give rise to the same structure whether we derive the characteristic function using the methods of von Neumann and Morgenstern or use the partition-function form.

EXERCISE

11.3. *Consider n traders trading in m + 1 goods, where the initial endowment of trader i is* $(a_1^i, a_2^i, \ldots, a_{m+1}^i)$, $a_j^i \geq 0$, *and* $\Sigma_j a_j^i > 0$. *Each trader has a utility function of the form* $\phi_i(x_1^i, x_2^i, \ldots, x_m^i) + x_{m+1}^i$, *where* $a_{m+1}^i \geq \max \phi_i(a_1, a_2, \ldots, a_m)$. *Show that the characteristic-function and partition-function forms are the same.*

This simplification, which also holds for simple games, does not hold for strategic market games (see section 10.2.2). When individuals are strategically bound by a market structure or other features of an economy or society, their own group or coalition S will be influenced by the coalition structure of the other players.

11.2.2 Ψ-Stability

Use of the partition-function form enlarges the coalitional possibilities available. The Ψ function of Luce (1955; Luce and Raiffa, 1957) offers a way to cut down on the possibilities, albeit in an ad hoc manner. The motivation behind this type of modeling is similar to that of Farquharson (1969) and Shubik (1959b): it may be costly to form some coalitions, or there may be reasons specific to a given problem which rule out certain coalitions. These considerations imply that the modeler is willing to give the coalitions some institutional properties.

Suppose we have a partition τ of the players in an n-person game. A rule Ψ is given such that $\Psi(\tau)$ specifies the admissible coalition changes from the partition τ. For example, one rule might be that given an existing partition, any coalition is permitted to add one player. Another might be that it can increase its membership by no more than k percent. This last rule might reflect dynamic properties of institutional growth, such that only a few new members can be absorbed within a given period.

Given the rule limiting collusive arrangements, Luce has suggested a cooperative side-payment solution consisting of pairs $[x, \tau]$, where the imputation x and partition τ are in equilibrium. The meaning of equilibrium here is illustrated below.

Let S be a coalition in $\Psi(\tau)$, that is, one of the coalitions that could form if the players were partitioned according to τ. For $[x, \tau]$ to be in equilibrium, we require that:

1. For every S in $\Psi(\tau)$, $v(S) \leq \Sigma_{i \in S} x_i$.
2. If $x_i = v(\{i\})$, then $\{i\}$ is in τ.

It follows that all points in the core are Ψ-stable.

There has been a certain amount of exploration of simple games, quota games, and symmetric games for Ψ-stable solutions, but, as in the case of its noncooperative-solution counterpart, the proliferation of cases is large, and no satisfactory systematic exploration of the solutions exists. Luce defined Ψ-stability for games with side payments. There is no formal difficulty in extending the definition to encompass no-side-payment games, and Nakamura (1973) has obtained some results. An application of Ψ-stability to congressional voting in a two-party system has been made by Luce and Rogow (1956).

Other modifications to the description of games studied by cooperative solution have been suggested. These include games in effectiveness form (Rosenthal, 1972), syndicates (Postlewaite and Rosenthal, 1974), unions (Charnes and Littlechild, 1975), and cooperative games with coalition structures (Aumann and Dreze, 1975; Wieczorek, 1976).

11.3 Concluding Remarks on Models and Solutions

11.3.1 Modeling and assumptions

Diverse considerations must be taken into account in the construction of models for game-theoretic analysis. A multitude of solution concepts have competing validity in applications to different problems.

This volume has provided an overview of modeling problems, preference-structure assumptions, the various formal descriptions of a game, presolution concepts, and solution concepts and also a formal but not deeply mathematical presentation of some of the properties of the solutions.

At this stage in the development of the social sciences the application of advanced mathematical methods must of necessity be somewhat ad hoc. It is premature to believe that the same type of modeling or analysis is going to be broadly applicable to economics, sociology, social psychology, and other disciplines unless care is taken to incorporate the differentiating features of the individual topics. Varying interests and concerns must be portrayed in the modeling and in the selection of appropriate presolutions and solutions.

For applications we have emphasized political economy. We do not pretend that our approach suits the study of diplomacy or picks up the nuances of sociopsychological differences that can be observed in simple experiments with two players. Chapter 12 outlines the domain of applicability of game theory. It must be stressed again, however, that considerable difficulties remain in modeling and in conceptualization. For example, there is as yet no satisfactory blending of game theory with learning theory; this poses a major hurdle to the application of game theory to some parts of social psychology.

In spite of the need to specialize in modeling and in the selection of solution concepts when applying game theory, it is important for the would-be user of this methodology to have a broad view of the topic so that he or she can see where difficulties may arise in application. Although our selection of examples in the preceding chapters has been somewhat biased to our intended application, the presentation of models and solutions has been general. The sociologist or psychologist who must consider personal differences and varying information conditions may blanch at this statement. We do not deny that our model of man has been made explicitly poor by our application of considerations of external symmetry (see section 2.1); however, our individual player is well defined, and his basic properties

Table 11.1
Summary of differences between game theory and behavioral theories

Game theory	Behavioral theories
Rules of the game	Laws and customs of society
External symmetry	Personal detail
No social conditioning	Socialization assumed
No role playing	Role playing
Fixed, well-defined payoffs	Payoffs difficult to define, may change
Perfect intelligence	Limited intelligence
No learning	Learning
No coding problems	Coding problems
Primarily static	Primarily dynamic

can be modified for specific application. The game-theoretic and the behavioral approaches are, in fact, complementary, not substitutes. They differ in the emphasis they apply in modeling. A brief listing of the major distinctions is given in table 11.1. They are discussed in some detail elsewhere (Shubik, 1975).

Given the assumptions on individuals, information, and other rules of the game, our presentation has been general. No attempt has been made to impose further specialized structures on the games. In application this step is critical. Prior to narrowing our consideration to the special classes of games suggested by the structure of political economy, then, we shall review the general aspects of modeling, presolutions, and solutions and note the interrelationships among the different approaches.

11.3.2 Models, presolutions, and solutions: A recapitulation

Models Three different representations of a finite game have been noted: *the extensive form* (section 3.3), *the strategic form* (section 3.4), and *the coalitional form* (section 6.2). The extensive form stresses the fine structure of the game, the details of moves and information. The strategic form suppresses much of the detail and highlights the strategic choices of the players, the details of the payoffs, and the possibilities for threats. The coalitional form suppresses strategic detail and highlights the joint gains that can be made by the formation of coalitions. A theory of games can be constructed starting with any one of these forms.

Starting with the extensive form, one can construct the strategic form in a straightforward manner. Since there is a considerable loss of information in this process, the reverse cannot be done.

Starting with the extensive or strategic form, one can construct several different representations of a game in cooperative form. For games with side payments, for example, the characteristic function can be constructed by the device of evaluating the worth of a coalition as though the set of players S were opposed by the remaining players $\bar{S}$ in a constant-sum game between S and $\bar{S}$. Even if this method is used to evaluate the characteristic function, though, there is no unique way to return to the strategic or extensive form, as much information is lost.

Presolutions The three representations noted above can be regarded as solutions to a game in and of themselves, in the sense that the construction of the extensive form, the payoff matrices, or a characteristic function already tells us a great deal about the phenomenon being studied. In general, we wish to know more. Frequently we want to predict a single outcome. This may not be possible without far more detailed knowledge than is usually available. However, a further narrowing of the possibilities may be obtained without the imposition of too many added conditions or a call for too much additional information.

Three presolutions (or weak solutions in the sense that they only weakly limit the outcomes) that apply to cooperative games are *Pareto optimality* (chapter 5), *the imputation space* (section 6.3), and *reasonable outcomes* (section 11.1.1). Pareto optimality limits the outcomes to a set with the property that it is not possible to improve the payoff to any member without lessening the payoff to at least one other member. The imputation space is a subset of the Pareto-optimal set obtained by adding the further condition of individual rationality; that is, any outcome in the Pareto set which offers an individual less than he can obtain by himself is excluded from the imputation space. The further restrictions on the set of reasonable outcomes have been discussed in section 11.1.1.

Modifications to basic models and presolutions Chapters 4 and 5 presented many different conditions on choice, utility functions, and preferences, any of which could serve as a basis for constructing a theory of games. The determining factors in one's choice must be realism, relevance, and tractability. Most of our examples and much of the subsequent analysis have been based on the assumption of the existence of individual utility scales, each measurable up to a linear transformation. This assumption appears to provide a reasonable approximation to reality and has the advantage of yielding tractable models.

Three different conditions on the transferability of wealth or chips or u-money merit the attention of the model builder: *no side payments, quasitransferability,* and *transferability.*

In a game with no side payments there is no good or medium that can be transferred among the players to serve as a common means of exchange or settlement. A game satisfies the conditions for quasi-transferability if all of the players possess some commodity that is used as a "money" or a means of exchange. The utility scale of each individual need not be linear in the chosen commodity. At various times and in different places, coconuts, cocoa beans, dried fish, salt bars, beaver pelts, and other items that were in sufficient but not abundant supply, were sufficiently durable and transportable, and tended to be held in some quantity by all members of the society have served as "money."

Full transferability requires the added condition that the money be a "u-money" in the sense that the utility of each person for the chosen commodity is linear. This assumption, while not true in general, provides a reasonable approximation in some instances and also yields considerable mathematical simplification.

It is important to reiterate that while transferability does not require that utility be extrinsically comparable, it usually leads to an intrinsic comparability of utility.

One may argue that a fully developed no-side-payment theory contains within it quasitransferability and transferability as special cases. The logic behind this position is undoubtedly true, but from the viewpoint of the development of game theory and its applications it can be misleading. Historically many developments have come from first studying the side-payment theory and then utilizing the insights obtained to generalize to the no-side-payment theory. It is tempting to believe that one might start with the no-side-payment theory and obtain the side-payment theory as a special case, but this does not appear to be the most fruitful approach. (An interesting unexplored problem is the development of a measure of "nosidepaymentness.")

Certain technical conditions arising from both mathematical and modeling considerations must be dealt with when we analyze the number of players, the number of strategies, and the length of play.

In particular, we may wish to study the behavior of solutions for games with indefinitely large numbers of players or with a continuum of strategies. The original theory of von Neumann and Morgenstern was designed to deal with games of finite duration, finite numbers of players, and finite pure-strategy sets for each player.

It is a well-known phenomenon in applied mathematics that a continuous model approximating many discrete particles may be far easier to analyze than a discrete model. This has led to several attempts to develop a game-theoretic analysis for games with infinitely many players and strategies (see chapter 3). Games in which a player can have infinitely many strategies can arise if the strategic variables are naturally continuous (for example, a strategy might involve the selection of a number in the interval 0–10, where the number is a distance to be traveled), or if the game is of indefinite length.

Solutions and representations Each of the three major representations of a game stresses different features of the phenomenon to be studied, and different solution concepts apply to each representation. The extensive form and the strategic form lend themselves best to noncooperative solutions, as discussed in chapters 9 and 10. The extensive form provides a natural structure from which one might try to construct explicitly dynamic solutions. The dynamics enters when we state the laws of motion for traveling down the game tree. Thus we may wish to consider as solutions behavioral rules for making decisions at each choice point.

The representation of a game in coalitional form is most appropriate for cooperative theories. These theories and their properties are sketched in section 11.3.4.

11.3.3 Experimental cooperative games

The characteristic-function form was designed essentially to bypass dynamics. There is nevertheless a growing literature on experimentation with cooperative games using the characteristic-function representation of a game. In fact several solution concepts have been motivated by experimentation. These include the *bargaining set* (Aumann and Maschler, 1964), the *equal-division core* (Selten, 1972), and the *competitive solution* for an n-person game (McKelvey, Ordeshook, and Winer, 1978). The bargaining set has already been defined. The equal-division core has the added condition that a payoff vector belongs to the equal-division core if no coalition can divide its value equally among its members to give more to each than they receive at the proposed payoff vector.

The competitive solution is somewhat in the same spirit as the no-side-payment bargaining sets. The basic idea is that coalitions must bid for their members in a competitive environment. If a player or set of players is critical to the formation of two coalitions, they should be indifferent between the offers of those coalitions. The formal definition of competitive solution is given by McKelvey, Ordeshook, and Winer (1978).

The experimentation with cooperative games is of interest but is extremely difficult to interpret in the sense that process (especially coalition formation) plays an important explicit role in the experiments but not in the theory.

One way to avoid process is to experiment with opinions. A series of "opinion experiments" has been run with several three-person games marked by somewhat different briefings and scenarios. Instead of having players actually play a game, opinions were solicited concerning how players should divide the proceeds from the game (Shubik, 1975b, 1978a, 1979). The main game had the following characteristic function:

$$v(1) = v(2) = v(3) = 0,$$
$$v(12) = 1, \quad v(13) = 2, \quad v(23) = 3,$$
$$v(123) = 4.$$

For a smaller but overlapping sample the following game was used:

$$v(1) = v(2) = v(3) = 0,$$
$$v(12) = 0, \quad v(13) = 4, \quad v(23) = 4,$$
$$v(123) = 4.$$

This game has a single-point core at (0, 0, 4). The main game has a large core, as shown by the trapezoid *ACDE* in figure 11.7.

The bargaining set consists of the core plus the line *BC*. As can be seen in figure 11.7, various points were suggested. In particular, the nucleolus is *N*, the value is *V*, and the equal-split point is at *U*.

Almost all of the replies to how the first game should be resolved were either points in the core or the equal-split point (4/3, 4/3, 4/3). The core seemed most attractive as a predictor in the first game, but this did not hold up in the second game. There was a tendency in the second game to give P_1 and P_2 the same amount, but more than zero; hence a point not in the core was most frequently selected.

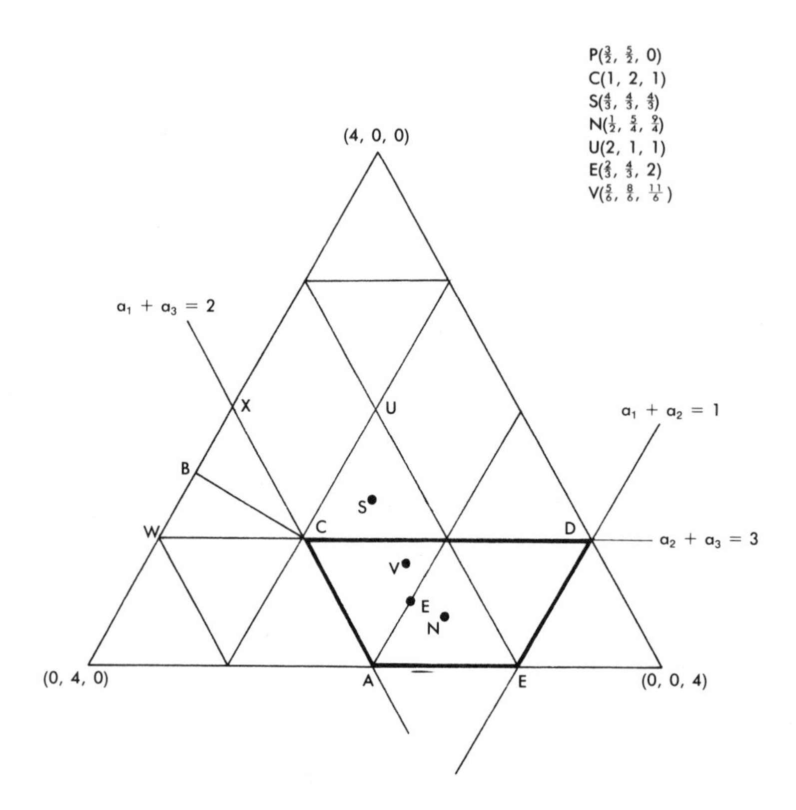

Figure 11.7
A three-person experimental game.

B. Roth's (1975) experiments found the results sensitive to different normalizations; Maschler's (1976) experiments found for the bargaining set. All of the solutions appear to have merit in special cases, but no general overall solution concept clearly dominates in all circumstances.

11.3.4 Intersolutional relationships

The most important division in the classification of solutions is between cooperative and noncooperative solutions. The latter are usually associated with games in strategic or extensive form. A listing of the properties of the noncooperative solution for games in general and strategic-market games and two-person zero-sum games in par-

Table 11.2
Properties of solutions to essential superadditive games in characteristic-function form

	Existence	Uniqueness	Maximum dimension	Dummy property	Composability; decomposability	Solution a convex set	Competitive-equilibrium convergence property
PAROPT	yes	yes	$n-1$	no*		yes	no
INDRAT	yes	yes	n	no		yes	no
IMPUTAT	yes	yes	$n-1$	no		yes	no
REASON	yes	yes	$n-1$	yes		yes	no
Core	yes if game is balanced; no otherwise	yes if it exists	$n-1$	yes	no	yes	yes
Weak ϵ-core	yes for ϵ suff. large	no	$n-1$	no	no	yes	yes as $\epsilon \to 0$
Strong ϵ-core	yes for ϵ suff. large	no	$n-1$	no	no	yes	yes as $\epsilon \to 0$
Least core	yes	yes	$n-2$			yes	yes
Nucleolus	yes	yes	0			yes	yes
Kernel	yes	yes	?			no*	yes
Bargaining set	yes	yes	$n-1$			no*	no*
Stable set	yes for simple (convex) games	no* (but yes for convex games)	$n-1$ if core exists; $n-2$ otherwise	yes	yes; no*	no*	no*
Value	yes	yes	0	yes	yes; yes	yes	yes

Presolutions: PAROPT = set of Pareto-optimal vectors; INDRAT = set of individually rational payoff vectors; IMPUTAT = set of imputations; REASON = set of reasonable payoff vectors. * Not in general.

Table 11.3
Hierarchies of solutions to essential superadditive games in characteristic-function form

	Contained in: PAROPT	INDRAT	IMPUTAT	SYM	REASON	Core	Weak ϵ-core	Strong ϵ-core	Least core	Kernel	Bargaining set	Stable set
PAROPT		no		no*								
INDRAT				no*								
IMPUTAT	yes	yes		no*		pure bargaining games					pure bargaining games	pure bargaining games
REASON	yes	yes	yes	no*								
Core	yes	yes	yes	no*	yes					yes in replicated markets	yes	yes
Weak ϵ-core	yes	if $\epsilon \le 0$	if $\epsilon \le 0$	no*	if $\epsilon \le 0$	if $\epsilon \le 0$	if $\epsilon' \le 0$	if $0 \le \epsilon \le \epsilon'$				
Strong ϵ-core	yes	if $\epsilon \le 0$	if $\epsilon \le 0$	no*	if $\epsilon \le 0$	if $\epsilon \le 0$	if $0 \le \epsilon \le \epsilon'/m$	if $\epsilon \le \epsilon'$				
Least core	yes			no*	yes	yes if core $\neq 0$		if it exists				
Nucleolus	yes	yes	yes	yes	yes	yes if core $\neq 0$		if it exists	yes	yes	yes	
Kernel	yes	yes	yes	yes	yes						yes	
Bargaining set	yes	yes	yes	no*		no*						no (unless identical)
Stable set	yes	yes	yes	no*	yes	yes for convex games						
Value	yes	yes	yes	yes	yes	no* (but yes for convex games)			symmetric games	symmetric games		

SYM = set of symmetric imputations (i.e., imputations where identical players receive identical amounts).
* Not in general.

ticular was given in chapter 10. In this section we confine ourselves to games in cooperative or coalitional form and to solutions related to them. Tables 11.2 and 11.3 summarize the relationships between solutions for side-payment games in characteristic-function form.

12 On the Applications of Game Theory

12.1 Introduction

The foregoing chapters have offered a general introduction to the methodology of model building and of game theory. In this chapter we complete the general survey with some observations on the applications of game theory to economics and other subjects. Some of the major misperceptions concerning the uses of game theory will be noted, as will the areas where attempts at application have encountered formidable difficulties. The application of any mathematical method to substantive problems calls for a great amount of hand-tailoring, and game theory is no exception to this rule.

Several books, articles, and bibliographies provide quick overviews of various aspects of the theory of games. In particular, there are the survey articles of Lucas (1971, 1972) and the excellent survey and lengthy bibliographies of Vorobyev (1970b,c, 1976, 1981). Other basic sources are the game-theoretic entries in Princeton's *Annals of Mathematics Studies* series (1950, 1953, 1957, 1959, 1964), the international conference reports (which unfortunately exist only in mimeographed form and are hard to come by), the *International Journal of Game Theory,* and the bibliographies of RAND publications on game theory (1971, 1977).

12.1.2 Two major misconceptions

Before applications to individual disciplines are considered, two major misconceptions should be addressed.

Misconception no. 1: "The maxmin strategies and the saddlepoint solution are central to the game-theoretic analysis of multiperson decision making"
Most expositions of game theory begin with a treatment of the two-person constant-sum game. The normative solution for games of this type calls for the players to employ maxmin strategies. Since their interests are diametrically opposed, the maxmin solution can be regarded as a reasonable extension of the concept of individual rational behavior to a two-person choice problem with an extremely special-

ized structure. Because each knows that the payoffs are completely negatively correlated, it is reasonable for each to expect that he can guarantee for himself no more than the largest amount the other is unable to prevent him from obtaining.

It is trivially easy to demonstrate that for a two-person non-constant-sum game the maxmin strategy can lead to unreasonable behavior. Even with two players, and certainly when there are more than two, communication and cooperation must be considered. Except for the two-person constant-sum game, it has long been recognized that there is no unique solution concept that can be regarded as *the* natural extension of individual rational behavior to multiperson decision making. This is one of the central problems of game theory. The current approach is to examine a multiplicity of solution concepts. Many of these have been discussed in chapters 6 to 11.

One source contributing to the overemphasis on the role of the maxmin solution is the formal mathematical relationship between the solution of a certain type of linear program and a two-person constant-sum game. This has led some authors of works on linear programming (e.g., Gale, 1960; Dantzig, 1964) to include a chapter on game theory, thus giving the impression that game-theoretic analysis is primarily concerned with a limited set of constant-sum matrix games.

It is important for the reader to appreciate that the definition of the characteristic function which is used heavily in the investigation of n-person non-constant-sum games does not necessarily depend on assumptions of maxmin behavior between opposing coalitions. This has been discussed in section 6.2 and again in section 11.2.

Misconception no. 2: "The application of the theory of games depends upon the assumption of the existence of transferable comparable utilities"

For ease of analysis von Neumann and Morgenstern presented their original work on n-person games in terms of games in which there exists a comparable and transferable utility or "money." Since their suggested applications were in the field of economics, they immediately drew criticism on the counts that utility in general may not be computed, it is not linearly transferable between individuals, and it may not even be possible to determine a utility scale unique up to a linear transformation for an individual. These criticisms were interpreted by many to imply that game theory is not applicable to economics. Leaving aside the merits of the three critical points, the methodology of game theory does not depend upon *any* of them. The mathematics is considerably simplified if the von Neumann–Morgenstern assumptions are fulfilled. Without them, however, the basic problem—the analysis of multiperson decision making—remains, and virtually all of the apparatus of game theory is applicable to it.

12.2 Game Theory and Economics

Periodically since the publication of *Theory of Games and Economic Behavior,* economists have asked what game theory has actually contributed to economics. Where have the applications been, and what if anything new has been learned?

This brief review is offered as a reference guide to the development of the applications of game theory to economics. It is not meant to be complete or completely self-contained. More exposition and references are given in the next volume, and several topics have been treated more mathematically by Aubin (1979).

12.2.1 Oligopolistic markets

The two major applications of game theory to economics have been to the study of various aspects of oligopolistic competition and collusion and to the study of the emergence of price in a closed economic system. A useful division of the study of oligopolistic markets is into:

1. duopoly
2. noncooperative and "quasicooperative" oligopoly
3. bilateral monopoly and bargaining
4. experimental gaming
5. auctions and bidding.

Duopoly Models of duopoly have always held a fascination for mathematically inclined economists. The literature is so vast that it merits a separate study. A good set of references up to around 1960 has been compiled by Chamberlin (1933, 1962). The first model that can be clearly identified as a well-defined mathematical description of competition in a duopolistic market is that of Cournot (1838). It is a game in strategic form in which the competitors, selling identical products, are each assumed to select a level of production in ignorance of the other's actions. The solution suggested is that of a noncooperative equilibrium. Bertrand's (1883) critique of Cournot was primarily directed toward the former's selection of production as the strategic variable. He suggested that price would be a more natural variable. Edgeworth's (1925) model of duopoly introduced rising costs, or the equivalent of a capacity restraint; Hotelling (1929), also working with a noncooperative model, introduced transportation costs as a form of product differentiation. Chamberlin's (1929, 1933, 1962) model introduced product differentiation in a more general manner.

A variety of duopoly models have been proposed and analyzed by authors such as Harrod (1934), Nichol (1934), Coase (1935), Kahn (1937), and Stigler (1940). These and many subsequent models cannot be regarded as explicitly game-theoretic inasmuch as the authors do not concern themselves with the detailed specification of the strategy spaces of the two firms. Indeed the careful mathematical statement of a duopoly model calls for considerable care and detail, as shown by Wald's (1951) analysis of equilibrium.

A series of explicitly game-theoretic models of duopoly have incorporated the effects of inventory carrying costs, fluctuations in demand, capacity constraints, and simultaneous decisions on both price and production. Other models have compared noncooperative solutions with other types of solution; stockout conditions and variation in information conditions have also been considered (Beckmann, 1965; Levitan and Shubik, 1971a,b, 1972, 1978; Mayberry, Nash, and Shubik, 1953; Shapley and Shubik, 1969c; Shubik, 1955a, 1968a, 1973c).

There is a growing literature on dynamic models of duopoly. Frequently, in teaching even the simple Cournot duopoly model, a dynamic process entailing action and reaction is sketched. Early work done by Stackelberg (1934; see also Heyward, 1941) suggested a series of quasidynamic models. More recent examples provide more mathematical game-theoretic and behavioral models of duopolistic behavior (Smithies and Savage, 1940; Shubik and Thompson, 1959; Cyert and DeGroot, 1973; Kirman and Sobel, 1974; Case, 1979).

Oligopoly One way of considering oligopolistic markets is to construct a duopoly model that can be studied and then enlarged in an appropriate manner as the number of competitors is increased. The assumption of symmetrically related firms facilitates comparisons among markets of different size. Cournot (1838) proceeds from an analysis of two similar firms to many, and Chamberlin (1933, 1962) considers small and large groups of competitors.

The investigation of oligopolistic markets calls for two if not three different skills. They are the skills of the economist at describing economic institutions and activities and selecting the relevant variables and relationships; the skills of the modeler in formulating a mathematical structure that reflects the pertinent aspects of the economic phenomena; and the skills of the analyst in deducing the properties of the mathematical model. Thus the work of Chamberlin may be regarded as a considerable step forward over that of Cournot in terms of its greater relevance and reality; however, it was no advance at all (and possibly a retrogression) in terms of rigorous mathematical formulation and analysis. Both the Cournot and Chamberlin large-group analyses are based on a noncooperative equilibrium analysis.

No amount of mathematical rigor can make up for lack of economic insight and understanding in the creation of the model to be analyzed. Thus the development of an adequate theory depends heavily upon verbal description and less upon fully formal models, as suggested by Stackelberg (1934) and Fellner (1949). Special variables must be considered. Brems (1951) introduces technological change, Bain (1956) considers entry, Baumol (1959), Shubik (1961), and Marris (1964) stress managerial structure, Levitan and Shubik (1971a) consider the role of inventories, and there are many other works dealing with important and special variables such as transportation, advertising, production change costs, multiple products, and financing.

What does the theory of games add beyond being a mathematical tidying-up device that translates the insights of others into a more heavily symbolic language? A partial answer is that the discipline required for specifying in detail the strategic options of individual actors leads to the discovery of gaps in the logic of less formally defined models. Many models of competition are "quasidynamic" in description. They describe competition in terms that gloss over the details of process and information conditions.

The Chamberlinian analysis of large-group behavior and the literature by Sweezy (1939), Stigler (1947, 1964), and many others on the "kinked oligopoly curve" provide examples of both the power and the danger of an informal mix of verbal and diagrammatic modeling. This is shown when one tries to formulate the structure of the market as a well-defined model. The kinked oligopoly curve has no objective existence; it is nothing more than a behavioral conjecture about the reactions of the competitors.

The arguments describing equilibrium or a tendency toward equilibrium using either the kinked oligopoly curve or Chamberlin's large-group analysis depend only upon the local properties of these subjective curves. Edgeworth's (1925) analysis of duopoly led him to conclude that no equilibrium need exist, but his results were obtained by considering the objective structure of oligopolistic demand over all regions of definition. In other words, the analysis requires that we be able to state the demand faced by the two firms for every pair of prices (p_1, p_2).

Shubik (1959b) suggested the term "contingent demand" to describe the demand faced by an individual firm given that the actions of competing firms are fixed. It can be shown that the contingent demand structure often depends on details of marketing that involve the manner in which individual demands are aggregated. Levitan (1964) related the description of oligopolistic demand to the theory of rationing. The study of the shape of contingent demand curves shows that the Chamberlin large-group equilibrium is easily destroyed for much the same reasons as indicated in Edgeworth's analysis (Levitan, 1964; Levitan and Shubik, 1971a; Roberts and Sonnenschein, 1977).

A full understanding of the problems posed by oligopoly requires that a clear distinction be made separating aspects of market structure, the intent of the firms, and the behavior of the firms (Shubik, 1956, 1957).

Perhaps the most important aspect of game-theoretic modeling for the study of oligopoly comes in describing information conditions and providing formal dynamic models that depend explicitly upon the information conditions. There is a growing interest in *state strategy models,* in which the system dynamics are dependent only upon the current state of the system. The work on games that model oligopolistic dynamics provides examples (Shubik, 1959b; Shubik and Thompson, 1959; Selten, 1960; Miyasawa, 1962; Morrill, 1966; Bell, 1968; Case, 1971, 1979; J. W. Friedman, 1971; Kuhn and Szegö, 1971; Liu, 1980).

The sensitivity of an oligopolistic market to changes in information has been studied (Shubik, 1973c). When information is relatively abundant, there is no strong reason to suspect that a few firms in an oligopolistic market will employ state strategies. Instead they may use *historical strategies,* in which previous history, threats, and counter-threats play an important role (Shubik, 1959b; Selten, 1973; Marschak and Selten, 1974).

Among the books devoted to a game-theoretic investigation of oligopoly are those of Shubik (1959), Jacot (1963), Sherman (1972), Telser (1972, 1979), J. W. Friedman (1977), and Shubik with Levitan (1980).

Bilateral monopoly and bargaining Whereas most of the models of oligopolistic behavior have either offered solutions based on the non-cooperative equilibrium or have sketched quasidynamic processes, the work on bilateral monopoly and bargaining has primarily stressed high levels of communication with a cooperative outcome, or a dynamic process leading to an optimal outcome. A few of the models suggest the possibility of nonoptimal outcomes such as strikes that materialize after threats are ignored or rejected.

Models of bargaining arise from a wide variety of institutional backgrounds. The major ones are bilateral trade among individual traders as characterized by Böhm-Bawerk's (1891) horse market or Bowley's (1928) model. Frequently, however, the model proposed refers to international trade or to laborer-employer bargaining. Edgeworth's (1881) famous initial model was cast in terms of the latter, as was the work of Zeuthen (1930).

The work of Edgeworth is clearly related to the game-theoretic solution of the core (Shubik, 1959a; Debreu and Scarf, 1963; Scarf, 1967). Böhm-Bawerk's analysis may be regarded as an exercise in determining the core and price in a market with indivisibilities (Shapley and Shubik, 1972a).

Zeuthen's analysis of bargaining is closely related to the various concepts of value as a solution (see chapter 7). This includes the work of Nash (1953), Shapley (1953e), Harsanyi (1956, 1959), Bishop (1963), Selten (1964), and Roth (1979).

A "solution" to an economic problem may be no more than a limiting of the feasible set of outcomes to a smaller set. No specific outcome may be predicted. The solution narrows down the set of outcomes but does not tell us exactly what will or should happen. The contract curve of Edgeworth and the core are solutions in this sense.

Other solutions may be used in an attempt to single out a solitary final outcome as the one that should or will emerge. In their static versions most of the various *value* solutions and other fair-division solutions that have been proposed may be regarded as normative in their suggestions and abstracted from any particular institutional background in their presentation (Steinhaus, 1949; Nash, 1953; Shapley, 1953e; Braithwaite, 1955; Harsanyi, 1956, 1959; Bishop, 1963; Kuhn, 1967; Kalai and Smorodinsky, 1975).

Still other solutions which may be used to select a single outcome are phrased in terms of a dynamics of the bargaining process (Zeuthen, 1930; Pen, 1952; Shubik, 1952; Raiffa, 1953; Harsanyi, 1956; Cross, 1969).

There is a danger that using mathematical models without a sufficient contextual reference can lead to false generalizations. This can come about if one fails to model the features that are important to a specific problem. It is our belief that game-theoretic modeling as applied to bilateral monopoly, bargaining, and fair-division problems will have considerable value in helping to sort out variables, to model information conditions, and in general to provide new analytical approaches using different solution concepts. Even so the modeler is well advised to check the context of the problem before proceeding far with a model or its analysis. Thus, for example, before applying a formal mathematical model to labor-management problems, one should consider the type of background provided by experts in collective bargaining such as Chamberlain (1951; see also Chamberlain and Kuhn, 1965).

Gaming One result of the development of game theory and the high-speed digital computer has been a growth of interest in using formal mathematical models of markets for teaching and experimental purposes. The earliest published article on an informal economic game experiment was by Chamberlin (1948). This, however, appeared in isolation from the rest of the literature. The first "business game" built primarily for training purposes was constructed by Bellman et al. (1957) several years later, and this was followed by many large-scale computerized business games. The use of these games has been broadly accepted in business schools and in some economics faculties.

Much of the earlier experimental work with games in economics did not use the computer. The games were presented in the form of matrices or diagrams. Siegel and Fouraker (1960), Fouraker and Siegel (1963), and Fouraker, Shubik, and Siegel (1961) were concerned with bilateral monopoly under various information conditions, duopoly, and triopoly. Stern (1966), Dolbear et al. (1968), and J. W. Friedman (1967, 1969) investigated the effect of symmetry and lack of symmetry in duopoly as well as several other aspects of oligopolistic markets. V. L. Smith (1965, 1967) considered the effect of market organization.

Experiments using computerized games have been run by Hoggatt (1959, 1967), McKenney (1962), Shubik, Wolf, and Lockhart (1971), Shubik, Wolf, and Eisenberg (1972), Shubik and Riese (1972), and Friedman and Hoggatt (1979). These games offer advantages in terms of control and ease of data processing over the noncomputer games.

Several of the games noted above can be and have been solved for various game-theoretic and other solutions. This means that in the duopoly games studied by Friedman, for example, it is possible to calculate the Pareto-optimal surface and the noncooperative equilibrium points. Similarly in many of the duopoly or oligopoly investigations it is usually possible to calculate the joint optimum, the noncooperative equilibria, and the competitive price system.

The game designed by Shubik and Levitan (1980) was specifically made amenable to game-theoretic analysis. Thus the joint maximum, the price noncooperative equilibrium, the quantity noncooperative equilibrium, the range of the Edgeworth cycle, the beat-the-average solution, and several other solutions have been calculated for this game.

It is probably too early to attempt a critical survey of the implications of all of the experimental work for oligopoly theory; however, a general pattern does seem to be emerging (see Sauermann, 1967, 1970, 1972; Shubik, 1975c; V. L. Smith, 1979). All other things being equal, an increase in the number of competitors does appear to lower price, as does an increase in cross-elasticities between products. With few competitors, however, information and communication conditions appear to be far more critical than a reading of oligopoly theory would indicate. Furthermore in duopoly experiments the role of symmetry appears as an important artifact.

An important development has been the growing interest in designing and doing experiments with self-policing systems, that is, systems that have been designed so that there are Pareto-optimal noncooperative equilibria (Groves and Ledyard, 1977; V. L. Smith, 1978).

In all of the experiments and in games for teaching, the importance of considering a richer behavioral model of the individual emerges when one observes the way in which the players attempt to deal with their environment. This observation is by no means counter to the game-theoretic approach; it is, rather, complementary to it. As yet no satisfactory dynamic oligopolistic solution has been provided by either standard economic theory or game theory. This appears to result from the difficulties involved in describing the role of information processing and communication.

A different set of games has been used for experimentation with a certain amount of economic content, but these are far less identified with an economic market than are the oligopoly games. These include simple bidding and bargaining exercises. In all of these instances there has been a specific interest in comparing the outcomes of the experiments with the predictions of various game-theoretic solutions (Kalisch et al., 1954; Flood, 1955; Stone, 1958; Maschler, 1965; Riker, 1967; Shubik, 1975a).

Auctions and bidding Auctions date back to at least Roman times. In many economies they still play an important role in financial markets and in commodity markets. Sealed bids are used in the letting of large systems contracts or in the sale of government property. Their history as economic market mechanisms is a fascinating subject by itself. Furthermore, since an auction or a bidding process is usually quite well defined by a set of formal rules (together, on occasions, with customs or other informal rules), it lends itself naturally to formal mathematical modeling.

The mathematical models of auctions and bidding fall into two major groups: those in which the role of competition is modeled by assuming a Bayesian mechanism (Lavalle, 1967), and those in which the model is solved as a game of strategy using the solution concept of the noncooperative equilibrium (Friedman, 1956; Vickrey, 1961; Griesmer and Shubik, 1963a,b,c; Griesmer, Shubik, and Levitan, 1967; Wilson, 1967; Amihud, 1976) or some other solution.

There is a considerable literature on problems encountered in various types of bidding (Beveridge, 1962; Cook, 1963; Christenson, 1965; Dean, 1965; Dean and Culkan, 1965; Loane, 1966; V. L. Smith, 1966) and on features such as problems in evaluation (Bremer, Hall, and Paulsen, 1957; Waggener and Suzuki, 1967; Feeney, 1968), risk minimization, and incentive systems (Haussmann and Rivett, 1959; Moriguti and Suganami, 1959; Rothkopf, 1969; Stark and Mayer, 1971).

The study of auctions and bidding falls into the zone between theory and application, as can be seen by observing the tendency of the publications to appear in journals such as the *Operations Research Quarterly* or *Management Science.* A useful bibliography on bidding has been supplied by Stark and Rothkopf (1979), and a survey has been provided by Engelbrecht-Wiggans (1980).

In general, game-theoretic work on auctions and bidding has been useful in two ways: descriptive and analytic. The careful specification of mathematical models has focused attention on the actual mechanisms, including informal rules and customs. The attempts at solution have shown that the models are extremely sensitive to information conditions and that many of the important features of auctions involve the individual's ability to evaluate what an object is worth to him and to others.

12.2.2 General-equilibrium models

Possibly the most important area of application of the theory of games to economic analysis has been to the closed general-equilibrium model of the economy. The solution concepts that have been explored are primarily cooperative solutions. Noncooperative solutions appear to be intimately related to monetary economies (see section 12.2.3).

There have been several stages of generalization in the application of game theory to economic analysis. These have dealt particularly with assumptions concerning preferences and utility and with the application of so-called limiting methods to study economies with many "small" agents. Von Neumann and Morgenstern originally modeled cooperative games in terms of a transferable utility. This restriction appears to be an unnecessarily strong and in many (but not all) instances unrealistic approximation of the economic world. It has, however, the virtue that it enormously simplifies the mathematical analysis required to investigate the various cooperative game solutions.

Shapley and Shubik (1953) suggested that the development of a theory of games did not require simplifications concerning utility, but it was not until the work of Aumann and Peleg (1960) that "no-side-payment solution theory" was adequately developed.

The earlier theory was referred to as side-payment theory because there was assumed to exist a special commodity, which we would call a "util" or a "u-money," such that if any trader effected a transfer of this commodity to another trader, there would be a constant marginal change in welfare for each.

In much economic literature it is claimed that the study of economics requires only the assumption of a preference ordering over the prospects faced by an individual (Hicks, 1938; Samuelson, 1948; Arrow and Debreu, 1954). Apart from the fact that such a strong assumption immediately rules out of economic consideration topics such as bargaining and fair division in which virtually no analysis can be made without stronger assumptions on the measurement of utility (see chapter 4), and even if we restrict our investigation to the free functioning of a price system, the assumption of only a preference ordering is not sufficient. One must at least restrict transformations to those which preserve the concavity of utility functions; otherwise markets involving gambles will emerge (Shubik, 1975b).

The investigation of game-theoretic solutions for application to economic problems has focused primarily on the core and secondarily on other cooperative solutions.

The core The first economist to consider bargaining and market stability in terms of the power of all feasible coalitions was Edgeworth (1881). He was dealing with a structure that can be described as a market game (Shapley and Shubik, 1969a), and his solution was, in essence, the core. Shubik (1959a) observed that the Edgeworth analysis was essentially an argument that could be described in terms of the core. He constructed a two-sided market model with side payments to demonstrate this and used the method of replication to illustrate the emergence of a price system. He conjectured that this result was true for no-side-payment games and proposed the problem to Shapley and Scarf.

Essentially the replication method boils down to considering an economy with thousands of butchers, bakers, and candlestick makers. Scarf (1967, 1971) and Debreu and Scarf (1963), also using the method of replication, were able to generalize the previous results considerably. They showed that under replication, in a market with any number of different traders, the core "shrinks down" (under the appropriate definition which takes into account the increasing dimensions) until a set of imputations that can be interpreted as price systems emerge as the limit of the core. Hildenbrand (1974) has generalized the technique of replication, doing away with the need for identical types. Further generalization is given by Anderson (1978).

A different way of considering markets with many traders is to imagine that we can "chop up" traders into finer and finer pieces. Using this approach we may consider a continuum of traders in which the individual trader whose strategic power is of no significance to the market is described as having a measure of zero. Aumann (1964b, 1966) first developed this approach.

In the past twenty years there has been a proliferation of the literature on the core of a market (Aumann, 1961b, 1964a,b, 1966; Debreu, 1963; Vind, 1964, 1965; Hildenbrand, 1968, 1970a,b, 1974; Dreze, Gabszewicz, and Gepts, 1969; Schmeidler, 1969a; Kannai, 1970; Gabszewicz and Mertens, 1971; Gabszewicz and Dreze, 1971; Hildenbrand and Kirman, 1976). Although much of this work has been devoted to the relationship between the core and the competitive equilibrium (for example, given a continuum of small traders of all types, the core and the competitive equilibrium can be proved to be identical), some has been directed to other problems. Thus Shapley and Shubik (1966) considered the effect of nonconvex preference sets, and Aumann (1973), Shitovitz (1973), and others (Gabszewicz, 1970; Gepts, 1970; Hansen and Gabszewicz, 1972; Telser, 1972) have been concerned with the economics of imperfect competition treated as a game with a continuum of traders. Caspi (1978) and Shubik (1973a, 1975b) also considered the effect of uncertainty on the core. Shapley (1975b) and Debreu (1975) and others have examined the rate of convergence of the core.

In summary, this work extends the concept of economic equilibrium and stability to many dimensions and raises fundamental questions concerning the role and the nature of coalitions in bringing about economic stability.

Other solutions The core may be regarded as characterizing the role of countervailing power among groups. There are other solution concepts which reflect other views for the determination of the production and distribution of resources. In particular, the family of solutions that can be described as the *value* of an n-person game stresses fair division, that is, division based both upon the needs or wants of the individual and upon his basic productivity and ownership claims.

The value There are a variety of differences among the value solutions that have been suggested. Three major factors are (1) whether there are two or more individuals, (2) whether or not a side-payment mechanism is present, and (3) whether threats play an important role and whether the status quo is difficult to determine. The work of Nash (1953), Shapley (1953a), Harsanyi (1956, 1959), and Selten (1964) covers the various cases (see also chapter 7).

Leaving aside the finer points, all of the major value solutions are based in one way or another on a symmetry axiom and an efficiency axiom. Describing them loosely, we can say that if individuals have equal claims, they should receive equal rewards, and the outcome should be Pareto-optimal.

These solutions appear to have no immediate relationship with a price system, yet using the method of replication one can show that under the appropriate conditions, as the number of individuals in a market increases, the value approaches the imputation selected by the price system. Markets with a continuum of traders have been investigated, and the coincidence of the value with the competitive equilibrium has been established (Shapley, 1964c,d; Shapley and Shubik, 1969d; Aumann and Shapley, 1974); applications to problems in taxation have been made (Aumann and Kurz, 1977a,b).

The bargaining set Rather than appeal to countervailing-power arguments or to considerations of fairness, one might try to delimit the outcomes by bargaining considerations (see chapter 11). Aumann and Maschler (1964) suggested a bargaining set, and Peleg (1967) established that such a set always exists. Many slight variations in the definition have been proposed. Shapley and Shubik (1972c) were able to show that under appropriate conditions the bargaining set lies within an arbitrarily small region of the imputation selected by the price system when the number of traders in an economy is large; this was, however, a somewhat restricted result.

The nucleolus A still different solution concept is the nucleolus, proposed by Schmeidler (1967). The nucleolus lies in the kernel and can be interpreted as the single point at which the maximum dissatisfaction of any coalition is minimized.

Consider any suggested imputation or division of the available wealth. Any coalition can compare what it could obtain by acting on its own with the amount it would obtain by accepting the suggested imputation. The difference between these two amounts is called the *excess,* which may be positive or negative. The nucleolus is the imputation at which the maximum excess is minimized. If the core exists, the nucleolus is located at its "center" (see chapter 11).

Since the nucleolus is in the core, under replication for large markets the nucleolus approaches the imputation selected by the price system. It is a natural concept to be considered for application to taxation and subsidy problems. It provides an immediate measure for the design of a taxation and subsidy program to minimize group claims or grievances.

The kernel There have been no significant applications to the social sciences in which the kernel plays a central role.

The stable set In applications to economics and the other social sciences the stable set has served more to suggest complications and difficulties in the study of strategic interaction than to offer a way of proceeding. The fact that the Lucas counterexample (see chapter 6), which showed that the stable set need not exist, can arise from a nonpathological market limits the applicability of the stable set to economics. Yet stable-set theory provides examples of phenomena, such as discriminatory solutions and the transfer of resources between apparently independent groups, that raise interesting questions in the modeling of economic, political, and social processes.

Solutions, market games, and the price system The class of market games (Shapley and Shubik, 1969b) provide a representation of a closed economic system for which a price system exists.

The study of market games with large numbers of participants has revealed a remarkable relationship between the imputations selected by a price system and the core, value, bargaining set, kernel, and nucleolus of large market games. An expository article by Shapley and Shubik (1967a) presents a nontechnical discussion of some of these relationships. Each solution concept models or picks up a different aspect of trading. The price system may be regarded as stressing decentralization (with efficiency); the core shows the force of countervailing power; the value offers a "fairness" criterion; the bargaining set and kernel suggest how the solution might be delimited by bargaining conditions; and the nucleolus provides a means to select a point at which dissatisfaction with relative tax loads or subsidies is minimized.

If for a large market economy these many different approaches call for the same imputation of resources, then we have what might be regarded as a nineteenth-century laissez-faire economist's dream. The imputation called for by the price system has virtues far beyond that of decentralization: it cannot be challenged by countervailing power, it is fair, and it satisfies certain bargaining conditions.

Unfortunately in most economies these euphoric conclusions do not hold for two important reasons. The first is that there are rarely if ever enough individuals of all types that oligopolistic elements are removed from all markets. The second is that the economies frequently contain elements that modify or destroy the conditions for the existence of an efficient price system. In particular, these include external economies and diseconomies, indivisibilities, and public goods. When the price system does not exist, we are forced to seek other solution concepts to provide alternatives.

12.2.3 Other applications to economics

The most-developed and best-known applications of game theory to economics have been noted above, but there is a growing literature in other areas.

Public goods, externalities, and welfare economics When we examine the literature on public goods, it is difficult to make a clear distinction between economic and political-science analysis. The basic nature of the problems is such that their investigation requires an approach based on political economy. More or less arbitrarily, even though it is related to welfare economics, we discuss the work on voting systems in the section on political science.

When externalities are present in an economy, an efficient price system may not exist. It may, however, be possible to design a tax and subsidy system that will allow an efficient price system to function.

A considerable literature now exists on game-theoretic approaches to public goods and taxation. No attempt will be made to cover it here (see Klevorick and Kramer, 1973; Aumann and Kurz, 1977a,b), but we shall return to the subject in some detail in the next volume.

Threats play an important role in studying many of the problems posed by externalities and public goods. Can you force an individual to share a public good? Can you prevent him from using a good unless he pays his share? Such questions allow us to differentiate many types of public goods, and a taxonomy based on these considerations has been suggested (Shubik, 1966). The difficulties involved in modeling economic or political activities by means of a characteristic function when threats are involved have been discussed in chapter 6.

Indivisibilities and other features that may cause the individual's consumption or production possibility sets to be nonconvex have been studied (Shapley and Shubik, 1966; Shubik, 1971a; Shapley and Scarf, 1974).

Externalities due to different ownership arrangements and pecuniary externalities caused by the presence of markets have also been studied. Although not primarily game-theoretic in content, the work of Buchanan and Tullock (1962), Davis and Whinston (1965), Zeckhauser (1973), and others is closely related to the game-theoretic approach to public goods and welfare economics.

Another aspect of welfare economics to which game-theoretic analysis is directly applicable involves the study of lump-sum taxation, subsidies, and compensation schemes (see Brams, Schotter, and Schwodiauer, 1979; Schleicher, 1979). These depend delicately on assumptions made concerning the availability of a side-payment mechanism and the relationship between social and economic prospects and the structure of individual preferences.

Money and financial institutions In recent years there has been considerable interest in the construction of an adequate microeconomic theory of money. Most of this work has taken as its basis the general-equilibrium nonstrategic model of the price system. Foley (1970a), Hahn (1971), Starr (1974), and others (Shubik, 1972a,b, 1978b; Shapley and Shubik, 1977; Shapley, 1976; Postlewaite and Schmeidler, 1978) have focused on static problems, while Grandmont (1977) provides a survey of the dynamics (see also Shubik and Whitt, 1973).

In contrast with the nonstrategic approaches, there has been a growth in explicitly strategic noncooperative game models of trading and production economies using a commodity or a fiat money (Shubik, 1976; Shubik and Wilson, 1977). A major theme of this work is that strategic modeling calls for the introduction of rudimentary structures and rules of the game which can be interpreted in terms of financial institutions, instruments, and laws. Thus in specifying the use of money, a distinction must be made between money and credit. Bankruptcy laws must be specified. The way in which money and credit enter the system must be noted. Bankers, government, and others may require modeling as separate players in the strategic market game (see chapter 10).

Information conditions clearly are of considerable importance in a mass economy. The sensitivity of economic models to the sequencing of market and financial moves, as well as the general problem of the sensitivity of market models to changes in information conditions, must be taken into account (Shubik, 1973c; Dubey and Shubik, 1977a,b).

Some results have been obtained using cooperative game-theoretic analysis. In particular, it has been shown that if trade is assumed to take place via markets, then pecuniary externalities are real (Shubik, 1971d). Other game-theoretic aspects of insurance have been considered by Borch (1968, 1979).

Macroeconomics There have been few applications of game theory to macroeconomic problems and international trade (Nyblen, 1951; Faxen, 1957; Munier, 1972). These have nevertheless suggested the possible uses of treating aggregated units as players in a game of strategy.

12.3 Game Theory and Political Science

Among the earliest attempts to apply game theory to political science are those of K. W. Deutsch (1954), Shapley and Shubik (1954), and Shubik (1954b). In these works the scope of application discussed was relatively limited when compared with what was to come. There are several topics in political science to which the application of game theory is *prima facie* appealing. These are voting, power, diplomacy, negotiation and bargaining behavior, coalition formation among political groups, and logrolling.

12.3.1 Voting and group preference

There is already a considerable body of literature using game theory or techniques closely related to game theory in application to political science. The relationship of individual preferences to group preference and the role of mechanisms for political choice, such as voting, are of interest to the political scientist.

In chapter 5 it was noted that much of the work on voting has left out the strategic aspects and concentrated on the "aggregation" of individual preferences via the vote. The assumptions usually made are that:

1. Individuals know their own preferences, and these are fixed.
2. They know and are able to evaluate all alternatives.
3. The rules of the game are known and understood by all.
4. Each individual is rational and suffers no information overload or computational problems in decision making.
5. It is possible to consider the social choice problem in a static context; that is, a static model serves as a reasonable approximation of a real-world social choice process such as an election.

A partial list of references is: Condorcet (1785), Dodgson (1873), Nanson (1882), Arrow (1951), Goodman and Markowitz (1952), Guilbaud (1952b), May (1952), Hildreth (1953), Buchanan (1954), Blau (1957), Black (1958), Vickrey (1960), Riker (1961), Sen (1964), Murakami (1966), Coleman (1966a), Plott (1967), Tullock (1967), Garman and Kamien (1968), Ledyard (1968), Fishburn (1969), Inada (1969), Rae (1969), DeMeyer and Plott (1970), Wilson (1972b), Kramer (1973).

The Condorcet voting paradox and the Arrow general possibility theorem referred to in chapter 5 are central to this approach. Various conditions on preference structures such as "single peaked preferences" have been investigated by Black and others.

The properties of different voting methods and different assumptions concerning the measurability and comparability of the intensity of individual preferences have been considered. Thus majority voting, weighted majority voting, various rank-ordering methods and rules for eliminating candidates, and other schemes have been studied (see Balinski and Young, 1982).

A different but highly related approach to problems of political choice takes into account the strategic aspect of voting. Explicit assumptions must now be made not only about what the individual knows about his own preferences but about what he knows of the preferences of others. In most of the early work either it was explicitly assumed that individuals were informed of all preferences or the information conditions were glossed over.

Studies of strategic voting can be divided into those using cooperative, noncooperative, and other solution concepts, and they can be further divided into those with a formal structure assumed (such as political parties) and those in which only individuals are considered in a setting with no institutional details specified explicitly.

A clear and concise application of the noncooperative equilibrium solution to strategic voting has been made by Farquharson (1969), who defined the conditions for sincere, straightforward, or sophisticated voting (see also Gibbard, 1973; Satterthwaite, 1975).

The theory of competition among political parties offers another important area of application for the theory of games. Some starts have been made (Downs, 1957; Chapman, 1967; Frey, 1968; Shubik, 1968c; Brams 1975, 1978; Kramer, 1977). These are based primarily on analogies between the economics of oligopolistic competition and noncooperative party struggles. The tendency has been to apply some form of noncooperative or mechanistic solution to the models. The classic work of Dahl (1961) describing political participation serves as a guide for those who wish to model different actors in political competition.

Some of the questions that can be examined by noncooperative game models are: How are the differences between economic and political competition manifested? What political mechanisms guarantee a Pareto-optimal distribution of resources, and under what conditions?

Although there are as yet no satisfactory models, it appears that it might be worthwhile to construct game-theoretic models of the political process in which power groups such as unions, large industries, and possibly several other organizations are distinguished along with the parties and the individual voters. With such models both cooperative and noncooperative solution concepts would merit examination.

Possibly the most important aspect of the application of game theory to economic analysis has been in providing formal mathematical models to study the effect of large numbers of participants in the economy. A well-known feature of the economic system is that mass markets exhibit characteristics highly different from those of markets with few participants. It is our belief that the study of game-theoretic models of political systems with large numbers of participants can cast some light on the role of numbers in the political process and can help to clarify some of the problems inherent in mass democracy. There are already a small number of works that contain formal mathematical models of voting processes in which the number of participants is a key variable. They deal primarily with extensions of the Voter Paradox (DeMeyer and Plott, 1970; Garman and Kamien, 1968) or with simple games (Shapley, 1962b; Wilson, 1972b).

Even with the simpler models of voting, extremely difficult problems are encountered if some of the assumptions are relaxed. For example, if individuals do not know each other's preferences, then in their strategic behavior they must evaluate the worth and consequences of deliberate misinformation. This becomes important, for example, in deciding when to disclose one's stand on tax bills or appropriations. There is a growing body of work dealing with the strategic aspects of incomplete information (Lumsden, 1966; Maschler, 1966a; Harsanyi, 1967, 1968a,b; Levine and Ponssard, 1977; Kalai and Rosenthal, 1978). This work has not been directed at political science in particular; hence any applications would call for more specific modeling.

It is important to stress that the applications to voting and group preferences noted above generally tend to play down the nonstrategic aspects of human behavior. After all, many people may vote more out of habit or affiliation than as the result of a carefully planned strategy. Even though this may be the case, game-theoretic and behavioral approaches are not incompatible. The problems of application lie at least as much in the model building as in the analysis. There is every indication, however, that a considerable amount of ad hoc modeling is required to reflect the intermix of habit, nonstrategic but conscious behavior, and strategic cooperative and noncooperative behavior that describes the actors in the political arena (Brams, 1975).

12.3.2 Coalitions and bargaining

Much of the writing on voting has dealt with a single vote. A special class of games, the *simple game* discussed in chapter 7, serves as a good representation of the single vote in isolation. Frequently, however, it is unreasonable to consider isolated issues. Attempts have been made to incorporate logrolling and the trading of votes into game-theoretic models (Shapley, 1953e; Shapley and Shubik, 1954; Mann and Shapley, 1964; Miller, 1973).

In designing new legislatures, or attempting to obtain an a priori feeling for how the voting power of individuals might change with changes in voting structure, the Shapley–Shubik and Banzhaf indices have been applied. They provide a measure of the importance of each individual.

Logrolling In many political processes the problem is not to pick one among a set of alternatives but to decide how many of a set of motions will be passed. Suppose that there are n individuals and m motions. The individuals each have one vote on each motion. The final outcome achieved by society is one of 2^m states. Since the individuals may have differing preferences for the outcomes, they may be in a position to trade votes among themselves. Thus the votes serve as a limited type of currency for making side payments (Bentley, 1949; Buchanan and Tullock, 1962; Wilson, 1969; Coleman, 1970; Shubik and Van der Heyden, 1978).

Given a voting method and preferences, several considerably different models may be specified. Three models are: (1) the voting process viewed as a market with prices for votes; (2) voting as a noncooperative game (this might also include a model of the agenda-selection process, which is frequently a key controlling factor in situations involving many alternatives); and (3) voting and logrolling as an explicit cooperative game.

We may wish to suppress the strategic aspects of individual choice and regard the votes of the individuals in much the same manner as the initial endowments of commodities are treated in models of an exchange economy. Assume that there is a price given for a vote on every issue. The initial vote endowment of each issue can be evaluated at the given prices, and a budget constraint can be calculated for each individual for buying and selling votes. It remains to establish the conditions under which an efficient price system will exist.

If we wish to consider the strategic aspects of logrolling, we might attempt to extend the noncooperative analysis suggested by Farquharson (1969). There are, however, some difficulties in modeling vote trading as a noncooperative game. A more natural approach is to consider logrolling as a cooperative game.

At least three cooperative solution concepts appear to be applicable to logrolling: the core, the value, and the bargaining set. The existence of the core is undoubtedly a necessary condition for the existence of a market for votes. Examples of this type of relationship are given by Klevorick and Kramer (1973) and by Wilson (1969, 1971a). Wilson has also considered the bargaining set. Shubik and Van der Heyden (1978) have shown the conditions for the core to be empty.

Noncooperative models of vote buying have also been considered (Shubik and Young, 1978; Young, 1978a,b).

Coalitions other than logrolling The applications of game theory to problems involving bargaining and coalition formation split into those with primary emphasis on statics and combinatorics and those which stress dynamics and process. The former tend to be parsimonious in the introduction of new variables, and as free as possible of personality and institutional factors; the latter are frequently far more descriptive and behavioristic.

The most direct and institution-free application of cooperative game theory to voting comes in the calculation of the value for simple games, where a simple game reflects the strategic structure of a voting procedure applied to a single issue.

When concern is directed toward voting on a single issue, many of the difficulties concerning the nature of preferences are not germane. When, furthermore, we assume that the voting process can be represented by a simple game, we have implicitly accepted a set of assumptions that limit the immediate applicability of any results yielded by the analysis. In particular (as is the case with all cooperative game-theoretic solutions and with the nonstrategic approaches of Arrow, Black, Plott, Rae, and others), the costs, details, and dynamics of the voting *process* are assumed away.

Thrown away with most formal game-theoretic and non-game-theoretic models of voting are important features such as the power and role of the chairman. The assumption of external symmetry (see chapter 2) is an important factor in most mathematical model building. It is valuable for many analytical purposes, but the price paid is high. Before interpreting any results obtained from a theory that makes such an assumption in terms of actual political processes, one must consider the sensitivity of the results to a variation of the assumption.

Even given the enormous simplifications implicit in the study of voting by means of simple game models, several interesting questions can be posed and answered. These concern the design of voting bodies and methods. Work on this topic includes that of Shapley and Shubik (1954) on the Security Council of the United Nations, Mann and Shapley (1964) on the Electoral College, Riker and Shapley (1968), Riker (1959), Banzhaf (1965, 1966), and Nozick (1968). Several of these papers deal with the problem of weighted voting and its relationship to the principle of "one-man, one-vote" in districts of unequal size. Further references are given by Dubey and Shapley (1979).

A step toward relaxing the assumptions made in applying game theory to models representing voting processes can be made without constructing a completely dynamic theory. This has been done by introducing some structure on coalitions. Luce (1954), Luce and Rogow (1956), Riker (1962, 1966), Leiserson (1968, 1970), Riker and Ordeshook (1968), and others have followed this approach. In particular, Luce (1954) has suggested a solution different from that of the value (see chapter 11), and Riker (1962) has stressed the importance of forming "minimal winning coalitions." This principle is in contradistinction to the analysis of Downs (1957), which suggests that political parties attempt to maximize their majorities. Riker argues that the payoffs to extra individuals not vital to forming a winning coalition may not be worth making (Riker and Ordeshook, 1973).

Riker's contribution is expressly based upon a game-theoretic approach to political coalitions, as contrasted with that of Downs, who provides an economic market model. Nevertheless Riker's proposal of a minimal size for winning coalitions (called his "size principle") is strongly dependent upon his description of a dynamics of coalition formation. Political scientists may find this description both fascinating and fruitful; but from the viewpoint of mathematical modeling it is difficult to formalize the appropriate game, characteristic function, and solution concept that are being suggested.

The approach of Luce and Rogow (1956) to the modeling of coalitions is somewhat more formal than that of Riker and simultaneously more static. They assume a priori that limitations on coalition formation are given as part of the description of the problem. The solution concept they use is that of Ψ-stability as proposed by Luce (see chapter 11).

Bargaining The literature on bargaining is considerable and varied. Several different domains of interest can be distinguished. Labor-management bargaining is almost a subject in itself. The economics of haggling and market bargaining between traders is an allied but somewhat different topic. The boundaries between economics and politics melt when international trade negotiations are considered. There is a large (and in general non-game-theoretic) literature on the social psychology of bargaining. In political science most of the literature on bargaining that has made use of game-theoretic concepts is concerned with party politics or with international negotiations.

The work in political science has to some extent been influenced by the studies of bargaining in economics or by economists and game theorists extending various mixtures of economic analysis and game-theoretic reasoning to bargaining problems (Edgeworth, 1881; Zeuthen, 1930; Pen, 1952; Shubik, 1952; Ellsberg, 1956; Harsanyi, 1956, 1959; Schelling, 1960b; Boulding, 1962; Bishop, 1963; Aumann and Maschler, 1964; Chamberlain and Kuhn, 1965; Cross, 1969, 1977; Stahl, 1972; A. E. Roth, 1979). This work has already been noted in section 12.2.

Among those who have made use of game theory in constructing analogies to international bargaining and conflict situations are Boulding (1962), M. Deutsch (1961), Ellsberg (1961), Harsanyi (1965), Iklé (1964), Midgaard (1970), Rapoport (1960, 1964), Sawyer and Guetzkow (1965), Schelling (1960a), Shubik (1963, 1968b), and Wohlstetter (1964).

Many of the problems that arise in attempting to apply game-theoretic reasoning to bargaining and negotiation are substantive. They involve the modeling of process. A brief survey such as this cannot do justice to these difficulties. The reader is referred to Iklé's (1964) *How Nations Negotiate* for a perceptive description of many aspects of the process of international negotiation and a useful but by now somewhat dated bibliographic note on literature relevant to international negotiation. *The Economics of Bargaining* by John Cross (1969) provides the political scientist as well as the economist with static and process models of bargaining in which stress is laid upon economic features such as the cost of the process. Frequently a "taxi meter" is running while bargaining is taking place.

Rapoport's (1960) *Fights, Games and Debates* stresses modeling problems in the application of game-theoretic reasoning. Schelling's (1960a) *The Strategy of Conflict* provides many provocative simple games and analogies to international conflict situations. However, there is a considerable danger in being misled if one tries to push analogies between extremely simple games and international bargaining too far. It is important to realize not only the usefulness but the limitations of simple game models as a didactic device.

12.3.3 Power

One of the key concerns of political science is the study of power. In our analysis of the various versions of a value solution (see chapter 7), we noted the relationship between these solutions and the concept of power. The immediate game-theoretic basis for the investigation of power is given in the writings of Nash (1953), Shapley (1953), Harsanyi (1962b,c), and Selten (1964). A paradoxical aspect of value solutions is that they were primarily motivated by a concern for fairness and equitable division. The relationship between fair division and power comes in the way in which threats enter into a consideration of how to evaluate the no-bargain point between any two coalitions. In essence, the status quo or no-bargain point may be determined by the power of the bargainers. The fair-division procedure is applied using this power-determined initial point as a basis for the settlement.

The various value solutions are essentially static and undoubtedly fail to portray adequately the interplay of power and influence in a dynamic system. Nevertheless they provide clear, well-defined concepts which can be subjected to scrutiny and compared with the theories of power proposed by political scientists and other social scientists. A brief set of references to writings on power which can fruitfully be contrasted with the game-theoretic value solutions might include: Goldhamer and Shils (1939), Lasswell and Kaplan (1950), H. A. Simon (1953), March (1955, 1966), Dahl (1957, 1961), Parsons (1957), Riker (1964), P. Blau (1964), Barber (1966), Coleman (1966a, 1970), Brams (1968), Nagel (1968), Wagner (1969), Axelrod (1970), Baldwin (1971), and Champlin (1971).

A comprehensive attempt to evaluate and reconcile the many approaches to the concept of power, including the game-theoretic approaches, is Nagel's (1975) *The Descriptive Analysis of Power.* This together with the papers of Harsanyi would serve as a sufficient guide for those who wish to pursue the investigation of the concept of power further.

12.3.4 Gaming associated with political science

There is a fair-sized literature on gaming associated with political science which varies somewhat in its direct association with game theory. A brief survey of the uses of gaming in this context is given by Shubik (1964, 1975c), and some extra references are included here to provide the interested reader with a lead into this literature:

Schelling (1960b), Gamson (1962), Joseph and Willis (1963), Riker and Niemi (1964), Barringer and Whaley (1965), Sawyer and Guetzkow (1965), Hermann (1967), Rapoport (1967), Riker (1967), Banks, Groom, and Oppenheim (1968), Friberg and Johnson (1968), Friedell (1968), Riker and Zavoina (1970).

12.4 Game Theory, Gaming, and Social Psychology

In the past two decades there has been an explosion of experimental work utilizing games. In 1948 Chamberlin ran what appears to be the first (somewhat informal) gaming experiment in economics. Flood (1955, 1958) in 1952 used six small experiments to explore the relationship between the predictions of game theory and the behavior of individuals in game or bargaining situations.

The greatest growth in the use of elementary games has, however, been in social psychology. The references to experiments involving 2×2 matrices now run into the thousands.

The predominant type of experiment has involved the two-person non-zero-sum 2×2 matrix game usually played repeatedly without face-to-face communication between the players. The most popular experimental game is the Prisoner's Dilemma; however, many other games have been used, and several matrix structures are referred to by special names, such as Chicken, or Battle of the Sexes:

	1	2
1	5, 5	−9, 10
2	10, −9	0, 0

Prisoner's Dilemma

	1	2
1	−10, −10	5, −5
2	−5, 5	0, 0

Chicken

	1	2
1	2, 1	0, 0
2	0, 0	1, 2

Battle of the Sexes

The name Chicken was suggested by analogy with a hotrod race in which two individuals drive toward each other, each with a pair of wheels on the line in the middle of the road; the individual who veers from the collision course is "chicken." The payoff matrix reflects the possibility of joint disaster, equal cowardice, or victory for one. Battle of the Sexes might describe the payoff to a husband and wife each wanting to go to a different movie together. The zero payoffs reflect not going at all.

Elaborate and complex games such as the Internation Simulation (INS) of Guetzkow (1963), the larger business games such as the Carnegie Tech Management Game, the Financial Allocation and Management Exercise of IBM, and the Harvard Business School Game (Levitan and Shubik, 1961; Cohen et al., 1964; McKenney, 1967; Shubik with Levitan, 1980) have been used for teaching, training, and experimentation. However, the direct testing of conjectures based on game theory has been relatively scarce with the larger games.

We have already noted in section 12.2 the work concerned with the testing of economic and game-theoretic hypotheses in environments that are by no means rich but may nevertheless be far more complex than the simple matrix experiments.

Before we discuss the different types of experimental games, we shall make some observations on the basic features of modeling in game theory as applied to experimental games.

12.4.1 Aspects of modeling for experimental gaming

Players and external symmetry One of the most important contrasts between the theory of games and most experimental gaming concerns the role of external symmetry. In most instances the game theorist assumes that unless otherwise specified all players are equally endowed with intelligence and all personality factors. They are a homogeneous population in all ways except those specifically noted in the description of the game. As can be seen from the literature on the Prisoner's Dilemma, though, a large part of sociopsychological investigation is concerned with examining how players who differ in many ways behave in symmetric games. The game-theoretic solutions applied to impersonal players then provide benchmarks for the measurement of the behavior of differentiated players.

Repeated plays and information When games are played repeatedly, the information aspects of the game become critical and can easily become extremely complex. The discussion in chapter 3 gives some indication of the nature of the problems and their treatment. Even simple solution concepts such as the noncooperative equilibrium (which serves as the basis for much experimental work) need extremely careful redefinition when repeated games are considered (see chapter 9).

The game theorist's usual assumptions explicitly ignore the coding and representational problems encountered when complex patterns of information must be displayed. Furthermore, although some work has been done under the explicit assumption of limited ability to process information, in general this is not assumed.

Most experimental games are run with repeated trials, and it is clear from the evidence that the static game-theoretic solutions are not adequate to explain behavior in repeated games. The concept of a strategy for a game in extensive form (see chapter 3) must be modified considerably if it is to reflect behavior in multiperiod games. Rapoport and Chammah (1965), Suppes and Carlsmith (1962), and others have suggested Markov models of player behavior in dynamic games.

Payoffs, utility, and side payments In chapters 3 and 4 we discussed the measurement of utility and interpersonal comparisons of utility. When experimental games are run, it is important that the experimenter distinguish between outcomes and the worth of outcomes.

We ask whether experimental evidence can help answer any of the following questions: Do individuals behave as though they had only preference orderings over the individual outcomes from a game? Do they behave as though their preferences could be represented by a utility function specified up to a linear transformation? Are interpersonal comparisons important? If, for example, the answer to the first of these questions is yes, we would expect that the three matrix games shown below would be played the same way (if no side payments are allowed):

	1	2
1	5, 5	−9, 10
2	10, −9	0, 0

	1	2
1	5, 50	−9, 100
2	10, −90	0, 0

	1	2
1	0, 0	−14, 5
2	5, −14	−5, −5

There is evidence that transformations on the outcomes do make a difference to the way players behave (Rapoport and Chammah, 1965; Rapoport, Guyer, and Gordon, 1976). Interpersonal comparisons and symmetry considerations appear to be applied frequently at the level of outcome and payoff. Thus individuals may argue that a dollar to one is different from a dollar to the other.

When games are played repeatedly, a new problem in the control of payoffs emerges. Should the players be paid after each trial, should they be required at least to add up score after each trial, or should all the book balancing and payments come at the end?

Many of society's decisions are made by individuals acting as fiduciaries for others. An important question that can be answered experimentally is how play is affected by having the payoffs made to individuals other than the players.

Most matrix-game experiments are run without allowing direct payments between the players. The numbers in the matrices represent the points or the money each will receive without any opportunity to make transfers. When the payoff matrices are nonsymmetric and side payments are not permitted, "equitable" outcomes may not be achievable. When payments can be made on the side, accounts can be evened up.

Symmetry, salience, and artifacts There is a tendency in applications of mathematical methods to the behavioral sciences to strive for a degree of generality that may be misleading. Thus it is easy to consider a set of prospects over which an individual has a complete preference order, without specifying what those prospects are. Unfortunately the individual's ability to evaluate and operate with sets of prospects may very well depend specifically upon what they are. A theory that fits consumer choice for staple foods may not be adequate for problems involving political or aesthetic choice.

It is my belief that the structure imposed on many economic problems and on gaming experiments by the use of goods, points, or money has a considerable effect upon individual decisions. Furthermore, as Schelling (1960) argued and experimental psychologists have learned, cues, points of salience, and symmetries manifested in the prospects or the set of outcomes can easily influence choice. The selection of "lucky numbers," the upper left or right cell of a matrix, the midpoint of an array, etc., all seem to be manifestations of complex human information-coding processes. Experiments with simple games have illustrated the important role played by symmetry, cues, salient points, and other artifacts (Stone, 1958; J. W. Friedman, 1967).

Communications An important distinction made in the construction of formal game models is between information flows specified as part of the rules of the game and communication and discussion that take place outside of the formal structure.

Many of the matrix-game experiments are run with no outside communication. The players may not see or know their competitors, and the only information they receive is numerical information concerning the outcomes of previous plays. In rare instances are players permitted to know the identity of the competitor, to see each other in action, and to talk to each other or otherwise communicate.

The modeling of verbal and other nonnumerical forms of communication poses deep problems for both the game theorist and the experimental gamer.

Verbal and other types of briefing The difficulties encountered in modeling or controlling communication among players are also all encountered in briefing. Briefings are usually given by means of written and/or spoken instructions. They are part of the structure of any game that is played, and their effect on the nature of play may be considerable (see R. I. Simon, 1967; Shubik, Wolf, and Poon, 1974), yet for even the simplest of games their role is not reflected in the formal description of the game.

Learning, experience, and aspiration Most of the work in game theory has been based upon the model of an intelligent individual with given preferences. Behavior may be encountered which resembles learning and change in aspiration; however, the current game-theoretic models of man do not provide us with adequate representations of these phenomena.

12.4.2 Simple matrix-game experiments

Although most experimental gaming has been done using 2×2 matrix games, in some instances larger matrices and other representations have been used. These experiments have been conducted under various conditions of briefing, communication, and repetition.

Some experiments have involved composite players consisting of more than one individual; in other words a player may be a team.

Other experiments have studied games with more than two players. Some have allowed for side payments; others have ruled out side payments. Still other games have had fairly complex and not rigidly defined sets of rules.

A natural division of the work in experimental gaming is into experiments that concentrate mainly on 2×2 matrix games and those that employ more complicated games.

Structure of the 2×2 matrix game We limit ourselves to four distinct outcomes for each player, consider only strong preference orderings, and adopt the convention that the first player's ordering of outcome preferences is given by $a_1 > b_1 > c_1 > d_1$ and the second player's by $a_2 > b_2 > c_2 > d_2$. Rapoport and Guyer (1966) have shown that in this case there exist 78 strategically different 2×2 matrices which can be generated by permuting the entries in the matrix shown below (taking into account the various symmetries):

	1	2
1	a_1, d_2	b_1, a_2
2	c_1, c_2	d_1, b_2

Further discussion of the classification of 2×2 games has been given by Guyer and Hamburger (1968), Harris (1969), and Moulin and Vial (1978).

Two-person strictly competitive games The maxmin strategy has been put forward as a strong normative recommendation for how an individual should play in a two-person constant-sum game. There exists a certain amount of evidence on how experimental subjects do in fact play. Many papers deal with either two-person constant-sum games or with games in which the goal of the players is to maximize the difference in their scores (Lieberman, 1960; Morin, 1960; Brayer, 1964; Littig, 1965; Malcolm and Lieberman, 1965; Gallo, 1966, 1969; Marwell, Ratcliff, and Schmitt, 1969; Shubik and Riese, 1972).

We noted in chapter 10 that a two-person non-constant-sum game is transformed into a zero-sum game when individuals strive to maximize score difference. The point or money payoffs are outcomes. When the stakes are small or merely symbolic, the goals pursued by the players may be better reflected in some aspects of status such as score difference.

Because of the difficulties in controlling motivation and in giving large rewards, it is important that in any non-constant-sum matrix-game experiment, player behavior be checked against the possibility that they are maximizing the difference in score.

The Prisoner's Dilemma The way in which most experimenters define the Prisoner's Dilemma implicitly assumes at least a measurable and usually an interpersonal comparison of utility, points, money, or payoffs. For example, Rapoport and Chammah (1965) define the game in terms of a symmetric matrix and impose the condition that $a + d < 2b$ (where dropping the subscripts indicates that the values are the same for each player):

	1	2
1	b, b	d, a
2	a, d	c, c

We can avoid interpersonal comparison, but we still need to utilize the measurability by assuming that, for $0 \leq r \leq 1$, $[ra_1 + (1 - r)d_1] < b_1$ and $[(1 - r)a_2 + rd_2] < b_2$. This disallows joint mixed strategies alternating between cells (2, 1) and (1, 2) that are better for both players than the payoff from playing (1, 1).

It is not our intention to provide a detailed discussion of the extensive literature on the Prisoner's Dilemma; an analysis and a large sample of references are given by Rapoport, Guyer, and Gordon (1976).

A series of studies has examined the effects of payoffs on play. These include the influence of the original stakes on behavior, money versus electric shock as rewards, large rewards, and systematic variations in the point score in the payoff matrices.

Differences in behavior among different types of players have been examined. Topics studied include the effect of race on strategy; the way prisoners play the Prisoner's Dilemma; experiments with high school students, with children, and with children compared to adults; the effect of sex differences and level of cooperation; and the use of performance in the Prisoner's Dilemma game as a measure of trust between schizophrenics.

One study, not experimental but empirical, has taken evidence from the courts on codefendant cases and plea bargaining. The results are not in accord with any simple noncooperative-equilibrium explanation (Forst and Lucianovic, 1977).

A "tournament" has been held in which game theorists, social psychologists, and others have suggested strategies for Prisoner's Dilemma (Axelrod, 1980): "tit-for-tat" and "an eye for an eye" appeared to fare well against all the other rules suggested.

Other 2 × 2 games Although Prisoner's Dilemma is the favorite experimental game, it is by no means the only one that has been used. Furthermore, whereas the social psychologist may be interested in how different types of individuals play the same game, the game theorist is interested in how the same individuals play different games.

We have already noted the problems and difficulties associated with the noncooperative-equilibrium solution. If more properties are imposed on the equilibrium point, though, a stronger case can be made for it as a predictor of behavior. In particular, some of the requirements might be: (1) uniqueness; (2) symmetry; (3) Pareto optimality; and (4) existence of dominant strategies. Any game that has a unique noncooperative equilibrium point that is Pareto-optimal and is formed by strategies that dominate other strategies is clearly one in which even a relatively cautious game theorist might venture a guess concerning player behavior. If the payoffs are symmetric as well, or if the players do not know each other's payoffs, this might even strengthen the prediction. Examples of two matrices, the first with properties 1, 3, and 4 and the second with all four properties, are shown below:

5, 6	−3, 4
−1, 2	−4, 3

5, 5	−3, −3
−3, −3	−4, −4

An experiment by Shubik (1962b) in which the players were not informed of their partners' payoffs provides some evidence to support as predictors of behavior outcomes that are not only noncooperative equilibrium points but have combinations of these other features. An excellent overall summary of experiments with the 2 × 2 game is given by Rapoport, Guyer, and Gordon (1976).

12.4.3 Other gaming experiments

Much of the other experimentation using gaming has been associated with economics and political science and has already been discussed in section 12.2.1. It remains to note a few of the variants from the 2×2 matrices that have been considered. Experiments have used matrices ranging from 3×3 up to 20×20 (Lieberman, 1960; Fouraker, Shubik, and Siegel, 1961), although the larger matrices have been generated from a regular and rather special structure.

Some two-person games that could easily be presented in matrix form have been described in other forms. Thus Stone (1958) used a payoff diagram, and Deutsch and Krauss (1962) utilized a scenario concerning truck driving complete with a map for a problem that could be abstractly represented by a matrix game.

One of the most important control variables in the study of two-person non-constant-sum games is the level of communication and interpersonal interaction between the players. Do they each know who their fellow player is? Do they know each other? Can they see each other? Can they talk with each other? A certain amount of work has been done controlling and varying these conditions (Hoggatt, 1969).

There is now a small but growing literature on games in characteristic-function form. Much of it deals with three-person games, but some work has involved larger numbers (Kalisch et al., 1954; Gamson, 1961b; Riker and Niemi, 1964; Shubik, 1975c). At least one solution concept, the bargaining set, was motivated by an attempt to explain the behavior observed in some experiments with games in characteristic-function form. A survey of this experimental literature is given elsewhere (Shubik, 1975c).

It is in general difficult to run experiments with more than two or three players or teams. Most such attempts have come in experiments designed to study the effect of numbers on economic competition, and these have been noted in section 12.2.1. Possibly the largest group studied in an experimental situation was a simulated stock market game with over 100 participants (Shubik, 1970c).

12.5 Game Theory and Operations Research

Until recently game theory has often been equated with the theory of two-person zero-sum games. *The* game-theoretic solution, for those who are only casually acquainted with the subject, is the maxmin solution. Game theory in the public mind is even today intimately associated with war, computers, and nuclear bombs as part of a Dr. Strangelove package. But two-person zero-sum games form an extremely small part of the theory of games. They are of little interest to the behavioral scientist. Nevertheless there is some basis for the popular view. In particular, there have been a number of applications of two-person zero-sum game theory to military problems.

In other parts of operations research a certain amount of interest in game theory arises from a formal mathematical analogy between a two-person zero-sum game and a linear program and its dual. There have been some applications of two-person zero-sum game theory to capital budgeting, and it is possible to model some tactical problems in economic warfare (such as the location of stores and the allocation of advertising expenditures) in terms of a zero-sum game.

One-person decision making under uncertainty can be formally treated as a two-person zero-sum game in which Nature is the opponent. [For the large related literature on statistical decision theory see Raiffa and Schlaiffer (1961).] We may note that in the operations research applications of game theory listed, the stress has been directly or indirectly on the two-person zero-sum game. This is not completely the case, but it is strongly so. Several exceptions are discussed below. There is an extensive literature on bidding, and several references have been given in section 12.2.1. Much of the work is based on models of the bidding process in terms of an n-person nonconstant-sum game. The solution concept used is predominantly the noncooperative equilibrium or some modification to be applied to repeated bids.

12.5.1 Military applications

Three types of zero-sum games have proved to be of military interest and have been used extensively in planning and in weapons evaluation. They are (1) allocation games, (2) duels, and (3) games of pursuit or search games. Thus in military operations research (as contrasted with the behavioral sciences) the emphasis is upon the application of two-person zero-sum theory and, for some problems, dynamics.

Allocation games A valuable summary of the work on allocation games has been provided by Dresher (1961). A popular example of this type of game, known as the Colonel Blotto game, has even been presented in the popular press (McDonald and Tukey, 1949). An example was given in section 8.2.4. This game is the prototype (see also Beale and Heselden, 1962; Blackett, 1954, 1958) of a series of games which have been used to study strategic air war, tactical air war, and target prediction.

EXERCISE

12.1 (formulated by Dresher). *There are n targets* $A_1, A_2, \ldots, A_n$. *They have values* $a_1, a_2, \ldots, a_n$, *respectively. Assume* $a_1 > a_2 > \ldots > a_n$. *Blue has one attacking unit that it can allocate to a target. If the target is undefended, it will be destroyed, and Blue gains* a_i *for target* A_i. *Red has one defending unit. If a defended target is attacked, the payoff is* pa_i, *where* $0 < p < 1$. *This amount is the expected damage to the target. Solve this game for optimal strategies.*

Another direct military application concerns the target assignment of a strike force against missile emplacements when the attacker does not know in which emplacements the missiles are stored.

Berkovitz and Dresher (1959, 1960a; Dresher, 1961) present a game-theoretic analysis of tactical air war involving tactical air forces in an array of theater air tasks. These include counterair activities, air defense, and ground support. Caywood and Thomas (1955) have studied fighter bomber combat, Fulkerson and Johnson (1957) other aspects of tactical air war, and Shubik and Weber (1981a) systems defense.

A general discussion of the military applicability of two-person zero-sum games has been provided by Haywood (1950, 1954). Other references of note include Beresford and Peston (1955), Johnson (1964), Dalkey (1965), Cohen (1966), Applegren (1967), and Suzdal (1976).

The literature on war gaming is covered by several bibliographies and is certainly related to the application of game theory to tactical studies (Riley and Young, 1957; Shubik, 1975b; Brewer and Shubik, 1979).

Duels: Lanchester equations or behavioral models Work on duels includes two bodies of somewhat different yet related literature. There is a considerable amount of writing on differential-equation models of combat characterized by variants of the Lanchester equations (Lanchester, 1916; Morse and Kimball, 1951; Engel, 1954; Driggs, 1956; Brackney, 1959; R. H. Brown, 1963; Helmbold, 1964; D. G. Smith, 1965; Taylor, 1974, 1979), an example of which is

$$\begin{aligned} dM/dt &= P - AN - CM, \\ dN/dt &= Q - BM - DN, \end{aligned} \qquad (12.1)$$

where M and N are the number of friendly and hostile troops at time t, P and Q are the rates at which new troops enter, A and B are combat loss rates, and C and D are operational loss rates (from accidents, weather, etc.).

These types of behavioristic damage exchange-rate models have been extended to non-zero-sum strategic combat models by L. F. Richardson (1960a,b). At this level of macromodeling it is easy to wax poetic and draw analogies that are attractive but unsubstantiated between the games, escalation, and the outbreak of war. Different views and interpretations of mathematical models of this type and of game theory in the study of conflict are given by Wohlstetter (1964) and Rapoport (1960). When escalation is considered, the underlying model is usually a non-constant-sum game. This is exemplified by the "dollar auction game" (Shubik, 1971b). Military models involving non-constant-sum games are discussed below.

Duels: Two-person zero-sum games

"The race is not always to the swiftest, nor the battle to the strong—but that's the way to lay your dough."—Damon Runyon

Virtually all of the game-theoretic work on duels has been devoted to two-person zero-sum games with nondenumerable sets of pure strategies for the duelists. An example of a simple duel was given in chapter 2.

In actual plane-to-plane, tank-to-tank, or man-to-man combat many important human factors undoubtedly play a large role in determining the outcome in spite of differences in weaponry. Nevertheless, assuming these factors to be equal on both sides, it is still important to be able to evaluate the relative effectiveness of the weapons. Hence the branch of game theory dealing with duels is directly applicable to weapons evaluation.

The study of duels can take into account differences in information. A simple example illustrates this. Consider two individuals walking toward each other with six-shooters; optimal strategies will vary depending upon whether they have silencers on their revolvers and whether one knows whether or not the other has a silencer. The literature distinguishes between silent and noisy duels (Belzer, 1948; Blackwell, 1948; Bellman and Girshick, 1949; Blackwell and Girshick, 1949; Glicksberg, 1950; Shapley, 1951a; Karlin, 1959; Matheson, 1967; D. G. Smith, 1967; Dresher, 1968; Fox and Kimeldorf, 1968, 1969, 1970; Sweat, 1969; Bhashyam, 1970; Ancker, 1975).

Duels may be regarded as belonging to a somewhat more general class of games which might be classified as "games of timing" (Shiffman, 1953; Dresher and Johnson, 1961).

Although there has been some work dealing with groupings of weapons and forces in which the emphasis is on allocation and an examination of doctrine, there has been little game-theoretic work designed to assist in the running of training or operational war games that deals with problems of doctrine or the study of communications or command and control.

Non-constant-sum games: Escalation and inspection We have argued in chapters 6–11 that no single solution concept can be applied successfully to non-constant-sum games. When one considers the complex dynamic problems that must be faced in analyzing escalation, nuclear weapons policy, disarmament agreements, or other aspects of top-level national military policy, it appears unlikely that mathematical models can offer more than some insight into various aspects of these problems. Dalkey (1965) has analyzed escalation as a special form of non-constant-sum game using a noncooperative solution.

Dresher (1962), Davis (1963b), Maschler (1966a), and several others have considered the inspection problem. The writings of H. Kahn (1960), Schelling (1960a), Wohlstetter (1964), and others are related to the type of strategic approach suggested by game theory but cannot strictly be classified as work in game theory.

Differential games: Search and pursuit games Possibly the earliest formulation of a differential game was that of Paxson in 1946, describing a destroyer firing at a maneuvering submarine.[1] This problem was shown to von Neumann, who not only sketched an approach but even did some machine coding for his as yet unborn computer. Since that time there has been considerable work on search-and-pursuit and evasion problems, including a large literature from the Soviet Union.

In chapter 2 an example of a game of pursuit was given. The literature includes lurid examples such as the homicidal driver and the pedestrian trying to avoid being killed (Isaacs, 1965), but more frequently the problems are posed abstractly or in terms of antisubmarine warfare or other specific military situations (Johnson, 1964; Pontryagin, 1966, 1970; Charnes and Schroeder, 1967; Cocklayne, 1967; Petrosjan and Murzov, 1967; Krasovskii, 1968, 1970; Krasovskii and Tretyakov, 1968; A. Friedman, 1970, 1971; Petrov, 1970; Nikolskii, 1971).

Related to these applications has been a growth of interest in differential games both in general and in connection with optimal control theory. We noted in chapter 3 and stress again here that the information assumptions and the implicit definition of strategy used in this work are far more specialized and restrictive than those usually employed in game-theoretic analysis. There are several books on differential games (Isaacs, 1965; Blaquiere, Gerard, and Leitman, 1969; A. Friedman, 1971; Kuhn and Szegö, 1971; Blaquiere, 1973; Zauberman, 1975), though none of these is confined to military applications.

12.5.2 Nonmilitary operations research

Although there have been far fewer direct applications of the theory of games to nonmilitary operations research than to military problems, there is a growing literature in the field. The topics that have received most attention have been bidding and contract submission (Shubik, 1955b, 1960; Stark and Rothkopf, 1979). This work has already been noted in section 12.2.1.

Two-person zero-sum games are of limited use in analyzing most industrial problems, but there are two classes of problems where they are of some use. The first consists of problems that can be modeled as one-person non-constant-sum games and hence treated as two-person zero-sum games if a maxmin criterion for a game against Nature is deemed acceptable. Analysis of this type is closely related to the decision-theoretic treatment of games against Nature. Among the earliest examples of application are Bennion's (1956) treatment of capital budgeting and Blackwell and Girshick's (1954) discussion of the use of maxmin in sampling.

The second class of problems are those which can be modeled to a good first approximation as actual two-person zero-sum games. (There are also several programming problems which can by formal mathematical analogy be considered equivalent to two-person zero-sum games.) Among the two-person game models is a pricing and quality problem studied by Boot (1965). Several advertising expenditure allocation and media mix problems have also been considered as zero-sum games (Gillman, 1950; L. Friedman, 1958; Shakun, 1965). This can be justified once the advertising budgets have been fixed. The non-zero-sum component of the problem is to decide the overall size of expenditures (Baligh and Richartz, 1967); the remaining allocation problem involves strict opposition. Several other problems have been considered, including location selection between competing banks or supermarkets. The optimum-assignment problem, which involves matching individuals of different talents with a set of different jobs (von Neumann, 1953a), provides an example of a programming problem that can be formally regarded as a game.

Several models of advertising and pricing have been based on non-constant-sum games (H. Mills, 1961; Bass et al., 1961; Shubik, 1962a; Eisenman, 1967; Bell, 1968).

Another group of applications have utilized the Shapley value (see chapter 7) to consider decentralization, joint costs, and incentive systems. The axioms used by Shapley have a natural interpretation in terms of accounting. This was first suggested by Shubik (1962a) with regard to the design of incentive-compatible internal corporate cost-accounting schemes. Since then applications using the value or the nucleolus have been made to joint cost and revenue problems such as air pollution regulation, internal telephone billing rates, public utility pricing, aircraft landing fees, water resource development, and pricing in public investments in general (Littlechild, 1970, 1974; Loehman and Whinston, 1971; Littlechild and Owen, 1973, 1976; Bird and Kortanek, 1974; Suzuki and Nakayama, 1976; Littlechild and Thompson, 1977; Billera, Heath, and Raanan, 1978).

The accounting literature on cost allocation gives more direct consideration to the use of game-theoretic methods including the core as well as the value (Jensen, 1977; Hamlen, Hamlen, and Tschirhart, 1977; Callen, 1978).

The value provides a means for evaluating the premium an individual should be willing to pay for a block of shares in a takeover bid or control fight; it is also useful in the study of voting and corporate directorships. Game-theoretic analysis has also been applied to merger agreements (Glasser, 1958) and to insurance problems (Borch, 1962, 1968).

The role of business games (Bellman et al., 1957; Bass, 1964; Jackson, 1968) has already been noted; the modeling techniques suggested by formal game theory have played an important part in their construction.

Game-theoretic thinking has been utilized in studies of regional cooperation (Gately, 1974), customs unions, OPEC, international agricultural planning, and cartels in general. Many of these working papers containing "scenario bundles" (a phrase suggested by Selten) are designed more to help formulate a problem and ask the right questions than to produce a formal answer. The mixture of qualitative and quantitative model building for competitive situations suggested by McDonald (1975) has applied strategic value in and of itself. The application of game theory does not always imply much formal calculation.

Relative to military operations research, to economics, and to political science, the direct applications of game theory to operations research are small, but they are considerably larger than suggested by H. Wagner (1969) in his survey of operations research. Furthermore the indirect effect of the introduction of new tools of analysis and new ways of thinking about multiperson decision problems is undoubtedly substantial.

12.6 Game Theory in Sociology, Anthropology, Biology, Other Behavioral Sciences, Law, and Ethics

12.6.1 Sociology and anthropology

In sociology and anthropology several problems appear to be amenable to game-theoretic analysis (Buchler and Nutini, 1968). Coalition formation among tribesmen has been studied (Barth, 1959). Games of status (R. E. Anderson, 1967; Shubik, 1971c; Laing and Morrison, 1973) offer a model to study pecking orders, competition for status, and phenomena such as potlatches. The roles of gifts (Mauss, 1954) and primitive money are undoubtedly closely related to the different conditions on transferability encountered in the theory of games. There has been some game-theoretic work on primitive economies (Salisbury, 1968).

Although the von Neumann–Morgenstern stable-set solution was phrased in terms of statics and does not appear to be a particularly desirable solution for economic application, it does suggest some interesting analogies and concepts relevant to the study of sociology. In particular, it stresses a totally different form of stability than is usually encountered in the physical sciences or in physical-science analogies to the behavioral sciences. The stability of the von Neumann–Morgenstern stable set comes about through an intricate balance of bargains and counterbargains implicit in the structure of the society. For example, even in the symmetric three-person constant-sum game there exist nonsymmetric or "discriminatory" solutions whose stability derives from the property that if any two players decide to enforce their social norm, the third cannot force them to deviate from it. Which two will enforce the norm cannot be predicted a priori in the game-theoretic formulation. Nonsymmetric features external to the model, such as race or religion, might determine which grouping is formed.

It is easy to give a broad and analogy-rich interpretation of a relatively limited mathematical model, but it is dangerous and misleading to do so without appropriate cautionary remarks. We must stress that although the interpretation in terms of discrimination is suggestive, we do not have a sufficiently developed dynamics to carry our analysis much beyond the level of making the analogy. The most important aspect of the model is its conceptualization of stability. Even without a fully developed dynamics, any attempt to discuss the effect of a "virtual displacement" shows that the stability being considered in a cooperative game is far different from that encountered in physical systems.

Even the classification of games in characteristic-function form provides a useful sociological insight. In particular, if the interrelatedness of a group of individuals can be described by an inessential game (see chapter 6), that group does not form a society. A game is inessential whenever its characteristic function is flat, and in this case there is nothing to be gained by communication. There is no need for language since the strategies directly determine the optimal outcome for each player. Of course, this does not mean that their fates are not related. For example, a two-person zero-sum game is an inessential game. Here, however, the interests of the players are diametrically opposed; in fact the best definition of the word "opponent" may be the status of one player with respect to the other in a two-person zero-sum game.

In an inessential game either players are strategically isolated or some (or all nonoverlapping) pairs of players are opponents.

Both in the definition of solution in a game-theoretic model and in experimental gaming, language and communication play an important role. Two-person zero-sum games are easier to experiment with than non-zero-sum games because language plays virtually no role (the only role being "gamesmanship"). Even for the simplest of non-constant-sum games the nature of the briefing and the possibilities for verbal communication have considerable influence.

The sociology of conflict as viewed by the theory of games has been discussed by Bernard (1950, 1954, 1965). Work on gaming, part of which has been discussed under social psychology, merges into sociology in the investigation of games with three or more parties and in the use of gaming models to study social interaction (Bond and Vinacke, 1961; Gamson, 1961a,b; J. Davis, 1964; Reichardt, 1968; Pruitt and Kimmel, 1977). It also raises questions concerning the relationship between game theory and learning (H. A. Simon, 1956; Bartos, 1964, 1967).

Other applications to sociology have included Gamson (1966) on game theory and administrative decision making and Sjoberg (1953) on social power. An interesting mix of economics and social institution building is presented by Schotter (1981).

12.6.2 Biology and zoology

In the 1960s and 1970s interest has grown in the application of the ideas of game theory to biology and zoology. A similar phenomenon has been a growth of interest by some economists in sociobiology. At least three aspects of game theory appear to be directly relevant: stochastic games, n-person non-constant-sum games, and the extensive form. Lewontin (1961) has suggested that stochastic games may be of use in the study of problems of evolution not adequately covered by the current theory of population (see also Hamilton, 1964; Trivers, 1971; Maynard Smith, 1974, 1976; J. S. Gale and Eaves, 1975; Selten, 1978). He notes that "the problem of choosing a proper numerical utility associated with a given outcome is a biological rather than a mathematical one." For any application the mathematics must be hand-tailored to fit the needs of the topic to be investigated. Turner and Rapoport (1971) consider the differences in the concepts of preference in economics and biology.

Marchi and Hansell (1973a,b) have considered evolutionary processes and zoological studies in terms of noncooperative game theory.

The Three-Person Duel discussed in chapter 2 raises several basic problems concerning the meaning of the survival of the fittest (Shubik, 1954a). When there are more than two animals involved in a fight for leadership of the group (herd, pack, pride, covey, etc.), how is the nature of the combat determined? A large variety of possibilities correspond to different solutions to n-person non-constant-sum games. The available empirical work suggests that there may be a considerable difference in behavior among species (K. Lorenz, private communication; Maynard Smith, 1974).

A nontechnical introduction to the possible ramifications of game-theoretic thinking in biology is provided by Dawkins (1976), while Selten (1978) develops the mathematical analysis of evolutionary stable strategies and noncooperative equilibria.

Another important feature of game-theoretic thinking as applied to the behavioral sciences in general, and to biology in particular, is the stress placed by the extensive form of the game on the historical process. We observed in chapter 3 that the extensive form, utilizing a tree representation, is extremely redundant for many purposes. For example, there are many different paths to the same position on a chessboard. To a good approximation Chess could be regarded as nonhistorical since the outcome of a game should depend upon the current array on the board and not upon the history of how it was arrived at. (In Chess as it is played, history may be important in that it allows opponents to sense each other's style of play. It is also true that some stalemate outcomes depend on the repetition of a position.) In the tree representation of Chess, though, even if two positions are the same, they have different positions if their histories are different.

A failure to recognize the importance of the difference between a historical process and a simple stochastic process can mislead a researcher into believing that state-strategy and control-theory models of dynamics are necessarily the correct models for biological or zoological processes.

The recent interest by economists in sociobiology is primarily focused on multigenerational models of the socioeconomic process and ideas concerning altruism (Hammond, 1975; Becker, 1976; Hirschleifer, 1978; Kurz, 1978; Shubik, 1981).

The applications of game-theoretic reasoning to biology, sociobiology, demography, and multigenerational models of the economy may eventually be extremely fruitful. Their fruitfulness, however, will depend on the empirical insights developed in these subjects and on the care that is exercised in ad hoc modeling. How much leeway the individual human has to trade among personal survival, survival of the species, individual quality of life, and societal quality of life, poses a series of deep empirical questions. Game-theoretic methods can be of help both in modeling and in exploring solutions, but their use must be judged in application. The class of games of social or economic survival (Shubik, 1959b; Shubik and Thompson, 1959) was designed specifically to reflect the intermix of individual desires to optimize and to survive. The interlinkage between generations, which is of great concern to sociobiologists and those interested in altruism, may be modeled in many different ways. The value of the various models will appear from investigations not yet undertaken and knowledge not yet developed.

12.6.3 Other applications

Applications of game theory to law have included studies of collusion, jury selection, and antitrust problems (Allen, 1956, Birmingham, 1968–69; Leff, 1970, 1976; Shubik, 1956). Beyond that, however, there is considerable scope for the application of game-theoretic reasoning to problems of property, equity, torts, and the whole array of situations involving extensions of the concept of ownership and the treatment of shared rights and commonly held goods. The strong relationship between legal definitions of equal treatment and game-theoretic analysis should allow another useful application in this area.

Although there have been no direct applications of game theory to psychiatry, a few points of contact merit noting. The game theorist and the psychiatrist may be regarded as focusing with different emphases on the same aspects of individual decision making under uncertainty in situations marked by conflicts of control or of goals. Whereas the psychiatrist may be interested in making explicit the nature of conflict that is not consciously perceived by the individual, the game theorist is interested in analyzing the aspects of conflict that must be present when an individual lives in an environment involving others.

Game-theoretic considerations can be used to construct games that highlight certain aspects of stress. For example, the Dollar Auction described in chapter 9 provides an example of a process similar to addiction.

In chapter 7 it was noted that the various value solutions can be viewed as formulations of fair-division principles. In chapter 10 fair-division games were discussed, and the references are not repeated here. There is a growing economic literature on fairness (Pazner and Schmeidler, 1974), which is only tangentially related to game theory. Several works in ethics have made use of game theory in the consideration of distribution. Braithwaite (1955) has specifically examined two-person fair division without interpersonal comparison.

The book by Rawls (1971) on the theory of justice developed a maxmin principle of distribution. Redistribution "should" take place as long as the level of the worst-off member of society can be raised. This measure, although *prima facie* appealing, unfortunately is not invariant under order-preserving transformations of individual utility functions. Harsanyi (1975a) offers a careful critique of Rawls.

There is a growing literature on strategy and ethics (Midgaard, 1970; Rapoport, 1964), and the relationship between man and God in the Bible has been analyzed in game-theoretic terms by Brams (1980).

Finally it must be noted that application is a two-way street. The application of game theory to a particular subject may not only contribute to substantive work but may also enrich the methodology itself.

Note to chapter 12

1. Part of Paxson's unpublished notes:

As far as I know there were few applications of game theory during World War II—not even by Johnny von N. As technical aide on the Applied Mathematics Panel of the OSRD, I was working with most of the country's leading mathematicians who were preoccupied, except for statistics, with rather elaborate but trivial applied work—as in aerial gunnery.

The one example is the airplane/submarine search game formulated and solved by nice heuristic methods by ASWORG. (See John McDonald's book, p. 121ff, or Bull. A.M.S. July 1948, Phil Morse.)

I bought my copy of TGEB in New York at Brentano's in April 1946. I was working at the Naval Ordnance Test Station, China Lake, California. Immediately fascinated, I formulated what was later to be called a differential game. This was a real problem, a duel between a destroyer firing at a maneuvering submarine, with allowance for denial of sonar coverage in the destroyer's wake.

In November of 1946, I spent a day in Princeton with Johnny von Neumann working on this problem. He sketched an approach, and even did some machine coding for his as yet unborn machine. I have saved his notes and flow charts—believed to be the first in the computing business. These have been blown up and shown at World Fairs and the IBM New York show.

In September of 1948, John Tukey showed at the Madison Symposium a three-D solution of the Colonel Blotto allocation game. I don't recall who originated the game itself.

Also in the 50's, Rand was studying the day by day allocation of air power to its various missions by opposing commanders. The movement of the battlefront on the ground was the campaign payoff. All of this was done manually on an analogue computer; i.e. the game was *played* many many times.

Berkovitz and Dresher finally gave an analytic *solution*. From the military point of view, these solutions tend to be extreme and the payoffs rather forced. A player may be told to concentrate all effort against the opposing air force and then shift over to concentrated close support of ground troops. This builds up points on the ground in the closing days of a campaign known to be of fixed length. No air commander could get away with such a strategy. Troops must be supported.

The point generalizes. Applications to military affairs use game formulations which can be solved. The mathematician is Procrustes. I myself am a confessed Áttic highwayman.

Military schools have always taught the matrix approach to match own courses of action against enemy courses of action. But Ollie Hayward then an Air Force colonel showed circa 1953 that Army doctrine for choice was too conservative, min-max rather than randomized saddle-point. This was also an early example of the use of ordinally ranked utilities. . . .

ABM developments fostered renewed interest in game theory. Selma Johnson worked for a long time on defense games in which the defender allocates a limited stock of interceptors to an incoming threat tube of re-entry vehicles and decoys, only probabilistically distinguishable.

These problems are probably unsolvable by current tools. It is a pity that the theoreticians haven't dug deeper in the behavior strategy mire.

About 1970, Mel Dresher solved a new type of duel. Opposing aircraft in air-to-air combat can fire either at the enemy aircraft or at the missiles already fired and coming at them (long range combat, of course). He only adumbrated the method used in his book. The method deserves wider exposure.

Currently, the Air Staff is again interested in tactical airground games. However, ground maneuver in considerable detail is desired. As presently formulated, I think a solution is impossible, without another rental of Procrustes' bed.

However, there is a class of tactical games which I would call Lanchester games worth further study. Assuming perfect information, these games can be formulated and solved in principle by Pontryagin's optimal control methods. The difficulty is that the underlying trajectory equations, embodying Lanchester attrition, are highly nonlinear. Pontryagin's beautiful examples are also beautifully linear.

A general theorem is needed in the field of tactical games. I carped earlier at payoffs maximizing (minimizing) the notion of a battle front line during a campaign of *fixed* length—a fact known to the players at the game's start.

It makes far more military sense to let Red choose an objective a given distance away and then try to minimize the time to achieve it. Blue tries to maximize this time.

Under what general conditions on the game description is it true that the campaign duration can be taken as parameter, the game solved for each value, and the answer for the objective payoff falls out for that duration value corresponding to the distance to the objective? This shouldn't be too hard.

Appendix A Some Details on Preference and Utility

A.1 Sets, Relations, and Orders

Before discussing the technical terminology of relations and orders, let us review the conventional language of elementary set theory, as employed throughout this book. *Sets* may be defined either directly, as

$\{a, b, \ldots\}$ is the set consisting of the things $a, b, \ldots,$

or through some stated property, as

$\{x: A(x)\}$ is the set of things x for which $A(x)$ is true.

The first notation is sometimes condensed with the aid of a vinculum (overscore) when there is no danger of confusion; thus $\overline{123}$ means $\{1,2,3\}$. The *empty set* is denoted $\varnothing$. The *cardinal number* (number of members) of a set S is denoted $|S|$. The *membership* symbols $\in$, $\notin$, $\ni$, $\not\ni$ mean "belongs to" (or "belonging to," etc.), "does not belong to," "has for a member," and "does not have for a member," respectively. The *inclusion* symbols $\subseteq$, $\subset$, $\supseteq$, $\supset$ (written between two sets) mean "is contained in," "is properly contained in," "contains," and "properly contains," respectively. If $S \subseteq T$, then S is a *subset* of T and T is a *superset* of S.

Operations on sets include:

Intersection:	$S \cap T = \{a: a \in S \quad \text{and} \quad a \in T\}$,
Union:	$S \cup T = \{a: a \in S \quad \text{or} \quad a \in T \quad \text{(or both)}\}$,
Difference:	$S - T = \{a: a \in S \quad \text{and} \quad a \notin T\}$,
Direct product:	$S \times T = \{(a, b): a \in S \quad \text{and} \quad b \in T\}$.

Thus the direct (or Cartesian) product of n sets is a set of ordered n-tuples. The *complement of S* (with respect to some designated superset N) is the set $N - S$.

A.1.1 Relations

A *binary relation* $\mathcal{R}$ is a set of ordered pairs (x, y) drawn from a domain $\mathcal{D}$; in other words, $\mathcal{R} \subseteq \mathcal{D} \times \mathcal{D}$. Rather than write $(x, y) \in \mathcal{R}$, however, we write $x \mathscr{R} y$ and say that x bears the relation $\mathscr{R}$ to y. The complementary relation $(\mathscr{D} \times \mathscr{D}) - \mathscr{R}$ is denoted $\not\mathscr{R}$. Relations can have the following properties:

Reflexivity: $x \mathscr{R} x$ for all x in $\mathscr{D}$,

Irreflexivity: $x \not\mathscr{R} x$ for all x in $\mathscr{D}$,

Symmetry: $x \mathscr{R} y \Rightarrow y \mathscr{R} x$ for all x,y in $\mathscr{D}$,

Antisymmetry: $x \mathscr{R} y \Rightarrow y \not\mathscr{R} x$ for all x,y in $\mathscr{D}$,

Completeness: $x \not\mathscr{R} y \Rightarrow y \mathscr{R} x$ for all x,y in $\mathscr{D}$,

Transitivity: $x \mathscr{R} y$ and $y \mathscr{R} z \Rightarrow x \mathscr{R} z$ for all x,y,z in $\mathscr{D}$,

p-Acyclicity: $x_1 \mathscr{R} x_2, x_2 \mathscr{R} x_3, \ldots, x_{p-1} \mathscr{R} x_p \Rightarrow$
$x_p \not\mathscr{R} x_1$ for all $x_1, x_2, \ldots, x_p$ in $\mathscr{D}$,

Acyclicity: $\mathscr{R}$ is p-acyclic for all positive integers p.

Note that a relation may be neither reflexive nor irreflexive, and may be neither symmetric nor antisymmetric. (The empty relation $\mathscr{R} = \varnothing$ is both symmetric and antisymmetric. It is also reflexive, transitive, and acyclic.) 1-Acyclicity is the same as irreflexivity, and 2-acyclicity the same as antisymmetry. A transitive, irreflexive relation is necessarily acyclic. A relation that is both symmetric and transitive (and hence reflexive) is an *equivalence.*

A.1.2 Preferences

A *preference relation* is an arbitrary irreflexive relation, denoted $>$, such that $x > y$ is interpreted to mean that x is preferred to y. There may be an *indifference relation,* denoted $\sim$, associated with $>$; we require that $\sim$ be an equivalence and that it hold only between members x and y of $\mathscr{D}$ that are *interchangeable* in $>$, in the sense that for every z in $\mathscr{D}$, $x > z \Leftrightarrow y > z$ and $z > x \Leftrightarrow z > y$. It follows in particular that $>$ and $\sim$ are disjoint.

Given $>$ and $\sim$, we say that x and y in $\mathscr{D}$ are *incomparable* if none of $x > y$, $y > x$, or $x \sim y$ holds. Unlike indifference, incomparability is not an equivalence. Indeed, though it is symmetric, it is not reflexive, nor is it transitive or even acyclic except for the trivial case where it is empty.

The symbol $\gtrsim$ denotes the union of $>$ and $\sim$, so $x \gtrsim y$ is interpreted to mean that x is preferred-or-indifferent (or *weakly preferred*) to y. Although many investigations take the relation $\gtrsim$ as the basic datum, it does not in general contain all the information conveyed by the pair of relations $>$ and $\sim$. Thus, if we try to recover the separate relations $\succ$ and $\sim$ from $\succsim$ by defining preference by $\mathscr{P} = \succsim - \precsim$ and indifference by $\mathscr{I} = \succsim \cap \precsim$, then we have in general only $\mathscr{P} \subseteq \succ$ and $\mathscr{I} \supseteq \sim$. (Here $\precsim$ is the transpose of $\succsim$; that is, $x \precsim y$ if and only if $y \succsim x$. The symbol $\prec$ may be defined similarly.) It can be shown that equality holds ($\mathscr{P} = \succ$ and $\mathscr{I} = \sim$) if and only if we started with an antisymmetric relation. It is even possible that $\mathscr{I}$, so defined, will not have the interchangeability property with respect to $\mathscr{P}$.

With transitivity the foregoing distinction vanishes. If we start with a transitive irreflexive relation $\succ$, then it is necessarily antisymmetric and there is no trouble. Conversely, if we start with a transitive reflexive relation $\mathscr{R}$, we can then define $\succ = \mathscr{R} - \mathscr{R}'$ and $\sim = \mathscr{R} \cap \mathscr{R}'$ (where $\mathscr{R}'$ denotes the transpose of $\mathscr{R}$) and show that these have all the properties required for preference and indifference and (obviously) yield $\succsim = \mathscr{R}$.

A.1.3 Orders

An *order* (or *ordering*) is a transitive preference relation $\succ$ with an associated indifference relation $\sim$. Equivalently we may take it to be a transitive reflexive relation $\succsim$. We shall use the following characterizations of an order:

it is *complete* if there are no incomparable pairs;

it is *partial* if it is not complete;

it is *strong* if there are no indifferent, unequal pairs;

it is *weak* if it is not strong.

Nomenclature in this area is far from standardized. Thus in the literature the definition of "partial" does not always exclude complete orders, and the definition of "weak" does not always exclude strong orders. "Complete" sometimes means our "strong complete," unless qualified by the adjective "weak." The terms "total" and "linear" are synonymous with either our "complete" or "strong complete," depending on the writer's taste. "Quasiorder" is one name for the $\succsim$ relation of an order; "preorder" is another. A "suborder," on the other hand, is an irreflexive acyclic relation and may not be an order at all. Fortunately most writers take care to define their terms explicitly.

Two other definitions may be of interest. An *interval order* is a strong order that satisfies

$x > y$ and $z > w \Rightarrow x > w$ or $z > y$,

for all x, y, z, w in $\mathscr{D}$.[1] A *semiorder* is an interval order that satisfies

$x > y$ and $y > z \Rightarrow x > w$ or $w > z$,

for all x, y, z, w in $\mathscr{D}$. Intuitively this last condition says that if there is at least one "discrimination level" between x and z, then no w can be incomparable to both x and z. A semiorder $>$ has a natural extension to a complete order $>'$, which may be defined as follows: $x >' y$ means that either $x > y$, or there is a z in $\mathscr{D}$ incomparable to x such that $z > y$, or there is a w in $\mathscr{D}$ incomparable to y such that $x > w$. (See the exercises at the end of section 4.2.2.) Under suitable conditions (e.g., if $\mathscr{D}$ is finite or countably infinite), a utility function can be defined for $>'$ with the property that

$x > y \Leftrightarrow u(x) > u(y) + 1$,

giving numerical expression to the concept of a "just noticeable difference" (see the references in section 4.2.1 and note 7 to chapter 4).

A.2 Rational Nonlinear Utility[2]

Utility is often thought of (misguidedly, I believe) as an ordering system, in essence. Here utility will be a choice operator, indicating the reaction of an organism to a given set of alternatives. This concept of utility can be applied not only to individuals but also to organizations or groups.

Let us suppose that the space of conceivable possibilities is a locally compact convex subset of a real vector space, where probability mixture (randomization) corresponds to taking a convex combination. And let us assume that the actual set of available alternatives is always a compact subset of this space.

The space of possibilities can be metrized, and the space of compact convex subsets of it also has a natural metrization. If S and T are two such subsets, we define

$$d(S, T) = \max_x \min_y d(x, y) \text{ subject to } \{x \in S, y \in T\}$$
$$= \max_y \min_x d(x, y) \text{ subject to } \{x \in S, y \in T\},$$

which is the Hausdorff metric for subsets.

Now let $C(S)$ be a choice operator that picks out from S a compact convex subset of preferred alternatives.

AXIOM 1. *$C(S)$ is a closed convex subset of S.*

AXIOM 2. *If $S_i \to S_0$ as $i \to \infty$, $x_i \in C(S_i)$ for each i, and $x_i \to x_0$ as $i \to \infty$, then $x_0 \in C(S_0)$.*

AXIOM 3. *If $S \supset T$ and $C(S) \cap T$ is nonempty, then $C(T) = C(S) \cap T$.*

An n-person noncooperative game can be quite satisfactorily discussed on the basis of this type of utility for the players.

Let each player have a space $\mathcal{S}_i$ of strategies s_i. An n-tuple $\mathfrak{s} = (s_1, s_2, \ldots, s_i, \ldots, s_n)$ is an equilibrium point if for each i the set $T_i(\mathfrak{s})$ of alternatives of the form $(s_1, s_2, \ldots, s_{i-1}, t_i, s_{i+1}, \ldots, s_n)$, where $t_i \in \mathcal{S}_i$, is such that $C_i[T_i(\mathfrak{s})]$ contains $\mathfrak{s}$, where C_i is the ith player's choice operator. Axioms 1 and 2 suffice for this.

In this fashion one can discuss noncooperative games between a coalition and an individual, or between coalitions.

A.3 Cardinal Utility from Intensity Comparisons

Let the domain of alternatives $\mathcal{D}$ be a nontrivial convex subset of the real line, that is, a finite interval, a half-line, or the whole line. Let preferences in $\mathcal{D}$ correspond to the natural ordering $\geq$ of the real numbers. Let the elements of $\mathcal{D} \times \mathcal{D}$ be related by a weak complete order $\succsim$ (see section A.1); the statement $(x, y) \succsim (z, w)$ will be interpreted to mean that a change from outcome y to outcome x is better than a change from outcome w to outcome z. The following three *intensity axioms* will be assumed for all x, y, z, w in $\mathcal{D}$:

I1: $(x, z) \succsim (y, z) \Leftrightarrow x \geq y$;

I2: $\{x, y, z, w \in \mathcal{D} : (x, y) \succsim (z, w)\}$ is closed in $\mathcal{D} \times \mathcal{D} \times \mathcal{D} \times \mathcal{D}$;

I3: $(x, y) \sim (z, w) \Leftrightarrow (x, z) \sim (y, w)$.

The first axiom ensures *consistency* between the intensity ordering and the preference ordering.[3] The second makes the ordering $\succsim$ *continuous*. The third we call the *crossover* property; it expresses the idea that differences are being compared.

THEOREM. *There exists a function u from $\mathcal{D}$ to the real numbers with the property that*

$(x, y) \succsim (z, w) \Leftrightarrow u(x) - u(y) \geq u(z) - u(w)$

for all x, y, z, w in $\mathscr{D}$. Moreover, this function is unique up to an order-preserving linear transformation.

A proof of this theorem has been given by Shapley (1975a).[4] The idea of the proof is to assign the utility values 0 and 1 arbitrarily to two chosen outcomes a_0 and $a_1 > a_0$, then to use the properties of $\gtrsim$ to determine outcomes $a_2, a_3, \ldots$ and $a_{-1}, a_{-2}, \ldots$ at "equal intensity" intervals, until all of $\mathscr{D}$ has been "staked out." (This may require countably many points.) Each a_n is assigned the utility n. Then a unique point $a_{1/2}$ between a_0 and a_1 is shown to exist such that $(a_1, a_{1/2}) \sim (a_{1/2}, a_0)$; it is assigned the utility 1/2. Continuing in this way, we extend the utility function u to a dense set of outcomes in $\mathscr{D}$ and finally, by continuity, to all of $\mathscr{D}$.

A.4 Cardinal Utility from Probability Mixtures

Possibly the most elegant of the several axiom systems that have been devised to establish a cardinal utility in the presence of risk was put forward by I. N. Herstein and John Milnor in 1953, not long after the original, pioneering work of von Neumann and Morgenstern (see the second [1947] and subsequent editions of *Theory of Games and Economic Behavior*). The following discussion is based closely on the Herstein–Milnor paper, but we have adopted a notational improvement due to Melvin Hausner (1954).

We call the domain $\mathscr{D}$ a *mixture set* if there is a way of associating an outcome $x\mu y$ in $\mathscr{D}$ to every ordered triple (x, μ, y), where x and y are in $\mathscr{D}$ and μ is in [0, 1] (the set of real numbers that lie between 0 and 1, inclusive), so that the following three *mixture axioms* are satisfied for every x, y in $\mathscr{D}$ and μ, ν in [0, 1]:

M1: $x1x = x$;

M2: $x\mu y = y(1 - \mu)x$;

M3: $[x\mu y]\nu y = x(\mu\nu)y$.

The intended interpretation is of course that $x\mu y$ should represent the prospect of having the outcome x occur with probability μ and the outcome y occur with probability $1 - \mu$.

An example of a mixture set is a convex set in a vector space, with $x\mu y$ defined to be the vector $\mu x + (1 - \mu)y$. Mixture sets are more general than convex sets, however, since axioms M1, M2, M3 are not sufficient to guarantee in $\mathscr{D}$ the full linear structure required for convexity.[5] Nevertheless the notion of a linear function still makes sense: f is defined to be *linear on* $\mathscr{D}$ if and only if

$$f(x\mu y) = \mu f(x) + (1 - \mu)f(y)$$

holds for all x, y in $\mathscr{D}$ and μ in $[0, 1]$.

Now let $>$ and $\sim$ be a preference relation and associated indifference relation on $\mathscr{D}$ (see section A.1), satisfying the following three *preference axioms* for all x, y, z in $\mathscr{D}$:

P1: $(>, \sim)$ is a complete order;

P2: $\{\mu: x\mu y > z\}$ and $\{\mu: z > x\mu y\}$ are closed;

P3: $x \sim y \Rightarrow x \frac{1}{2} z \sim y \frac{1}{2} z$.

We shall call a function u a *linear utility for* $(>, \sim)$ if it is linear on $\mathscr{D}$, as above, and satisfies

$$x > y \Leftrightarrow u(x) > u(y)$$

for all x, y in $\mathscr{D}$. (This at once implies $x \sim y \Leftrightarrow u(x) = u(y)$ because of P1.)

Axiom P1 speaks for itself. Axiom P2 is a way of stating that the preferences are continuous with respect to probabilities. Axiom P3 is a "rule of substitution between indifferent alternatives" (Marshak, 1950).[6] It turns out that the conclusion $x\mu z \sim y\mu z$ needs to be assumed only for one intermediate value of μ, such as $\mu = 1/2$; it then follows for all others with the aid of M1, M2, M3 and a limiting argument using P2.

THEOREM (Herstein and Milnor). *If $\mathscr{D}$ and $(>, \sim)$ satisfy M1, M2, M3 and P1, P2, P3, then a linear utility for $(>, \sim)$ exists and is unique up to an order-preserving linear transformation.*

The general idea of the proof is not unlike that of section A.3: take two elements a_0 and $a_1 > a_0$ of $\mathscr{D}$, arbitrarily assign them utilities 0 and 1, respectively, and then show that a sort of linear interpolation and extrapolation can be carried out that will determine uniquely the utilities of all the other elements of $\mathscr{D}$.

Notes to appendix A

1. Jamison and Lau (1973, p. 902) unaccountably give the name *semitransitivity* to this property, although it obviously implies transitivity for any irreflexive relation $>$ (*Proof*: set $y = z$).

2. This section by John F. Nash, Jr., was written as an informal note dated August 8, 1950; it is reproduced here with the permission of the author.

3. The symmetric form of I1, namely

$$\text{I1}': \quad (x, y) \gtrsim (x, z) \Leftrightarrow y \leq z,$$

has been omitted for logical economy, since it follows from I1, I2, and I3. From the standpoint of economic logic, however, I1 and I1′ are equally acceptable, and the proof of the main theorem can be shortened if we take both as axioms (see Shapley, 1975a).

4. Related theorems have been proved by Suppes and Winet (1955), Debreu (1959b), Pfanzagl (1959), Scott (1964), and others; but the crossover axiom I3 appears to be new in this context.

5. For example, neither the associative law for probability mixes,

$$\text{M3}': \quad [x\mu y]\nu z = x(\mu\nu)\left[y\left(\frac{\nu - \mu\nu}{1 - \mu\nu}\right)z\right] \quad \text{if } \mu\nu < 1,$$

nor the cancellation law,

$$\text{M4}: \quad x\mu y = x\mu z \Rightarrow y = z \quad \text{if } \mu < 1,$$

can be derived from M1, M2, M3. Adding M3′ and M4 to the list of axioms makes $\mathcal{D}$ what Hausner called a mixture *space,* which, as he proved, is always representable as a convex set in some linear space, possibly infinite-dimensional. [Rather than use the system M1, M2, M3, M3′, M4, Hausner, who was not interested in mixture *sets,* used the equivalent but simpler system M2, M3, M4, M5, where M5 is the obviously desirable property

$$\text{M5}: \quad x\mu x = x.$$

The point is that M1, M2, M3 suffice to prove M5, but M1, M2, M3′, M4 do not; this was the only reason for retaining M3 after adding the apparently strong M3′. Including M5 then makes M1 redundant as well.]

6. This axiom has also been called the "strong independence axiom" and the "sure-thing principle" (see Rothenberg, 1961, pp. 234–250).

Appendix B Bibliography on Stable-Set Solutions

More than 100 articles and books dealing with stable sets are listed in this appendix, in chronological order and with coded descriptors. The latter may be used to generate more specialized bibliographies for four-person games, finite solutions, political applications, and so forth. A few unpublished items are included, primarily for historical reasons. The listing goes through 1973. The following abbreviations are used:

1. Game classifications

3prs	three-person
4prs	four-person, etc.
nprs	also general finite-person games
infp	infinite-person
comp	composite or compound
conv	convex
mark	market
$n - 1$	coalitions of less than $n - 1$ players lose
quot	quota (or m-quota)
simp	simple (or semisimple, extreme)
symm	symmetric (or partially symmetric)
notr	nontransferable utility
part	partition-function form
abst	abstract (nongame)

2. Solution classifications

barg	bargaining curves
disc	discriminatory
fin	finite (main simple, etc.)
inf	also infinite solutions
noex	nonexistent
path	pathological
prod	product
symm	symmetric
uniq	unique

3. Other classifications

book	book
thes	thesis
surv	survey
bset	relationship to bargaining set
calc	calculation
card	cardinality of solutions
core	relationship to core
defs	definitions extending the theory
exam	example or counterexample
exp	experimental games

kern	relationship to kernel	vari	variant of classical theory
loc	location of solutions	econ	application to economics
noco	relationship to noncooperative theory	poli	application to politics or voting
solv	solvable classes of games		

	Descriptors		
Authors and Dates	Games	Solutions	Other
von Neumann–Morgenstern (1944)	3prs simp	fin symm	book solv
	4prs symm nprs mark comp abst	inf barg disc	surv vari defs econ
Copeland (1945)			surv
M. Richardson (1946)	abst		solv
Wald (1947)			surv econ
McKinsey (1952a)			book surv
(1952b)			surv exam
Milnor (1952)			loc
Shapley (1952a)			vari defs
Bott (1953)	simp symm	symm	
Gillies (1953a)	simp symm	disc barg	
(1953b)			loc
(1953c)	comp abst simp	disc uniq	thes loc defs solv core vari
M. Richardson (1953a)	abst		solv
(1953b)	abst		solv
Shapley (1953a)	4prs quot nprs	barg	solv
(1953b)			card loc core
Shapley–Shubik (1953)	notr		defs
Kalisch et al. (1954)	4prs		exp

Authors and Dates	Descriptors		
	Games	Solutions	Other
W. H. Mills (1954)	4prs		calc
Isbell (1955)	simp	fin	
M. Richardson (1955)	abst		solv
Shapley (1955b)	mark symm	barg	econ
M. Richardson (1956)	simp symm	fin symm	
Berge (1957b)	nprs abst		book loc core
Luce–Raiffa (1957)			book surv
Berge (1958)	abst		book
Burger (1959)	mark		book surv
Gelbaum (1959)	symm	symm	
Gillies (1959)	comp abst simp	disc uniq	defs solv core vari loc
Griesmer (1959)	simp	fin disc	
Gurk (1959)	5prs simp	disc	
Gurk–Isbell (1959)	simp	fin	
Harary–Richardson (1959)	abst		solv calc
Isbell (1959)	notr 3prs	fin	vari defs
Kalisch (1959)	quot		
Kalisch–Nering (1959)	infp	noex	defs exam
W. H. Mills (1959)	4prs		calc loc
Nering (1959)	4prs symm	symm	
Peleg (1959)			solv
Shapley (1959a)	simp	path	
(1959b)	mark symm	barg uniq	core loc
Shubik (1959a)	mark	barg	econ
Vickrey (1959)	simp 4prs		vari defs
Aumann–Peleg (1960)	notr		defs core
Hoffman–Richardson (1961)	simp symm	fin	
von Neumann–Morgenstern (1961)	simp $n-1$ symm	symm	loc calc

Authors and Dates	Descriptors		
	Games	Solutions	Other
Bondareva (1962)		uniq	core solv
M. Davis (1962)	infp symm simp	symm noex	defs exam
Riker (1962)			poli
Thrall (1962)	part		defs
Bondareva (1963)	quot nprs	uniq	core solv
Lucas (1963)	part 4prs 3prs $n-1$		thes
Peleg (1963a)	notr nprs 3prs		defs solv core
Thrall–Lucas (1963)	part nprs 3prs		defs
Eisenman (1964)	part		thes defs vari
Galmarino (1964)	4prs	fin inf	card
Hebert (1964)	4prs	disc	
Owen (1964)	comp simp	barg prod	core loc
Shapley (1964a)	comp simp	barg prod	
Shubik (1964)			surv
Stearns (1964a)	notr 3prs		calc core
(1964b)	notr nprs 6prs	noex	defs exam
Lucas (1965)	part 4prs	uniq	
Peleg (1965b)	comp		kern
(1965c)	comp		
Stearns (1965)	notr 7prs	noex	exam
Eisenman (1966)	part		defs vari
Lucas (1966a)	part		surv
(1966b)	$n-1$	symm	
Owen (1966)	4prs nprs	disc	solv
Parthasarathy (1966)	comp simp	prod	
Aumann (1967)	notr		surv
Lucas (1967a)	part $n-1$	uniq	
(1967b)	5prs	path uniq	exam core

Authors and Dates	Descriptors		
	Games	Solutions	Other
Shapley (1967b)		disc	solv
Skala (1967)	abst		core
Guilbaud (1968)			book surv
Leiserson (1968)	simp	fin	poli
Lucas (1968a)	part 11prs	noex	exam
(1968b)	10prs	noex	exam
(1968c)	5prs 8prs	path symm uniq	exam
Owen (1968a)	nprs infp	noex	book exam surv
(1968b)	$n-1$	disc	
Shubik (1968a)	mark	barg	econ core
Wilson (1968)	simp notr	fin	defs poli
Bondareva (1969)	comp 4prs nprs 5prs	uniq	core
Kulakovskaya (1969)		uniq	core
Lucas (1969a)	8prs	path uniq	exam
(1969b)	10prs	noex	exam calc
Parthasarathy (1969a)	comp simp	prod	
(1969b)	comp simp	prod	
Shapley–Shubik (1969a)	mark	path noex	core exam
Bondareva (1970)	3prs mark nprs	barg	core solv
Leiserson (1970)			surv poli
Rosenthal (1970)	simp	fin	thes bset vari defs
Vorobyev (1970a)			book surv
(1970b)			surv
Wilson (1970)	abst		econ
Kulakovskaya (1971)		uniq	core
Parthasarathy–Raghavan (1971)	comp simp	prod	book surv
Shapley (1971a)	conv	uniq	core solv
Skérus–Jačiauskas (1971)	3prs		noco

Authors and Dates	Descriptors		
	Games	Solutions	Other
Sokolina–Bondareva (1971)	comp	uniq	core
Wilson (1971a)	simp notr	fin	bset poli
(1971b)	simp notr	fin	bset poli calc
Yasuda (1971)	mark		econ
Wilson (1972a)	abst		econ
Lucas (1972)			surv
Vilkov (1972)	notr	uniq	core
Harsanyi (1973)		uniq	noco defs core
Hart (1973a)	symm	symm	econ
(1973b)	notr infp mark nprs	prod	thes econ
Weber (1973a)	$n - 1$	disc	
(1973b)	$n - 2$	disc	

Appendix C Games against Nature

A classical statistical problem can be modeled as a one-person non-constant-sum game, which in turn can be regarded formally as a two-person game against a strategic opponent without a utility function called "Nature." How do we describe the behavior of Nature? What model of his strategies and his behavior does the statistician have? There is a large body of literature on this subject (see Luce and Raiffa, 1957; Raiffa, 1968). Even now, however, the perceptive survey of Milnor (1954) provides insight into the problem and several proposed solutions.

Consider a game in which player P_1 is the row player and player N is Nature, the column player. We are given a payoff matrix for P_1. There are a number of possibilities for setting up the game:

Laplace criterion: If Nature's probabilities are unknown, assume that they are all equal.

Wald's minimax: Assume that Nature's payoffs are the negative of P_1's. Then P_1 should play as though he were in a two-person zero-sum game with Nature out to get him.

Hurwicz optimism criterion: Select some parameter α, $0 \leq \alpha \leq 1$. For each row or probability mixture of rows let m be the smallest and M the largest component. Then P_1 should select the row or mixture of rows for which $\alpha M + (1 - \alpha)m$ is a maximum. When $\alpha = 0$, this is the Wald criterion.

Savage's minimax regret: Savage has suggested the formation of a "regret" matrix obtained by measuring the maximum of any column minus any other entry. The regret is the difference between what P_1 obtains and what he might have obtained given Nature's choice.

Milnor considers the following ten axioms that assign a preference relation $\gtrsim$ between pairs of rows of the matrix (a_{ij}). He then establishes which axioms are required by each of the criteria and shows which are compatible. Luce and Raiffa give a relatively detailed discussion of each axiom.

The first five axioms are compatible with all four decision criteria:

1. *Ordering*. The relation $\gtrsim$ is a complete ordering of the rows. That is, it is a transitive relation such that for any two rows r, r' either $r \gtrsim r'$ or $r' \gtrsim r$.

2. *Symmetry*. The ordering is independent of the numbering of the rows and columns. (Thus we are not considering situations where there is any reason to expect one state of Nature more than another.)

3. *Strong domination*. If each component of r is greater than the corresponding component of r', then $r > r'$ (shorthand for $r \gtrsim r'$ but not $r' \gtrsim r$).

4. *Continuity*. If the matrices a_{ij}^k converge to a_{ij}, and if $r^k > r_1^k$ for each k, then the limit rows r and r_1 satisfy $r \gtrsim r_1$.

5. *Linearity*. The ordering relation is not changed if the matrix (a_{ij}) is replaced by (a'_{ij}), where $a'_{ij} = \lambda a_{ij} + \mu$, $\lambda > 0$ and μ being constants.

The following four axioms serve to distinguish between the four criteria:

6. *Row adjunction*. The ordering between the old rows is not changed by the adjunction of a new row.

7. *Column linearity*. The ordering is not changed if a constant is added to a column.

8. *Column duplication*. The ordering is not changed if a new column, identical with some old column, is adjoined to the matrix. (Thus we are only interested in what states of Nature are possible, and not in how often each state may have been counted in the formation of the matrix.)

9. *Convexity*. If row r is equal to the average $(r' + r'')/2$ of two equivalent rows, then $r \gtrsim r'$. (Two rows are *equivalent*, $r' \sim r''$, if $r' \gtrsim r''$ and $r'' \gtrsim r'$. This axiom asserts that the player is not prejudiced against randomizing: if two rows are equally favorable, then he does not mind tossing a coin to decide between them.)

Finally we will need a modified form of axiom 6 that is compatible with all four criteria:

10. *Special row adjunction*. The ordering between the old rows is not changed by the adjunction of a new row, providing that no component of this new row is greater than the corresponding components of all old rows.

Table C.1

	Laplace	Wald	Hurwicz	Savage
1. Ordering	+	+	+	+
2. Symmetry	+	+	+	+
3. Strong domination	+	+	+	+
4. Continuity	×	+	+	+
5. Linearity	×	×	+	×
6. Row adjunction	+	+	+	
7. Column linearity	+			+
8. Column duplication		+	+	+
9. Convexity	×	+		+
10. Special row adjunction	×	×	×	+

Milnor's principal results are shown in table C.1, which describes the relations between the ten axioms and the four criteria. The symbol × indicates that the corresponding axiom and criterion are compatible. Each criterion is characterized by the axioms marked +. No criterion satisfies all of the properties. Furthermore, as the matrix below indicates, the advice offered by the proponents of the various criteria may call for divergent actions:

2	2	0	1	Laplace
1	1	1	1	Wald
0	4	0	0	Hurwicz (for $\alpha > 1/4$)
1	3	0	0	Savage

The compilation of these axioms and the construction of criteria all raise the same fundamental questions concerning how we do or should construct priors about the unknown. In most applications—to bargaining, oil drilling, or betting on the next California earthquake—pure mathematics may be of help in structuring our beliefs, but the essential step lies in the modeling of behavior, in the choice of the axioms rather than in their manipulation.

Bibliography

Abbreviations used

AMS = American Mathematical Society

AnMS = *Annals of Mathematics Studies* series (Princeton, NJ: Princeton University Press)

APSR = *American Political Science Review*

CNRS = Centre National de la Recherche Scientifique

IEEE = Institute of Electrical and Electronics Engineers

IJGT = *International Journal of Game Theory*

JORSA = *Journal of the Operations Research Society of America*

NRLQ = *Naval Research Logistics Quarterly*

PNAS = *Proceedings of the National Academy of Sciences of the United States*

RAND = The RAND Corporation, Santa Monica, CA

SIAM = Society for Industrial and Applied Mathematics

Adams, E. W., 1965. Elements of a theory of inexact measurement. *Philosophy of Science* 32:205–228.

Ahrens, J., 1962. Zur Gesamtheit der symmetrischen *n*-Personen-Spiele. *Metrika* 5:81–95.

Allais, M., 1953. Le comportement de l'homme rationnel devant le risque: Critique des postulats et axioms de l'école Américaine. *Econometrica* 21:503–546.

Allen, L., 1956. Games bargaining: A proposed application of the theory of games to collective bargaining. *Yale Law Review* 65:660–693.

Allingham, M. G., 1975. Economic power and values of games. *Zeitschrift für Nationalökonomie* 35:293–299.

Amihud, Y., ed., 1976. *Bidding and Auctioning for Procurement and Allocation*. New York: New York University Press.

Ancker, C. H., Jr., 1975. Stochastic duels with round dependent hit probabilities. *NRLQ* 22:575–583.

Anderson, R. E., 1967. Status structures in coalition bargaining games. *Sociometry* 30:393–403.

Anderson, R. M., 1978. An elementary core equivalence theorem. *Econometrica* 46:1483–1487.

Annals of Mathematics Studies (*AnMS*), 1950, 1953, 1957, 1959, 1964. Volumes 24, 28, 39, 40, 52 of the series. Princeton, NJ: Princeton University Press.

Applegren, L., 1967. An attrition game. *Operations Research* 15:11–31.

Armstrong, W. E., 1939. The determinateness of the utility function. *Economic Journal* 49:453–467.

Arrow, K. J., 1951. *Social Choice and Individual Values,* 1st ed. New York: John Wiley (2nd ed., 1963).

———, 1973. Rawls's principle of just saving. *Swedish Journal of Economics* 75:323–335.

Arrow, K. J., and G. Debreu, 1954. Existence of an equilibrium for a competitive economy. *Econometrica* 22:265–290.

Asscher, N., 1975. Bargaining set for 3-person games without side payments. Publication RM-11, Center for Research in Game Theory and Mathematical Economics, Hebrew University, Jerusalem.

———, 1976. An ordinal bargaining set for games without sidepayments. *Mathematics of Operations Research* 1:381–389.

———, 1977. A cardinal bargaining set for games without sidepayments. *IJGT* 6:87–114.

Aubin, J. P., 1979. *Mathematical Methods of Game and Economic Theory.* Amsterdam: North-Holland.

———, 1981. Cooperative fuzzy games. *Mathematics of Operations Research* 6:1–13.

Aumann, R. J., 1959. Acceptable points in general cooperative n-person games. *AnMS* 40:287–324.

———, 1961a. Almost strictly competitive games. *SIAM Journal* 9:544–550.

———, 1961b. The core of a cooperative game without side payments. *Transactions of the AMS* 98:539–552.

———, 1962. Utility theory without the completeness axiom. *Econometrica* 30:445–462.

———, 1964a. Introduction to G. Jentzsch's "Some thoughts on the theory of cooperative games." *AnMS* 52:407–409.

———, 1964b. Markets with a continuum of traders. *Econometrica* 32:39–50.

———, 1964c. Mixed and behavior strategies in infinite extensive games. *AnMS* 52:627–650.

———, 1964d. Subjective programming. In *Human Judgments and Optimality,* eds. M. W. Shelly II and G. L. Bryan. New York: John Wiley, pp. 217–242.

———, 1964e. Utility theory without the completeness axiom: A correction. *Econometrica* 32:210–212.

———, 1966. Existence of competitive equilibria in markets with a continuum of traders. *Econometrica* 34:1–17.

———, 1967. A survey of cooperative games without side payments. In *Essays in Mathematical Economics,* ed. M. Shubik. Princeton, NJ: Princeton University Press, pp. 3–27.

———, 1973. Disadvantageous monopolies. *Journal of Economic Theory* 6:1–11.

———, 1974. Subjectivity and correlation in randomized strategies. *Journal of Mathematical Economics* 1:67–96.

———, 1975. Values of markets with a continuum of traders. *Econometrica* 43:907–912.

Aumann, R. J., and J. H. Dreze, 1975. Cooperative games with coalition structures. *IJGT* 3:217–237.

Aumann, R. J., and M. Kurz, 1977a. Power and taxes. *Econometrica* 45:1137–1161.

———, 1977b. Power and taxes in a multicommodity economy. *Israel Journal of Mathematics* 27:185–234.

Aumann, R. J., and M. Maschler, 1964. The bargaining set for cooperative games. In *Advances in Game Theory,* eds. M. Dresher, L. S. Shapley, and A. W. Tucker. Princeton, NJ: Princeton University Press, pp. 443–447.

———, 1967. Repeated games with incomplete information: A survey of recent results. In *Report to the U.S. Arms Control and Disarmament Agency.* Mathematica Policy Research, Inc., Princeton, NJ, pp. 287–403.

———, 1972. Some thoughts on the minimax principle. *Management Science* 18 (part 2):54–63.

Aumann, R. J., and B. Peleg, 1960. von Neumann–Morgenstern solutions to cooperative games without side payments. *Bulletin of the AMS* 66:173–179.

Aumann, R. J., and L. S. Shapley, 1968. Values of nonatomic games, I: The axiomatic approach. RAND Publication RM-5468-PR.

———, 1969. Values of nonatomic games, II: The random ordering approach. RAND Publication RM-5842-PR.

———, 1970a. Values of nonatomic games, III: Values and derivatives. RAND Publication RM-6216-PR.

———, 1970b. Values of nonatomic games, IV: The value and the core. RAND Publication RM-6260.

———, 1971. Values of nonatomic games, V: Monetary economies. RAND Publication R-843-RC.

———, 1974. *Values of Non-Atomic Games.* Princeton, NJ: Princeton University Press.

Axelrod, R., 1970. *Conflict of Interest.* Chicago, IL: Markham.

———, 1980. Effective choice in the prisoner's dilemma. *Journal of Conflict Resolution* 24:3–26.

Bachet de Meziriac, 1612. *Problèmes plaisants et délectables, qui ses font par les nombres.* Lyon.

Bain, J., 1956. *Barriers to New Competition.* Cambridge, MA: Harvard University Press.

Baker, K. A., P. C. Fishburn, and F. S. Roberts, 1972. Partial orders of dimension 2. *Networks*: 11–28.

Baldwin, D. A., 1971. The costs of power. *Journal of Conflict Resolution* 15:145–155.

Baligh, H. H., and L. E. Richartz, 1967. Variable sum game models of marketing problems. *Journal of Marketing Research* 4:173–183.

Balinski, M. L., and H. P. Young, 1982. *Fair Representation.* New Haven, CT: Yale University Press.

Banks, M. H., O. J. R. Groom, and A. N. Oppenheim, 1968. Gaming and simulation in international relations. *Political Studies* 16:1–17.

Banzhaf, J. F., 1965. Weighted voting doesn't work: A mathematical analysis. *Rutgers Law Review* 19:317–343.

———, 1966. Multi-member electoral districts: Do they violate the "one man, one vote" principle? *Yale Law Journal* 75:1309.

———, 1968. One man, 3.312 votes: A mathematical analysis of the electoral college. *Villanova Law Review* 13:304–332.

Barber, J. D., 1966. *Power in Committees.* Chicago, IL: Rand McNally.

Barringer, R. E., and B. S. Whaley, 1965. The M.I.T. political-military gaming experience. *Orbis* 9:437–458.

Barth, F., 1959. Segmentary opposition and the theory of games: A study of pathan organization. *Journal of the Royal Anthropological Society* (London) 89:5–21.

Bartos, O. J., 1964. A model of negotiation and the recency effect. *Sociometry* 27:311–326.

———, 1967. *Simple Models of Group Behavior.* New York: Columbia University Press.

Basar, T., and Y. Ho, 1974. Informational properties of Nash solutions of two stochastic nonzero sum games. *Journal of Economic Theory* 7:370–387.

Bass, B. M., 1964. Business gaming for organization research. *Management Science* 10:545–556.

Bass, B. M., R. D. Buzzel, M. R. Greene, W. Lozer, E. A. Pessemier, D. L. Showver, A. Suchman, C. A. Theodore, and G. W. Wilson, eds., 1961. *Mathematical Models and Methods in Marketing.* Homewood, IL: Irwin.

Baudier, E., 1969. Un critère de choix collectif dans un jeu à *n* personnes et à somme constante. In *La Décision: Agrégation et dynamique des ordres de préférence.* Paris: Editions du CNRS, pp. 47–52.

———, 1973. Competitive equilibrium in a game. *Econometrica* 41:1049–1068.

Baumol, W. J., 1959. *Business Behavior Value and Growth.* New York: Macmillan.

Beale, E. M. L., and G. P. M. Heselden, 1962. An approximate method of solving Blotto games. *NRLQ* 9:65–79.

Becker, G., 1976. Altruism, egoism, and genetic fitness economics and sociobiology. *Journal of Economic Literature* 14:817–826.

Beckmann, M. J., with the assistance of D. Hochstadter, 1965. Edgeworth-Bertrand duopoly revisited. In *Operations Research–Verfahren III,* ed. R. H. Sonderdruck. Meisenheim: Verlag Anton Hain, pp. 55–68.

Bell, C. E., 1968. The *n* days of Christmas. *Management Science* 14:525–535.

Bellman, R., 1954. Decision-making in the face of uncertainty, II. *NRLQ* 1:327–332.

———, 1957. *Dynamic Programming.* Princeton, NJ: Princeton University Press.

Bellman, R., C. E. Clark, D. G. Malcolm, C. J. Craft and F. M. Ricciardi, 1957. On the construction of a multi-stage multi-person business game. *Journal of Operations Research* 5:469–503.

Bellman, R., and M. A. Girshick, 1949. An extension of results on duels with two opponents, one bullet each, silent guns, equal accuracy. RAND Publication D-403.

Belzer, R. L., 1948. Silent duels, specified accuracies, one bullet each. RAND Publication RAD(L)-301.

Beniest, W., 1963. Jeux stochastiques totalement cooperatifs arbitres. *Cahiers du Centre d'Etude de Recherche Opérationelle* (Brussels) 5:124–138.

Bennion, E. G., 1956. Capital budgeting and game theory. *Harvard Business Review* 34:115–123.

Bentley, A. F., 1949. *The Process of Government,* 3rd ed. San Antonio, TX: Principia Press.

Beresford, R. S., and M. H. Peston, 1955. A mixed strategy in action. *Operational Research Quarterly* 6:173–175.

Berge, C., 1957a. Topological games with perfect information. *AnMS* 39:165–178.

———, 1957b. *Théorie générale des jeux à n personnes.* Paris: Gauthier-Villars.

———, 1958. *Théorie des graphes et ses applications.* Paris: Dunod.

Bergson, A., 1938. A reformulation of certain aspects of welfare economics. *Quarterly Journal of Economics* 52:310–334.

Berkovitz, L. D., and M. Dresher, 1959. A game-theory analysis of tactical air war. *Operations Research* 7:599–620.

———, 1960a. Allocation of two types of aircraft in tactical air war: A game-theoretic analysis. *Operations Research* 8:694–706.

———, 1960b. A multimove infinite game with linear payoff. *Pacific Journal of Mathematics* 10:743–765.

Bernard, J., 1950. Where is the modern sociology of conflict? *American Journal of Sociology* 56:11–16.

———, 1954. The theory of games of strategy as a modern sociology of conflict. *American Journal of Sociology* 59:411–424.

———, 1965. Some current conceptualizations in the field of conflict. *American Journal of Sociology* 70:442–454.

Bernoulli, D., 1954. Exposition of a new theory on the measurement of risk (translated from Latin; original, 1738). *Econometrica* 22:23–36.

Bertrand, J., 1883. Théorie mathématique de la richesse sociale (review). *Journal des Savants* (Paris): 499–508.

Beveridge, J. M., 1962. To bid or not to bid. *Aerospace Management* 5:24–28.

Bewley, T., and E. Kohlberg, 1978. On stochastic games with stationary optimal strategies. *Mathematics of Operations Research* 32:104–125.

Bhashyam, N., 1970. Stochastic duels with nonrepairable weapons. *NRLQ* 17:121–129.

Billera, L. J., 1970a. Some theorems on the core of an *n*-person game without side payments. *SIAM Journal of Applied Mathematics* 18:567–579.

———, 1970b. Clutter decomposition and monotonic Boolean functions. *Annals, New York Academy of Sciences* 175:41–48.

———, 1970c. Existence of general bargaining sets for cooperative games without side payments. *Bulletin of the AMS* 76:375–379.

———, 1971a. On the composition and decomposition of clutters. *Journal of Combinatorial Theory* 11:234–245.

———, 1971b. Some recent results in *n*-person game theory. *Mathematical Programming* 1:58–67.

———, 1972a. A note on the kernel and the core of games without side payments. Technical Report 152, Department of Operations Research, Cornell University.

———, 1972b. Global stability in *n*-person games. *Transactions of the AMS* 172:45–56.

———, 1974. On games without side payments arising from a general class of markets. *Journal of Mathematical Economics* 1:129–139.

Billera, L. J., and R. E. Bixby. 1973. A characterization of polyhedral market games. *IJGT* 2:253–261.

———, 1974. Market representations of *n*-person games. *Bulletin of the AMS* 80:522–526.

Billera, L. J., D. C. Heath, and J. Raanan, 1978. Internal telephone billing rates: A novel application of nonatomic game theory. *Operations Research* 26:956–965.

Bird, C. G., and K. O. Kortanek, 1974. Game theoretic approaches to some air pollution regulation problems. *Socio-Economic Planning Sciences* 8:141–147.

Birmingham, R. L., 1968–69. Legal and moral duty in game theory: Common law contract and Chinese analogies. *Buffalo Law Review* 18:99–117.

Birnbaum, Z. W., and J. D. Esary, 1965. Modules of coherent binary systems. *SIAM Journal of Applied Mathematics* 13:442–462.

Birnbaum, Z. W., J. D. Esary, and S. C. Saunders, 1961. Multicomponent systems and structures and their reliability. *Technometrics* 3:55–77.

Birnbaum, Z. W., and F. Proschan, 1963. Coherent structures of nonidentical components. *Technometrics* 5:191–209.

Bishop, R. L., 1963. Game theoretic analysis of bargaining. *Quarterly Journal of Economics* 77:559–602.

Bixby, R., 1971. On the length-width inequality for compound clutters. *Journal of Combinatorial Theory* 11:246–248.

Black, D., 1958. *The Theory of Committees and Elections.* Cambridge, England: Cambridge University Press.

Blackett, D. W., 1954. Some Blotto games. *NRLQ* 1:55–60.

———, 1958. Pure strategy solutions of Blotto games. *NRLQ* 5:107–109.

Blackwell, D., 1948. The silent duel, one bullet each, arbitrary accuracy. RAND Publication RM-302.

———, 1949. The noisy duel, one bullet each, arbitrary nonmonotone accuracy. RAND Publication RM-131.

———, 1954. On multi-component attrition games. *NRLQ* 1:210–216.

———, 1955. The prediction of sequences. RAND Publication RM-1570.

———, 1956. An analogy of the minimax theorem for vector payoffs. *Pacific Journal of Mathematics* 6:1–8.

———, 1962. Discrete dynamic programming. *Annals of Mathematical Statistics* 33:719–726.

———, 1965. Discounted dynamic programming. *Annals of Mathematical Statistics* 36:226–235.

Blackwell, D., and T. S. Ferguson, 1968. The big match. *Annals of Mathematical Statistics* 39:159–163.

Blackwell, D., and M. A. Girshick, 1949. A loud duel with equal accuracy where each duelist has only a probability of possessing a bullet. RAND Publication RM-219.

———, 1954. *Theory of Games and Statistical Decisions.* New York: John Wiley.

Blaquiere, A., 1973. *Topics in Differential Games.* Amsterdam: North-Holland.

Blaquiere, A., F. Gerard, and G. Leitmann, 1969. *Quantitative and Qualitative Games.* New York: Academic Press.

Blau, J. H., 1957. The existence of social welfare functions. *Econometrica* 25:302–313.

Blau, J., and D. Brown, 1981. The structure of neutral monotonic social functions. *Review of Economic Studies* (forthcoming).

Blau, P., 1964. *Exchange and Power in Social Life.* New York: John Wiley.

Bloomfield, S., and R. Wilson, 1972. The postulates of game theory. *Journal of Mathematical Sociology* 2:221–234.

Böhm-Bawerk, E. von, 1923. *Positive Theory of Capital* (translated from German; original, 1891). New York: G. E. Steckert.

Bohnenblust, H. M., S. Karlin, and L. S. Shapley, 1950. Solutions of discrete two person games. *AnMS* 24:51–72.

Bonacheck, P., 1968. Sin and error. Ph.D. dissertation, Harvard University.

Bond, J. R., and W. E. Vinacke, 1961. Coalitions in mixed-sex triads. *Sociometry* 24:61–75.

Bondareva, O. N., 1962. Theory of the core in the *n*-person game (in Russian). *Vestnik Leningradskii Universitet* 13:141–142.

———, 1963. Some applications of linear programming methods to the theory of cooperative games (in Russian). *Problemy Kibernetiki* 10:119–139.

———, 1969. Solutions for a class of games with empty core (in Russian). *Doklady Akademii Nauk SSSR* 185:247–250.

———, 1970. A theorem on externally stable sets (in Russian). *Doklady Academii Nauk SSSR* 192:259–261.

Boot, J. C. G., 1965. Price determination based on quality: An application of minimax. *Statistica Neerlandica* 19:41–53.

———, 1967. *Mathematical Reasoning in Economics and Management Science.* Englewood Cliffs, NJ: Prentice-Hall.

Borch, K., 1962. Application of game theory to some problems in automobile insurance. *The Astin Bulletin* 2 (part 2):208–221.

———, 1968. *The Economics of Uncertainty.* Princeton, NJ: Princeton University Press.

———, 1979. Mathematical models for marine insurance. *Scandinavian Actuarial Journal*: 25–36.

Borel, E., 1921. La théorie du jeu et les équations intégrales à noyau symétrique. *Comptes Rendus de l'Académie des Sciences* (Paris) 173:1304–1308.

———, 1924. Sur les jeux on interviennent l'hasard et l'habilité des joueurs. In *Théorie des probabilités,* ed. J. Herman. Paris: Librairie Scientifique, pp. 204–224.

———, 1938. *Traité du calcul des probabilités et de ses applications,* vol. 4. Paris: Gauthier-Villars.

———, 1953. On games that involve chance and the skill of the players (translated from French; original, 1924). *Econometrica* 21:101–115.

Bott, R., 1953. Symmetric solutions to majority games. *AnMS* 28:319–323.

Boulding, K. E., 1962. *Conflict and Defense: A General Theory.* New York: Harper & Row.

Bowley, A. L., 1928. Bilateral monopoly. *Economic Journal* 38:651–659.

Brackney, H., 1959. The dynamics of military combat. *JORSA* 7:30.

Bradley, G., and M. Shubik, 1974. A note on the shape of the Pareto optimal surface. *Journal of Economic Theory* 81:530–538.

Braithwaite, R. B., 1955. *Theory of Games as a Tool for the Moral Philosopher.* Cambridge, MA: Harvard University Press.

Brams, S. J., 1968. Measuring the concentration of power in political systems. *APSR* 62:461–475.

———, 1975. *Game Theory and Politics.* New York: Free Press.

———, 1978. *The Presidential Election Game.* New Haven, CT: Yale University Press.

———, 1980. *Biblical Games.* Cambridge, MA: MIT Press.

Brams, S. J., A. Schotter, and G. Schwodiauer, 1979. *Applied Game Theory.* Würzburg: Physica Verlag.

Brayer, A. R., 1964. An experimental analysis of some variables of minimax theory. *Behavioral Science* 9:33–44.

Bremer, H., W. Hall, and M. Paulsen, 1957. Experiences with bid evaluation problems. *NRLQ* 4:27–30.

Brems, H., 1951. *Product Equilibrium under Monopolistic Competition.* Cambridge, MA: Harvard University Press.

Brewer, G., and M. Shubik, 1979. *The War Game.* Cambridge, MA: Harvard University Press.

Brown, D., 1974. An approximate solution to Arrow's problem. *Journal of Economic Theory* 9:375–383.

———, 1975. Aggregation of preferences. *Quarterly Journal of Economics* 89:456–469.

Brown, G. W., 1951. Iterative solutions of games by fictitious play. In *Activity Analysis of Production and Allocation,* ed. T. C. Koopmans. New York: John Wiley, pp. 374–376.

Brown, R. H., 1963. Theory of combat: The probability of winning. *JORSA* 11:418.

Buchanan, J. M., 1954. Individual choice in voting and the market. *Journal of Political Economy* 62:334–343.

Buchanan, J. M., and G. Tullock, 1962. *The Calculus of Consent.* Ann Arbor, MI: University of Michigan Press.

Buchler, I. R., and H. G. Nutini, 1968. *Game Theory in the Behavioral Sciences.* Pittsburgh, PA: University of Pittsburgh Press.

Burger, E., 1959. *Einführung in die Theorie der Spiele.* Berlin: Walter de Gruyter. English translation, *Introduction to the Theory of Games* (Englewood Cliffs, NJ: Prentice-Hall, 1963.)

Burgin, G. H., 1969. On playing 2-person zero-sum against non-minimax players. *IEEE Transactions on Systems Science and Cybernetics* SSC-5:369–370.

Butterworth, R. W., 1972. A set-theoretic treatment of coherent systems. *SIAM Journal of Applied Mathematics* 22:590–598.

Callen, J. L., 1978. Financial cost allocations: A game theoretic approach. *The Accounting Review* 53:303–308.

Camacho, A., 1974. Societies and social decision functions. In *Developments in the Methodology of Social Sciences,* eds. W. Leinfellner and E. Köhler. Dordrecht: Riedel.

Case, J. H., 1971. Applications of the theory of differential games to economic problems. In *Differential Games and Related Topics,* eds. H. W. Kuhn and G. P. Szegö. Amsterdam: North-Holland.

———, 1979. *Economics and the Competitive Process.* New York: New York University Press.

Caspi, Y., 1978. A limit theorem on the core of an economy with individual risks. *Economic Studies* 45:267–271.

Cassady, R., 1967. *Auctions and Auctioneering.* Berkeley, CA: University of California Press.

Caywood, T. E., and C. J. Thomas, 1955. Applications of game theory in fighter versus bomber combat. *JORSA* 3:402–411.

Chamberlain, N., 1951. *Collective Bargaining.* New York: McGraw-Hill.

Chamberlain, N., and J. W. Kuhn, 1965. *Collective Bargaining.* New York: McGraw-Hill.

Chamberlin, E. H., 1929. Duopoly: Value where sellers are few. *Quarterly Journal of Economics* 44:63.

———, 1933. *The Theory of Monopolistic Competition,* 6th ed. Cambridge, MA: Harvard University Press.

———, 1948. An experimental imperfect market. *Journal of Political Economy* 56:95–108.

———, 1962. *Monopolistic Competition,* 8th ed. Cambridge, MA: Harvard University Press.

Champlin, J., 1971. On the study of power. *Politics and Society* 1:91–111.

Chapman, D., 1967. Models of the working of a two party electoral system, I. *Papers on Non-Market Decision Making* 3:19–37.

Charnes, A., and K. Kortanek, 1967. On balanced sets, cores, and linear programming. *Cahiers du Centre d'Etude de Recherche Opérationelle* (Brussels) 9:32–43.

Charnes, A., and S. C. Littlechild, 1975. Unions in *n*-person games. *Journal of Economic Theory* 10:386–403.

Charnes, A., and R. G. Schroeder, 1967. On some stochastic tactical antisubmarine games. *NRLQ* 14:291–312.

Chow, C. K., 1961. On the characterization of threshold functions. In *Switching Circuit Theory and Logical Design,* ed. R. S. Ledley. New York: American Institute of Electrical Engineers.

Christenson, C., 1965. *Strategic Aspects of Competitive Bidding for Corporate Securities.* Cambridge, MA: Division of Research, Graduate School of Business Administration, Harvard University.

Church, R., 1940. Numerical analysis of certain free distributive structures. *Duke Mathematical Journal* 6:732–734.

Coase, R. H., 1935. The problem of duopoly reconsidered. *Review of Economic Studies* 2:137.

Cocklayne, E., 1967. Plane pursuit with coordinate constraints. *SIAM Journal of Applied Mathematics* 15:1511–1516.

Cohen, K. J., W. Dill, A. Kuehn, and P. Winters, 1964. *The Carnegie Tech Management Game: An Experiment in Business Education.* Homewood, IL: Irwin.

Cohen, N. D., 1966. An attack defense game with matrix strategies. RAND Publication RM-4274-PR.

Cole, S. G., 1969. An examination of the power-inversion effect in three-person mixed-motive games. *Journal of Personality and Social Psychology* 11:50.

Cole, S. G., and J. L. Phillips, 1967. The propensity to attack others as a function of the distribution of resources in a three-person game. *Psychonomic Science* 9:239–240.

Coleman, J. S., 1966a. Foundations for a theory of collective decisions. *American Journal of Sociology* 71:615–627.

———, 1966b. The possibility of a social welfare function. *American Economic Review* 56:1105–1122.

———, 1967. The possibility of a social welfare function: Reply [to comments by Park and Mueller]. *American Economic Review* 57:1311–1317.

———, 1970. Political money. *APSR* 64:1074–1087.

———, 1971. Control of collectivities and the power of a collectivity to act. In *Social Choice,* ed. B. Lieberman. New York: Gordon & Breach, pp. 260–300.

Condorcet, Marquis de, 1785. *Essai sur l'application de l'analyse à la probabilité des décisions rendues à la pluralité des voix.* Paris.

Cook, P. W., Jr., 1963. Fact and fancy on identical bids. *Harvard Business Review* 41:67–72.

Copeland, A. H., 1945. Review of *Theory of Games and Economic Behavior. Bulletin of the AMS* 51:498–504.

Cornwall, R. R., 1969. The use of prices to characterize the core of an economy. *Journal of Economic Theory* 1:353–373.

Cournot, A. A., 1897. *Researches into the Mathematical Principles of the Theory of Wealth* (translated from French; original, 1838). New York: Macmillan.

Cross, J. G., 1969. *The Economics of Bargaining.* New York: Basic Books.

———, 1977. Negotiations as a learning process. *Journal of Conflict Resolution* 21:581–606.

Cyert, R. M., and H. M. DeGroot, 1973. An analysis of cooperation and learning in a duopoly context. *American Economic Review* 63:24–37.

Cyert, R. M., and J. G. March, 1963. *Behavioral Theory of the Firm.* Englewood Cliffs, NJ: Prentice-Hall.

Dahl, R. A., 1957. The concept of power. *Behavioral Science* 2:201–215.

———, 1961. *Who Governs? Democracy and Power in an American City.* New Haven, CT: Yale University Press.

Dalkey, N., 1953. Equivalence of information patterns and essentially determinate games. *AnMS* 28:217–243.

———, 1965. Solvable nuclear war models. *Management Science* 11:783–791.

Danskin, J. M., and L. Gillman, 1953. A game over function space. *Rivista Di Mathematica* (Universita di Parma) 4:83–94.

Dantzig, G., 1951. A proof of the equivalence of the programming problem and the game problem. In *Activity Analysis of Production and Allocation,* ed. T. C. Koopmans. New York: John Wiley, pp. 330–335.

———, 1956. Constructive proof of the min max. *Pacific Journal of Mathematics* 6:25–33.

———, 1964. *Linear Programming and Its Extensions.* Princeton, NJ: Princeton University Press.

David, P. T., R. M. Goldman, and R. C. Bain. 1960. *The Politics of National Party Conventions.* Washington, DC: Brookings Institution.

Davis, J., 1964. Passatella: An economic game. *British Journal of Sociology* 15:191–206.

Davis, M., 1962. Symmetric solutions to symmetric games with a continuum of players. In *Recent Advances in Game Theory.* Princeton, NJ: Princeton University Conferences, pp. 119–126.

———, 1963a. A bargaining procedure leading to the Shapley value. Research Memorandum 61, Economics Research Program, Princeton University.

———, 1963b. Verification of disarmament by inspection: A game theoretic model. Research Memorandum 62, Economics Research Program, Princeton University.

———, 1964. Infinite games of perfect information. *AnMS* 52:85–101.

———, 1974. Some further thoughts on the minimax principal. *Management Science* 20:1305–1310.

Davis, M., and M. Maschler, 1965. The kernel of a cooperative game. *NRLQ* 12:223–259.

Davis, O. A., and A. B. Whinston, 1965. Welfare economics and the theory of second best. *Review of Economic Studies* 32:1–14.

Dawkins, R., 1976. *The Selfish Gene.* New York: Oxford University Press.

Dean, B. V., 1965. Contract award and bidding strategies. *IRE Transactions on Engineering Management* 12:53–59.

Dean, B. V., and R. H. Culkan. 1965. Contract research proposal preparation strategies. *Management Science* 11:B187–199.

Debreu, G., 1952. A social existence theorem. *PNAS* 38:886–893.

———, 1954. Representation of a preference ordering by a numerical function. In *Decision Processes,* eds. R. M. Thrall, C. H. Coombs, and R. L. Davis. New York: John Wiley, pp. 159–165.

———, 1959a. *Theory of Value.* New York: John Wiley.

———, 1959b. Cardinal utility for even-chance mixtures of pairs of sure prospects. *Review of Economic Studies* 26:174–177.

———, 1960. Topological methods in cardinal utility theory. In *Mathematical Methods in the Social Sciences,* eds. K. J. Arrow, S. Karlin, and P. Suppes. Stanford, CA: Stanford University Press, pp. 16–26.

———, 1963. On a theorem of Scarf. *Review of Economics Studies* 30:178–180.

———, 1967. Preference functions on measure spaces of economic agents. *Econometrica* 35:111–122.

———, 1969. Neighboring economic agents. In *La Décision: Agrégation et dynamique des ordres de préférence.* Paris: Editions du CNRS, pp. 85–90.

———, 1970. Economies with a finite set of equilibria. *Econometrica* 38:387–392.

———, 1975. The rate of convergence of the core of an economy. *Journal of Mathematical Economics* 2:1–7.

Debreu, G., and H. Scarf. 1963. A limit theorem on the core of an economy. *International Economic Review* 4:235–246.

Dedekind, R., 1897. Uber Zerlegungen von Zahlen durch ihre grossten gemeinsamen Teiler. In *Werke,* vol. 2 (Braunschweig, 1931), pp. 103–148.

DeMeyer, F., and C. Plott, 1970. The probability of a cyclical majority. *Econometrica* 38:345–354.

Denardo, E., 1981. *Dynamic Programming and Allied Topics: Definite Planning Horizons.* Englewood Cliffs, NJ: Prentice-Hall.

Deutsch, K. W., 1954. Game theory and politics: Some problems of application. *Canadian Journal of Economics and Political Science* 20:76–83.

Deutsch, M., 1961. The face of bargaining. *Operations Research* 9:886–897.

Deutsch, M., and R. M. Krauss, 1962. Studies in interpersonal bargaining. *Journal of Conflict Resolution* 6:52–76.

Djubin, G. N., 1968. On games on the unit square with a "roof" payoff (in Russian). *Teoriya Veroyastnostei i ee Primeneniya* 13:138–149.

———, 1969. A set of games on the unit square with unique solution. *Soviet Mathematics-Doklady* 10:51–53.

Dodgson, C. L., 1873. A discussion of the various methods of procedure in conducting elections, Oxford. Parrish Collection, Princeton University Library.

Dolbear, F. T., L. A. Lave, G. Bowman, A. Lieberman, E. Prescott, F. Reuter, and R. Shepman. 1968. Collusion in oligopoly: An experiment on the effect of numbers and information. *Quarterly Journal of Economics* 82:240–259.

Downs, A., 1957. *An Economic Theory of Democracy*. New York: Harper & Row.

Dresher, M., 1961. *Games of Strategy: Theory and Applications*. Englewood Cliffs, NJ: Prentice-Hall.

———, 1962. A sampling inspection problem in arms control agreements. RAND Publication RM-2972.

———, 1968. An armed AWACS duel. RAND Publication RM-5916.

———, 1970. Probability of a pure equilibrium point in n-person games. *Journal of Combinatorial Theory* 8:134–145.

Dresher, M., and S. Johnson, 1961. Optimal timing in missile launching. RAND Publication RM-2723.

Dresher, M., S. Karlin, and L. S. Shapley, 1950. Polynomial games. *AnMS* 24:161–180.

Dreyfus, S. E., 1965. *Dynamic programming and the Calculus of Variations*. New York: Academic Press.

Dreze, J. H., J. Gabszewicz, and S. Gepts, 1969. On cores and competitive equilibria. In *La Décision: Agrégation et dynamique des ordres de préférence*. Paris: Editions du CNRS, pp. 91–114.

Dreze, J. H., J. Gabszewicz, D. Schmeidler, and K. Vind, 1972. Cores and prices in an exchange economy with an atomless sector. *Econometrica* 40:1091.

Driggs, I. H., 1956. A Monte Carlo model of Lanchester's square law. *JORSA* 4:148–151.

Dubey, P., 1975. On the uniqueness of the Shapley value. *IJGT* 4:131–139.

———, 1976. Probabilistic generalizations of the Shapley value. Discussion Paper 440, Cowles Foundation, Yale University.

———, 1978. Finiteness and inefficiency of Nash equilibria. Discussion Paper 508, Cowles Foundation, Yale University.

———, 1980. Nash equilibria of market games: Finiteness and inefficiency. *Journal of Economic Theory* 22:363.

Dubey, P., A. Mas-Collel, and M. Shubik, 1980. Efficiency properties of strategic market games: An axiomatic approach. *Journal of Economic Theory* 22:339–362.

Dubey, P., A. Neyman, and R. J. Weber. 1981. Value theory without efficiency. *Mathematics of Operations Research* 6:122–128.

Dubey, P., and L. S. Shapley, 1976. Noncooperative exchange with a continuum of traders. Discussion Paper 447, Cowles Foundation, Yale University.

———, 1979. Mathematical properties of the Banzhaf power index. *Mathematics of Operations Research* 4:99–131.

Dubey, P., and M. Shubik, 1977a. Trade and prices in a closed economy with exogenous uncertainty, different levels of information, money and compound futures markets. *Econometrica* 45:1657–1680.

———, 1977b. A closed economic system with production and exchange modelled as a game of strategy. *Journal of Mathematical Economics* 4:253–287.

———, 1978a. The noncooperative equilibria of a closed trading economy with market supply and bidding strategies. *Journal of Economic Theory* 17:1–20.

———, 1978b. Strategic market games and market mechanisms. Preliminary Paper CF80415, Cowles Foundation, Yale University.

———, 1980. A strategic market game with price and quantity strategies. *Zeitschrift fur Nationalökonomie* 40:25–34.

———, 1981. Information conditions, communication and general equilibrium. *Mathematics of Operations Research* (forthcoming).

Dubins, L., 1957. A discrete evasion game. *AnMS* 39:231–255.

Dubins, L., and E. H. Spanier. 1961. How to cut a cake fairly. *American Mathematical Monthly* 68:1–17.

Dushnik, B., and E. W. Miller, 1941. Partially ordered sets. *American Journal of Mathematics* 63:600–610.

Duveen, J. H., 1935. *Art Treasures and Intrigue.* New York: Doubleday.

Dyckman, T. R., and S. Smidt, 1970. An axiomatic development of cardinal utility theory using decision theory. *Decision Sciences* 1:245–257.

Dynkin. E. B., 1969. Game variant of a problem on optimal stopping. *Soviet Mathematics-Doklady* 10:270–274.

Edgeworth, F. Y., 1881. *Mathematical Psychics.* London: Kegan Paul.

———, 1925. *Papers Relating to Political Economy, I.* London: Macmillan, pp. 111–142.

Edmonds, J., and D. R. Fulkerson, 1970. Bottleneck extrema. *Journal of Combinatorial Theory* 8:299–306.

Eilenberg, S., 1941. Ordered topological spaces. *American Journal of Mathematics* 63:39–45.

Eisenberg, E., 1961. Aggregation of utility functions. *Management Science* 7:337–350.

Eisenman, R. L., 1964. On solutions of alliance games. Ph.D. dissertation, University of Michigan, Ann Arbor.

———, 1966. Alliance games of n-persons. *NRLQ* 13:403–411.

———, 1967. A profit-sharing interpretation of Shapley values for n-person games. *Behavioral Science* 12:396–398.

Ellsberg, D., 1956. Theory of the reluctant duelist. *American Economic Review* 46:909–923.

———, 1961. The crude analysis of strategic choices. *American Economic Review* 51:472–478.

Elliot, R. J., and N. J. Kalton, 1972. Values in differential games. *Bulletin of the AMS* 78:427–431.

Engel, J. H., 1954. A verification of Lanchester's law. *JORSA* 2:163–171.

Engelbrecht-Wiggans, R., 1980. Auctions and bidding models: A survey. *Management Science* 26:119–142.

Everett, H., 1957. Recursive games. *AnMS* 39:47–78.

Fan, K., 1952. Fixed point and minimax theorems in locally convex topological linear spaces. *PNAS* 39:42–47.

Farquhar, P. H., 1974. Fractional hypercube decompositions of multiattribute utility functions. Technical Report 222, Department of Operations Research, Cornell University.

Farquharson, R., 1969. *Theory of Voting.* New Haven, CT: Yale University Press.

Faxen, K. O., 1957. *Monetary and Fiscal Policy under Uncertainty.* Stockholm: Almqvist and Wiksell.

Fechner, G. T., 1850. *Elements der Psychophysik.* Leipzig: Breitkopf und Hartel.

Feeney, G. J., 1968. Risk-aversion in incentive contracting. RAND Publication RM-4231-PR.

Fellner, W., 1949. *Competition Among the Few.* New York: Knopf.

———, 1965. *Probability and Profit: A Study of Economic Behavior Along Bayesian Lines.* Homewood, IL: Irwin.

Ferguson, T. S., 1967. On discrete evasion games with a two-move information lag. *Proceedings, Fifth Berkeley Symposium on Mathematical Statistics and Probability,* vol. 1, pp. 453–462.

Fishburn, P. C., 1968. Utility theory. *Management Science* 14:335–378.

———, 1969. Preference summation and social welfare. *Management Science* 15:179–186.

———, 1970a. *Utility Theory for Decision Making.* New York: Wiley-Interscience.

———, 1970b. Suborders on commodity spaces. *Journal of Economic Theory* 2:1–7.

———, 1970c. Comments on Hansson's "group preferences." *Econometrica* 38:933–935.

———, 1971. A study of lexicographic expected utility. *Management Science* 17:672–678.

Fleming, M., 1952. A cardinal concept of welfare. *Quarterly Journal of Economics* 66:366–384.

Flerov, Yu. A., 1969. Multilevel dynamic games. *Soviet Mathematics-Doklady* 10:994–996.

Flood, M. M., 1955. A group preference experiment. In *Mathematical Models of Human Behavior,* ed. J. W. Dunlop. Stamford, CT: Dunlop and Associates.

———, 1958. Some experimental games. *Management Science* 5:5–26.

Foley, D. K., 1970a. Economic equilibrium with costly marketing. *Journal of Economic Theory* 2:276–281.

———, 1970b. Lindahl's solution and the core of an economy with public goods. *Econometrica* 38:66–72.

Forst, B., and J. Lucianovic, 1977. The prisoner's dilemma: Theory and reality. *Journal of Criminal Justice* 5:55–64.

Fouraker, L. E., M. Shubik, and S. Siegel, 1961. Oligopoly bargaining: The quantity adjuster models. Publication RB 20, Pennsylvania State University.

Fouraker, L. E., and S. Siegel, 1963. *Bargaining Behavior.* New York: McGraw-Hill.

Fox, M., and G. Kimeldorf, 1968. Strategies and values in noisy duels. *Proceedings of the 1968 U.S. Army Operations Research Symposium,* pp. 27–34.

———, 1969. Noisy duels. *SIAM Journal of Applied Mathematics* 17:353–361.

———, 1970. Values and shooting times in noisy duels. *Journal of the American Statistical Association* 65: 422–430.

Frechet, M., 1953. Commentary on the three notes of Emile Borel. *Econometrica* 21:118–127.

Frey, B., 1968. A general model of resource allocation in a democracy. *General Systems* 13:157–163.

Friberg, M., and D. Johnson, 1968. A simple war and armament game. *Journal of Peace Research* 3:233–247.

Friedell, M. F., 1968. A laboratory experiment on retaliation. *Journal of Conflict Resolution* 12:357–378.

Friedman, A., 1970. Existence of value and saddle points for differential games of pursuit and evasion. *Journal of Differential Equations* 7:92–110.

———, 1971. *Differential Games.* New York: Wiley-Interscience.

Friedman, J. W., 1967. An experimental study of cooperative duopoly. *Econometrica* 35:379–397.

———, 1969. On experimental research in oligopoly. *Review of Economic Studies* 36:399–415.

———, 1971. A noncooperative view of oligopoly. *International Economic Review* 12:106–122.

———, 1977. *Oligopoly and the Theory of Games.* Amsterdam: North-Holland.

Friedman, J. W., and A. C. Hoggatt, 1979. *An Experiment in Noncooperative Oligopoly.* Greenwich, CT: JAI Press.

Friedman, L., 1956. A competitive bidding strategy. *Operations Research* 4:104–112.

———, 1958. Game theory models in the allocation of advertising expenditures. *Operations Research* 6:699–709.

Friedman, M., and L. J. Savage, 1948. The utility analysis of choices involving risk. *Journal of Political Economy* 56:279–304.

Frisch, H., 1971. On Lerner's theorem of equal income distribution. *Zeitschrift fur Nationalökonomie* 31:395–404.

Frisch, R., 1926. Sur une problème d'économie pure. *Norsk Matematisk Forenings Skrifter* 16 (série I): 1–40.

———, 1932. *New Methods of Measuring Marginal Utility.* Tübingen: Mohr.

———, 1937. General choice-field theory. In *Report of Third Annual Research Conference on Economics and Statistics,* Cowles Commission, University of Chicago, pp. 64–69.

———, 1964. Dynamic utility. *Econometrica* 32:418–424.

Fulkerson, D. R., 1968. Networks, frames, blocking systems. In *Mathematics of the Decision Sciences, I.* Providence, RI: AMS, pp. 303–334.

Fulkerson, D. R., and S. M. Johnson. 1957. A tactical air game. *JORSA* 5:704–712.

Gabszewicz, J., 1970. Théories du noyau et de la concurrence imparfaite. *Recherches Economiques de Louvain* 36:21–37.

Gabszewicz, J., and J. H. Dreze. 1971. Syndicates of traders in an exchange economy. In *Differential Games and Related Topics,* eds. H. W. Kuhn and G. P. Szegö. Amsterdam: North-Holland, pp. 399–414.

Gabszewicz, J., and J. F. Mertens, 1971. An equivalence theorem for the core of an economy whose atoms are not "too" big. *Econometrica* 39:713–721.

Gale, D., 1960. *The Theory of Linear Economic Models.* New York: McGraw-Hill.

Gale, D., and O. A. Gross, 1958. A note on polynomial and separable games. *Pacific Journal of Mathematics* 8:735–741.

Gale, D., H. W. Kuhn, and A. W. Tucker, 1951. Linear programming and the theory of games. In *Activity Analysis of Production and Allocation,* ed. T. C. Koopmans. New York: John Wiley, pp. 317–329.

Gale, D., and A. Mas-Colell, 1974. A short proof of existence of equilibrium without ordered preferences. Working Paper IP-207, Center for Research in Management Science, University of California, Berkeley.

Gale, D., and L. S. Shapley, 1962. College admissions and the stability of marriage. *American Mathematics Monthly* 69:9–15.

Gale, D., and F. M. Stewart, 1953. Infinite games with perfect information. *AnMS* 28:245–266.

Gale, J. S., and L. J. Eaves, 1975. Logic of animal conflict. *Nature* 254:463–464.

Gallo, P. S., Jr., 1966. The effects of score feedback and strategy of the other on cooperative behavior in a maximizing difference game. *Psychonomic Science* 5:401–402.

———, 1969. Personality impression formation in a maximizing difference game. *Journal of Conflict Resolution* 13:118–122.

Galmarino, A. R., 1964. On the cardinality of solutions of four-person constant-sum games. *AnMS* 52:327–344.

Gamson, W. A., 1961a. A theory of coalition formation. *American Sociological Review* 26:373–382.

———, 1961b. An experimental test of a theory of coalition formation. *American Sociological Review* 26:265–273.

———, 1962. Coalition formation at presidential nominating conventions. *American Journal of Sociology* 68:157–171.

———, 1966. Game theory and administrative decision-making. In *Empathy and Ideology: Knowledge for Administrative Innovation,* eds. B. Press and A. Adrian. Chicago, IL: Rand McNally.

Garcia, C. B., C. E. Lemke, and H. Leuthi, 1973. Simplicial approximations of an equilibrium point for noncooperative *N*-person games. In *Mathematical Programming,* eds. T. C. Ho and S. M. Robinson. New York: Academic Press, pp. 227–260.

Gardner, M., 1961. *The Second Scientific American Book of Mathematical Puzzles and Diversions.* New York: Simon and Schuster.

Garman, M., and M. Kamien, 1968. The paradox of voting: Probability calculations. *Behavioral Science* 13:306–316.

Gately, D., 1974. Sharing the gains from regional cooperation: A game theoretic application to planning investment in electric power. *International Economic Review* 15:195–208.

Geanakoplos, J., 1978. The bargaining set and nonstandard analysis. Publication TR-1, Center on Decision and Conflict in Complex Organizations, Harvard University.

Gelbaum, B. R., 1959. Symmetric zero-sum *n*-person games. *AnMS* 40:95–109.

Gepts, S., 1970. De Kern van een Riuleconomie. *Tijdschrift voor Economie* 15:50–64.

Gerber, H. U., 1972. Games of economic survival with discrete- and continuous-income processes. *Operations Research* 20:37–45.

Gibbard, A., 1973. Manipulation of voting schemes: A general result. *Econometrica* 41:587–601.

Gilbert, E. N., 1954. Lattice theoretic properties of frontal switching functions. *Journal of Mathematical Physics* 33:55–67.

Gillette, D., 1957. Stochastic games with zero stop probabilities. *AnMS* 39:179–187.

Gillies, D. B., 1953a. Discriminatory and bargaining solutions to a class of symmetric *n*-person games. *AnMS* 28:325–342.

———, 1953b. Location of solutions. In *Report of an Informal Conference on the Theory of n-Person Games,* Department of Mathematics, Princeton University, mimeographed, pp. 11–12.

———, 1953c. Some theorems on n-person games. Ph.D. dissertation, Department of Mathematics, Princeton University.

———, 1959. Solutions to general non-zero-sum games. *AnMS* 40:47–85.

Gillman, L., 1950. Operations analysis and the theory of games: An advertising example. *Journal of the American Statistical Association* 45:541–545.

Glasser, G. J., 1958. Game theory and cumulative voting for corporate directors. *Management Science* 5:151–156.

Glicksberg, I., 1950. Noisy duel, one bullet each, with simultaneous fire and unequal worths. RAND Publication RM-474.

———, 1952. A further generalization of Kakutani's fixed point theorem. *Proceedings of the AMS* 3:170–174.

Glicksberg, I., and O. A. Gross, 1952. Solution sets for games on the square. RAND Publication RM-901.

———, 1953. Notes on games over the square. *AnMS* 28:173–182.

Goldhamer, H., and E. Shils, 1939. Types of power and status. *American Journal of Sociology* 45:171–182.

Goldman, A. J., 1957. The probability of a saddle point. *American Mathematical Monthly* 64:729–730.

Goodman, L. A., and H. Markowitz, 1952. Social welfare functions based on individual rankings. *American Journal of Sociology* 58:257–262.

Gorman, W. M., 1968a. Conditions for additive separability. *Econometrica* 36:605–608.

———, 1968b. The structure of utility functions. *Review of Economic Studies* 35:367–390.

Grandmont, J. M., 1977. Temporary general equilibrium theory. *Econometrica* 45:535–572.

Granger, C. G., 1956. *La mathématique social du Marquis de Condorcet.* Paris: Presses Universitaires de France.

Griesmer, J. H., 1959. Extreme games with three values. *AnMS* 40:189–212.

Griesmer, J. H., and M. Shubik, 1963a. Toward a study of bidding processes: Some constant sum games. *NRLQ* 10:11–21.

———, 1963b. Toward a study of bidding processes, 2: Games with capacity limitations. *NRLQ* 10:151–173.

———, 1963c. Toward a study of bidding processes, 3: Some special models. *NRLQ* 10:199–217.

Griesmer, J. H., M. Shubik, and R. E. Levitan, 1967. Toward a study of bidding processes, 4: Games with unknown costs. *NRLQ* 14:415–433.

Gross, O. A., 1956. Games with payoff discontinuities at discrete points. RAND Publication RM-1755.

———, 1957. A rational game on the square. *AnMS* 39:307–311.

Groves, T., and J. Ledyard, 1977. Optimal allocation of public goods: A solution to the "free rider" problem. *Econometrica* 45:783–809.

Guetzkow, H., C. F. Alger, R. A. Brady, R. C. Noel, and R. C. Snyder, 1963. *Simulation in International Relations.* Englewood Cliffs, NJ: Prentice-Hall.

Guilbaud, G. Th., 1952a. Les problèmes du partage matériaux pour une enquête sur les algèbres et les arithmétiques de la répartition. *Economie Appliquée* 5:93–137.

———, 1952b. Les théories de l'intérêt général et le problème logique de l'agrégation. *Economie Appliquée* 5:501–584.

———, 1968. *Eléments de la théorie mathématique des jeux.* Paris: Dunod.

Gurk, H. M., 1959. Five-person, constant-sum, extreme games. *AnMS* 40:179–188.

Gurk, H. M., and J. R. Isbell, 1959. Simple solutions. *AnMS* 40:247–265.

Guyer, M., and H. Hamburger, 1968. A note on the enumeration of all 2 × 2 games. *General Systems* 13:205–208.

Hahn, F. H., 1971. Equilibrium with transactions costs. *Econometrica* 39:417–439.

Halphen, E., 1955. La notion de vraisemblance. *Publications de l'Institut de Statistique de l'Université de Paris* 4:41–92.

Hamilton, W. D., 1964. The genetical evaluation of social behavior. *Journal of Theoretical Biology* 7:1–52.

Hamlen, S. S., W. W. Hamlen, and J. T. Tschirhart, 1977. The use of core theory in evaluating joint cost allocation schemes. *The Accounting Review* 52:616–626.

Hammond, P., 1975. Charity: Altruism or cooperative egoism. In *Altruism, Morality and Economic Theory,* ed. N. Phelps. New York: Russell Sage.

Hanani, H., 1960. A generalization of the Banach and Mazur game. *Transactions of the AMS* 94:86–102.

Hanisch, H., P. J. Hilton, and W. M. Hirsch, 1969. Algebraic and combinatorial aspects of coherent structures. *Transactions, New York Academy of Sciences* 31:1024–1037.

Hanner, O., 1959. Mean play of sums of positional games. *Pacific Journal of Mathematics* 9:81–99.

Hansel, G., 1966. Sur le nombre des fonctions Booléennes monotones de *n* variables. *Comptes Rendus de l'Académie des Sciences* (Paris) 262:1088–1090.

Hansen, T., and J. Gabszewicz, 1972. Collusion of factor owners and distribution of social output. *Journal of Economic Theory* 4:1–18.

Hansson, B., 1969. Group preferences. *Econometrica* 37:50–54.

Harary, F., and M. Richardson, 1959. A matrix algorithm for solutions and r-bases of a finite irreflexive relation. *NRLQ* 6:307–314.

Hardin, G., 1968. The tragedy of the commons. *Science* 162:1243–1248.

Harris, R. J., 1969. Note on "optimal policies for the prisoner's dilemma." *Psychological Review* 76:363–375.

Harrod, R. F., 1934. The equilibrium of duopoly. *Economic Journal* 44:335.

Harsanyi, J. C., 1955. Cardinal welfare, individualistic ethics, and interpersonal comparisons of utility. *Journal of Political Economy* 63:302–321.

———, 1956. Approaches to the bargaining problem before and after the theory of games: A critical discussion of Zeuthen's, Hick's and Nash's theories. *Econometrica* 24:144–157.

———, 1959. A bargaining model for the cooperative *n*-person game. In *Contributions to the Theory of Games,* vol. 4, eds. A. W. Tucker and D. R. Luce. Princeton, NJ: Princeton University Press, pp. 324–356.

———, 1962a. Bargaining in ignorance of the opponent's utility function. *Journal of Conflict Resolution* 6:29–38.

———, 1962b. Measurement of social power, opportunity costs and the theory of two-person bargaining games. *Behavioral Science* 7:67–80.

———, 1962c. Measurement of social power in *n*-person reciprocal power situations. *Behavioral Science* 7:81–92.

———, 1962d. Rationality postulates for bargaining solutions in cooperative and in non-cooperative games. *Management Science* 9:141–153.

———, 1963. A simplified bargaining model for the *n*-person cooperative game. *International Economic Review* 4:194–220.

———, 1964. A general solution for finite noncooperative games based on risk-dominance. *AnMS* 52:651–679.

———, 1965. Bargaining and the conflict situation in the light of a new approach to game theory. *American Economic Review* 55:447–457.

———, 1967. Games with incomplete information played by "Bayesian" players, I: The basic model. *Management Science* 14:159–182.

———, 1968a. Games with incomplete information played by "Bayesian" players, II: Bayesian equilibrium points. *Management Science* 14:320–334.

———, 1968b. Games with incomplete information played by "Bayesian" players, III: The basic probability distribution of the game. *Management Science* 14:486–502.

———, 1973. Oddness of the number of equilibrium points: A new proof. *IJGT* 2:235–250.

———, 1974. An equilibrium-point interpretation of stable sets and a proposed alternative definition. *Management Science* 20:1472–1495.

———, 1975a. Can the maxmin principle serve as a basis for morality? A critique of John Rawls's theory. *APSR* 69:594–606.

———, 1975b. The tracing procedure: A Bayesian approach to defining a solution for *n*-person games. *IJGT* 4:61–94.

———, 1976. A solution concept for n-person noncooperative games. *IJGT* 5:211–225.

———, 1979. A new general solution concept for both cooperative and noncooperative games. In *Papers of the Rhineland-Westphalian Academy of Sciences,* no. 287, Opladen: Westdeutscher Verlag, pp. 7–28.

Harsanyi, J. C., and R. Selten, 1972. A generalized Nash solution for two-person bargaining games with incomplete information. *Management Science* 18:P80–P106.

Hart, S., 1971. Values of mixed games. Masters thesis, Department of Mathematical Sciences, Tel Aviv University.

———, 1973a. Symmetric solutions of some production economies. *IJGT* 2:53–62.

———, 1973b. The formation of cartels in large markets. Mimeographed paper, Department of Mathematical Sciences, Tel Aviv University.

Hartman, E. A., 1971. Development and test of a model of conflict in a truel. Report 71-1, Cooperation/Conflict Research Group, Michigan State University, East Lansing.

Haussmann, F., and B. H. P. Rivett, 1959. Competitive bidding. *Operational Research Quarterly* 16:49–55.

Hausner, M., 1952a. Games of survival. RAND Publication RM-776.

———, 1952b. Optimal strategies in games of survival. RAND Publication RM-777.

———, 1954. Multidimensional utilities. In *Decision Processes,* eds. R. M. Thrall, C. H. Coombs, and R. L. Davis. New York: John Wiley, pp. 167–180.

Hausner, M., and J. G. Wendel, 1952. Ordered vector spaces. *Proceedings of the AMS* 3:977–987.

Haywood, O. G., Jr., 1950. Military decision and the mathematical theory of games. *Air University Quarterly* 4:17–30.

———, 1954. Military decision and game theory. *JORSA* 2:365–385.

Hebert, M. H., 1964. The doubly discriminatory solutions of the four person constant-sum game. *AnMS* 52:345–375.

Helmbold, R. L., 1964. Some observations of the use of Lanchester's theory for prediction. *Operations Research* 12:778–781.

Henry, C., 1972. Market games with indivisible commodities and non-convex preferences. *Review of Economic Studies* 39:73–76.

Henshaw, R. C., and J. R. Jackson, 1966. *The Executive Game.* Homewood, IL: Irwin.

Hermann, C. F., 1967. Validation problems in games and simulations with special reference to models of international politics. *Behavioral Science* 12:216–231.

Herstein, I. N., and J. W. Milnor, 1953. An axiomatic approach to measurable utility. *Econometrica* 21:291–297.

Heyward, E. J. R., 1941. H. von Stackelberg's work on duopoly. *Economic Record* 17:99–106.

Hicks, J. R., 1938. *Value and Capital.* London: Oxford University Press.

Hildenbrand, W., 1968. On the core of an economy with a measure space of economic agents. *Review of Economic Studies* 35:443–452.

———, 1969. Pareto optimality for a measure space of economic agents. *International Economic Review* 10:363–372.

———, 1970a. On economies with many agents. *Journal of Economic Theory* 2:161–188.

———, 1970b. Existence of equilibria for economies with production and a measure space of consumers. *Econometrica* 38:608–623.

———, 1974. *Core and Equilibria of a Large Economy.* Princeton, NJ: Princeton University Press.

Hildenbrand, W., and A. P. Kirman, 1976. *Introduction to Equilibrium Analysis.* Amsterdam: North-Holland.

Hildreth, C., 1953. Alternative conditions for social orderings. *Econometrica* 21:81–94.

Hirschleifer, J., 1978. Natural economy versus political economy. *Journal of Social and Biological Structures* 1:319–337.

Hochberg, M., and W. M. Hirsch, 1970. Sperner families, s-systems, and a theorem of Meshalkin. *Annals, New York Academy of Sciences* 175:224–237.

Hoffman, A. J., and R. M. Karp, 1966. On nonterminating stochastic games. *Management Science* 12(series A):359–370.

Hoffman, A. J., and M. Richardson, 1961. Block design games. *Canadian Journal of Mathematics* 13:110–128.

Hoggatt, A. C., 1959. An experimental business game. *Behavioral Science* 4:192–203.

———, 1967. Measuring the cooperativeness of behavior in quantity variation duopoly games. *Behavioral Science* 12:109–121.

———, 1969. Response of paid student subjects to differential behavior of robots in bifurcated duopoly games. *Review of Economic Studies* 36:417–432.

Holladay, J. C., 1957. Cartesian products of termination games. *AnMS* 39:189–200.

Hotelling, H., 1929. Stability of competition. *Economic Journal* 39:41–57.

Houthakker, H. S., 1960. Additive preferences. *Econometrica* 28:244–257.

Howard, N., 1966. The theory of meta-games. *General Systems* 2:167–186.

———, 1971. *Paradoxes of Rationality.* Cambridge, MA: MIT Press.

Hurwicz, L., 1977. On the interaction between information and incentives in organizations. In *Communication and Control,* ed. K. Klippendorf. New York: Gordon & Breach.

Iklé, F., 1964. *How Nations Negotiate.* New York: Harper & Row.

Inada, K., 1969. On the simple majority decision rule. *Econometrica* 37:490–506.

Isaacs, R., 1952. A pursuit game with incomplete information. RAND Publication RM-791.

———, 1965. *Differential Games: A Mathematical Theory with Applications to Warfare and Pursuit, Control and Optimization.* New York: John Wiley.

Isaacs, R., and S. Karlin, 1954. A game of aiming and evasion. RAND Publication RM-1316.

Isbell, J. R., 1955. A class of game solutions. *Proceedings of the AMS* 6:346–348.

———, 1956. A class of majority games. *Quarterly Journal of Mathematics* 7:183–187.

———, 1957a. Finitary games. *AnMS* 39:79–96.

———, 1957b. Homogeneous games. *Mathematics Student* 25:123–128.

———, 1958. A class of simple games. *Duke Mathematical Journal* 25:423–439.

———, 1959. Absolute games. *AnMS* 40:357–396.

———, 1960a. Homogeneous games, II. *Proceedings of the AMS* 11:159–161.

———, 1960b. A modification of Harsanyi's bargaining model. *Bulletin of the AMS* 66:70–73.

———, 1964. Homogeneous games, III. *AnMS* 52:255–265.

Isbell, J. R., and W. H. Marlow, 1956. Attrition games. *NRLQ* 3:71–94.

Jackson, J. R., 1968. On decision theory under competition. *Management Science* 15:12–32.

Jacot, S.-P., 1963. *Stratégie et Concurrence.* Paris: SEDES.

Jamison, D. T., and L. J. Lau, 1973. Semiorders and the theory of choice. *Econometrica* 41:901–912.

Jamison, D. T., and E. Luce, 1972. Social homogeneity and the probability of intransitive majority rule. *Journal of Economic Theory* 5:79–87.

Jensen, D. L., 1977. A class of mutually satisfactory allocations. *The Accounting Review* 52:842–850.

Jentzsch, G., 1964. Some thoughts on the theory of cooperative games. *AnMS* 52:407–442.

Johnson, S. M., 1964. A search game. In *Advances in Game Theory,* eds. L. S. Shapley and A. W. Tucker. Princeton, NJ: Princeton University Press, pp. 39–48.

Jones, B., M. Steele, J. Gahagan, and J. Tedeschi, 1963. Matrix values and cooperative behavior in the prisoner's dilemma game. *Journal of Personality and Social Psychology* 8:148–153.

Joseph, M. J., and R. H. Willis, 1963. An experimental analog to two-party bargaining. *Behavioral Science* 8:117–127.

Jumarie, G., 1969. Differential games with delayed information (in French). *Comptes Rendus de l'Académie des Sciences* (Paris) 268:1040.

Kahn, H., 1960. *On Thermonuclear War.* Princeton, NJ: Princeton University Press.

Kahn, R. F., 1937. The problem of duopoly. *Economic Journal* 47:1–20.

Kalai, E., 1972. Cooperative non-sidepayment games: Extensions of sidepayment game solutions, metrics, and representative functions. Ph.D. dissertation, Cornell University.

Kalai, E., M. Maschler, and G. Owen, 1975. Asymptotic stability and other properties of trajectories and transfer sequences leading to the bargaining sets. *IJGT* 4:193–213.

Kalai, E., and R. W. Rosenthal, 1978. Arbitration of two-party disputes under ignorance. *IJGT* 7:65–72.

Kalai, E., and M. Smorodinsky, 1975. Other solutions to Nash's bargaining problem. *Econometrica* 43:513–518.

Kalisch, G. K., 1959. Generalized quota solutions of n-person games. *AnMS* 40:163–177.

Kalisch, G. K., J. W. Milnor, J. F. Nash, and E. D. Nering, 1954. Some experimental n-person games. In *Decision Processes,* eds. R. M. Thrall, C. H. Coombs, and R. L. Davis. New York: John Wiley, pp. 301–327.

Kalisch, G. K., and E. D. Nering, 1959. Countably infinitely many person games. *AnMS* 40:43–45.

Kannai, Y., 1963. Existence of a utility in infinite dimensional partially ordered spaces. *Israel Journal of Mathematics* 1:229–234.

———, 1966. Values of games with a continuum of players. *Israel Journal of Mathematics* 4:54–58.

———, 1968. Continuity properties of the core of a market. Publication RM-34, Department of Mathematics, Hebrew University, Jerusalem.

———, 1969. Countably additive measures in cores of games. *Journal of Mathematical Analysis and Applications* 27:227–240.

———, 1970. Continuity properties of the core of a market. *Econometrica* 38:791–815.

———, 1971. Continuity properties of the core of a market: A correction. Publication RM-67, Department of Mathematics, Hebrew University, Jerusalem.

Kaplansky, I., 1945. A contribution to von Neumann's theory of games. *Annals of Mathematics* 46:474–479.

Karlin, S., 1950. Operator treatment of the minmax principle. *AnMS* 24:133–154.

———, 1953a. Reduction of certain classes of games to integral equations. *AnMS* 28:125–158.

———, 1953b. On a class of games. *AnMS* 28:159–171.

———, 1953c. The theory of infinite games. *AnMS* 58:371–401.

———, 1957a. An infinite move game with a lag. *AnMS* 39:257–272.

———, 1957b. On games described by bell-shaped kernels. *AnMS* 39:365–391.

———, 1959. *Mathematical Methods and Theory in Games, Programming and Economics,* 2 volumes. Reading, MA: Addison-Wesley.

Keeler, E., 1971. Bridge as a game. RAND Publication P-4647.

Keeney, R. L., 1971. Utility independence and preferences for multiattributed consequences. *Operations Research* 19:875–893.

———, 1972. Utility functions for multiattributed consequences. *Management Science* 18:276–287.

Kemeny, J. G., O. Morgenstern, and G. L. Thompson, 1956. A generalization of the von Neumann model of an expanding economy. *Econometrica* 24:115–135.

Kemeny, J. G., and G. L. Thompson, 1957. The effect of psychological attitudes on the outcomes of games. *AnMS* 39:273–298.

Kifer, Y. I., 1969. Optimal strategy in games with an unbounded sequence of moves. *Theory of Probability and Its Applications* 14:279.

Kikuta, K., 1978. A lower bound on an imputation of a game. *Journal of the Operations Research Society of Japan* 21:457–468.

Kinnard, C., 1946. *Encyclopedia of Puzzles and Pastimes.* Secaucus, NJ: Citadel.

Kirman, A. P., and M. Sobel, 1974. Dynamic oligopoly with inventories. *Econometrica* 42:279–287.

Kirman, A. P., and D. Sonderman, 1972. Arrow's theorem, many agents, and invisible dictators. *Journal of Economic Theory* 5:267–277.

Kisi, T., 1961. Some silent duels. *Memoirs of the Defense Academy* (Yokosuka, Japan) 2:118–138.

Kleitman, D., 1969. On Dedekind's problem: The number of monotone Boolean functions. *Proceedings of the AMS* 21:677–682.

Klevorick, A. K., and G. H. Kramer, 1973. Social choice on pollution management: The *Genossenschaften. Journal of Public Economics* 2:101–146.

Knight, F. H., 1933. *Risk, Uncertainty and Profit.* London: School Reprints of Scarce Works.

Kohlberg, E., 1970. On non-atomic games: Conditions for f pNA. Publication RM-34, Department of Mathematics, Hebrew University, Jerusalem.

———, 1971. On the nucleolus of a characteristic function game. *SIAM Journal of Applied Mathematics* 20:62–66.

Koopmans, T. C., 1972a. Representations of preference orderings with independent components of consumption. In *Decision and Organization,* eds. C. B. McGuire and R. Radner. Amsterdam: North-Holland, pp. 57–78.

———, 1972b. Representations of preference orderings over time. In *Decision and Organization,* eds. C. B. McGuire and R. Radner. Amsterdam: North-Holland, pp. 79–100.

Kopelowitz, A., 1967. Computation of the kernels of simple games and the nucleolus of n-person games. Publication RM-31, Department of Mathematics, Hebrew University, Jerusalem.

Korobkov, V. K., 1965. On monotonic functions of the algebra of logic (in Russian). *Problemy Kibernetiki* 13:5–28.

Kramer, G., 1973. On a class of equilibrium conditions for majority rule. *Econometrica* 41:285–297.

———, 1977. A dynamical model of political equilibrium. *Journal of Economic Theory* 16:310–334.

Krasovskii, N. N., 1968. On capture in differential games. *Soviet Mathematics-Doklady* 9:988–991.

———, 1970. On a problem of pursuit. *Soviet Mathematics-Doklady* 11:343–346.

Krasovskii, N. N., and V. E. Tretyakov, 1968. Regularizing a problem on the encounter of motions of game theory (in Russian). *Prikladnaia Matematiki I Makhanika* 32:3–14.

Krentnel, W. D., J. C. C. McKinsey, and W. V. Quine, 1951. A simplification of games in extensive form. *Duke Mathematical Journal* 18:885–900.

Kreps, D. M., and R. Wilson, 1980. Sequential equilibria. Mimeographed paper, Graduate School of Business, Stanford University.

Krinskii, V. I., and V. A. Ponomarev, 1964. Playing blind games (in Russian). *Biofizika* 9:372–375.

Kuhn, H. W., 1950a. A simplified two-person poker. *AnMS* 24:97–103.

———, 1950b. Extensive games. *PNAS* 36:570–576.

———, 1953. Extensive games and the problem of information. In *Contributions to the Theory of Games, II.* eds. H. W. Kuhn and A. W. Tucker. Princeton, NJ: Princeton University Press.

———, 1961. An algorithm for equilibrium points in bimatrix games. *PNAS* 47:1657–1662.

———, 1967. On games of fair division. In *Essays in Mathematical Economics,* ed. M. Shubik. Princeton, NJ: Princeton University Press, pp. 27–37.

Kuhn, H. W., and G. P. Szegö, eds., 1971. *Differential Games and Related Topics.* Amsterdam: North-Holland.

Kulakovskaya, T. E., 1969. Sufficient conditions for the coincidence of the core and solution in a cooperative game (in Russian). *Lietuvos Mathematikos Rinkinys* 9:424–425.

———, 1971. Necessary and sufficient conditions for the coincidence of core and solution in a classical cooperative game. *Soviet Mathematics-Doklady* 12:1231–1234.

Kurz, M., 1978. Altruism as an outcome of social interaction. *American Economic Review* 68:216–222.

Kushner, H. J., and S. G. Chamberlain, 1969. Finite state stochastic games: Existence theorems and computational procedures. *IEEE Transactions on Automatic Control* 14:248–255.

Laing, J. D., and R. J. Morrison, 1973. Coalitions and payoffs in three-person sequential games. *Journal of Mathematical Sociology* 3:1–23.

Lanchester, T. W., 1916. *Aircraft in Warfare: The Dawn of the Fourth Arm.* London: Constable.

Lange, O., 1934. The determinateness of the utility function. *Review of Economic Studies* 1:218–225.

Lapidot, E., 1972. The counting vector of a simple game. *Proceedings of the AMS* 31:228–231.

Larsen, H. D., 1948. A dart game. *American Mathematical Monthly* 55:248, 640–641.

Lasswell, H., and A. Kaplan, 1950. *Power and Society.* New Haven, CT: Yale University Press.

Lavalle, I. H., 1967. A Bayesian approach to an individual player's choice of bid in competition sealed auctions. *Management Science* 13:A584–A597.

Ledyard, J. O., 1968. Resource allocation in unselfish environments. *American Economic Review* 58:227–237.

Leff, A. A., 1970. Injury, ignorance and spite: The dynamics of collection. *Yale Law Journal* 80:1–46.

———, 1976. *Swindling and Selling.* New York: The Free Press.

Leiserson, M., 1968. Factions and coalitions in one-party Japan: An interpretation based on the theory of games. *APSR* 62:770–787.

———, 1970. Game theory and the study of coalitional behavior. In *The Study of Coalitional Behavior,* eds. S. Groennings et al. New York: Holt, Rinehart & Winston, pp. 255–272.

Lemke, C. E., and J. T. Howson, 1964. Equilibrium points of bimatrix games. *SIAM Journal of Applied Mathematics* 12:413–423.

Leontief, W., 1947a. A note on the interrelationship of subsets of independent variables of a continuous function with continuous first derivatives. *Bulletin of the AMS* 53:343–350.

———, 1947b. Introduction to a theory of the internal structure of functional relationships. *Econometrica* 15:361–373.

Lerner, A. P., 1944. *The Economics of Control.* New York: Augustus M. Kelley.

———, 1970. Distributional equality and aggregate utility: Reply. *American Economic Reivew* 60:442–443.

Levine, E., and D. Lubell, 1970. Sperner collections on sets of real variables. *Annals, New York Academy of Sciences* 175:272–276.

Levine, P., and J. P. Ponssard, 1977. The values of information in some nonzero sum games. *IJGT* 6:221–229.

Levitan, R. E., 1964. Demand in an oligopolistic market and the theory of rationing. Publication RC-1545, IBM Corporation, Yorktown Heights.

Levitan, R. E., and M. Shubik, 1961. Financial allocation and marketing executive game (FAME). T. J. Watson Laboratory, IBM Corporation.

———, 1971a. Price variation duopoly with differentiated products and random demand. *Journal of Economic Theory* 3:23–39.

———, 1971b. Noncooperative equilibria and strategy spaces in an oligopolistic market. In *Mathematical Methods of Action and Reaction,* eds. H. W. Kuhn and G. P. Szegö. Amsterdam: North-Holland, pp. 429–447.

———, 1972. Price duopoly and capacity constraints. *International Economic Review* 13:111–122.

———, 1978. Duopoly with price and quantity as strategic variables. *IJGT* 7:1–11.

Lewontin, R. C., 1961. Evolution and the theory of games. *Journal of Theoretical Biology* 10:382–403.

Lieberman, B., 1960. Human behavior in a strictly determined 3 × 3 matrix game. *Behavioral Science* 5:317–322.

Liggett, T. M., and S. A. Lippman, 1970. Stochastic games with perfect information and time average payoff. *SIAM Review* 11:604–607.

Lindahl. E., 1958. Just taxation: A positive solution. Reprinted in part in *Classics in the Theory of Public Finance,* eds. R. A. Musgrave and A. E. Peacock. London: Macmillan.

Littig, L. W., 1965. Behavior in certain zero sum two person games. *Journal of Social Psychology* 66:113–125.

Little, I. M. D., 1957. *A Critique of Welfare Economics,* 2nd ed. New York: Oxford University Press.

Littlechild, S. C., 1970. A game theoretic approach to public utility pricing. *Western Economic Journal* 8:162–166.

———, 1974. A simple expression for the nucleolus in a special case. *IJGT* 3:21–29.

Littlechild, S. C., and G. Owen, 1973. Simple expression for the Shapley value in a special case. *Management Science* W3:370–373.

———, 1976. A further note on the nucleolus of the "airport game." *IJGT* 5:91–95.

Littlechild, S. C., and G. F. Thompson, 1977. Aircraft landing fees: A game theory approach. *Bell Journal of Economics* 8:186–204.

Liu, P. T., ed., 1980. *Dynamic Optimization and Mathematical Economics.* New York: Plenum Press.

Loane, E. P., 1966. Note on level-debt-service municipal bidding. *Management Science* 13:291–293.

Loehman, E., and A. Whinston, 1971. A new theory of pricing and decisionmaking for public investment. *Bell Journal of Economics* 2:606–628.

Loomis, L. H., 1946. On a theorem of von Neumann. *PNAS* 32:213–215.

Lorimer, P., 1967. A note on orderings. *Econometrica* 35:537–539.

Lubell, D., 1966. A short proof of Sperner's lemma. *Journal of Combinatorial Theory* 1:299.

Lucas, W. F., 1963. On solutions to n-person games in partition function form. Ph.D. dissertation. University of Michigan, Ann Arbor.

———, 1965. Solutions for four-person games in partition function form. *SIAM Journal* 13:118–128.

———, 1966a. Solution theory for n-person games in partition-function form. In *Theory of Games: Techniques and Applications,* ed. A. Mensch. New York: American Elsevier, pp. 131–134.

———, 1966b. n-person games with only 1, $n - 1$, and n-person coalitions. *Zeitschrift für Wahrscheinlichkeitstheorie und verwandte Gebiete* 6:287–292.

———, 1967a. Solutions for a class of n-person games in partition function form. *NRLQ* 14:15–21.

———, 1967b. A counterexample in game theory. *Management Science* 13:766–767.

———, 1968a. A game in partition function form with no solution. *SIAM Journal of Applied Mathematics* 16:582–585.

———, 1968b. A game with no solution. *Bulletin of the AMS* 74:237–239.

———, 1968c. On solutions for n-person games. RAND Publication RM-5567-PR.

———, 1969a. Games with unique solutions that are nonconvex. *Pacific Journal of Mathematics* 28:599–602.

———, 1969b. The proof that a game may not have a solution. *Transactions of the AMS* 137:219–229.

———, 1971. Some recent developments in n-person game theory. *SIAM Review* 13:491–523.

———, 1972. An overview of the mathematical theory of games. *Management Science* 18:3–19.

———, 1974. Measuring power in weighted voting systems. Publication TR-227, Department of Operations Research, Cornell University.

———, 1978. On the existence of stable sets. In Report on Fourth International Workshop on Game Theory (mimeographed), Cornell University, p. 20.

Lucas, W. F., and J. C. Maceli, 1977. Discrete partition function games. Publication TR-344, School of Operations Research and Industrial Engineering, Cornell University.

Lucas, W. F., and M. Rabie, 1980. Games with no solutions and empty cores. Technical Report 474, School of Operations Research, Cornell University.

Luce, R. D., 1954. A definition of stability for *N*-person games. *Annals of Mathematics* 59:357–366.

———, 1955. Ψ-stability: A new equilibrium concept for *n*-person game theory. In *Mathematical Models of Human Behavior,* ed. J. W. Dunlop. Stamford, CT: Dunlop and Associates, pp. 32–44.

———, 1956. Semiorders and a theory of utility discrimination. *Econometrica* 24:178–191.

———, 1959a. *Individual Choice Behavior.* New York: John Wiley.

———, 1959b. On the possible psychophysical laws. *Psychological Review* 66:81–95.

Luce, R. D., and W. Edwards, 1958. The derivation of subjective scales from just noticeable differences. *Psychological Review* 65:222–237.

Luce, R. D., and H. Raiffa, 1957. *Games and Decisions.* New York: John Wiley.

Luce, R. D., and A. A. Rogow, 1956. A game theoretic analysis of congressional power distribution for a stable two-party system. *Behavioral Science* 1:83–95.

Lumsden, H., 1966. Perception and information in strategic thinking. *Journal of Peace Research* 3:257–277.

Maitra, A., and T. Parthasarathy, 1970. On stochastic games. *Journal of Optimization Theory and Application* 5:289–300.

Malcolm, D., and B. Lieberman, 1965. The behavior of responsive individuals playing a two person zero sum game requiring the use of mixed strategies. *Psychonomic Science* 2:373–374.

Mangasarian, O. L., 1964. Equilibrium points of bimatrix games. *SIAM Journal* 12:778–780.

Mangasarian, O. L., and H. Stone, 1964. Two person nonzero sum games and quadratic programming. *Journal of Mathematical Analysis and Applications* 9:348–355.

Mann, I., and L. Shapley, 1964. The a priori voting strength of the electoral college. In *Game Theory and Related Approaches to Social Behavior,* ed. M. Shubik. New York: John Wiley, pp. 151–164.

March, J. G., 1955. An introduction to the theory and measurement of influence. *APSR* 49:431–451.

———, 1966. The power of power. In *Varieties of Political Theory,* ed. D. Easton. Englewood Cliffs, NJ: Prentice-Hall, pp. 39–70.

Marchi, E., and R. I. C. Hansell, 1973a. A framework for systematic zoological studies with game theory. *Mathematical Biosciences* 16:31–58.

———, 1973b. Generalizations on the parsimony question in evolution. *Mathematical Biosciences* 17:11–34.

Marris, R., 1964. *The Economic Theory of Managerial Capitalism.* New York: Free Press.

Marschak, J., 1950. Rational behavior, uncertain prospects, and measurable utility. *Econometrica* 18:111–141.

Marschak, T., and R. Selten, 1974. *General Equilibrium with Price Making Firms.* Berlin: Springer-Verlag.

Marwell, G., K. Ratcliff, and D. R. Schmitt, 1969. Minimizing differences in a maximizing difference game. *Journal of Personality and Social Psychology* 12:158–183.

Maschler, M., 1963. The power of a coalition. *Management Science* 10:8–29.

———, 1965. Playing an n-person game: An experiment. Publication RN-73, Economic Research Program, Princeton, NJ.

———, 1966a. A price leadership method for solving the inspector's nonconstant-sum game. *NRLQ* 13:11–33.

———, 1966b. The inequalities that determine the bargaining set $\mathcal{M}_1^{(i)}$. *Israel Journal of Mathematics* 4:127–133.

———, 1976. An advantage of the bargaining set over the core. *Journal of Economic Theory* 13:184–192.

Maschler, M., B. Peleg, and L. S. Shapley, 1970. The kernel and nucleolus of a cooperative game as locuses in the strong ϵ-core. Publication RM-60, Research Program in Game Theory and Mathematical Economics, Hebrew University, Jerusalem.

———, 1972. The kernel and bargaining set for convex games. *IJGT* 1:73–93.

Maschler, M., and M. Perles, 1978. The superadditive solution for the Nash bargaining game. In Report on Fourth International Workshop On Game Theory (mimeographed), Cornell University. (Submitted to *IJGT.*)

Mas-Colell, A., 1974. An equilibrium existence theorem without complete or transitive preferences. Working Paper IP-195, Center for Research in Management Science, University of California, Berkeley.

Maskin, E., 1977. On game forms with efficient Nash equilibria. Paper presented at Stanford University, mimeographed.

Matheson, J. D., 1967. Preferential strategies with imperfect weapons. Publication AR 67-1, April 1967, AD 813 915, Analytic Services, Inc.

Matula, D. W., 1966. Games of sequence prediction. Publication ORC 66-3, Operations Research Center, University of California, Berkeley.

Mauss, M., 1954. *The Gift.* Glencoe, IL: Free Press.

May, K. O., 1952. A set of independent, necessary and sufficient conditions for simple majority decision. *Econometrica* 20:680–684.

———, 1954. Transitivity, utility, and aggregation in preference patterns. *Econometrica* 22:1–13.

Mayberry, J., J. F. Nash, Jr., and M. Shubik, 1953. A comparison of treatments of a duopoly situation. *Econometrica* 21:141–155.

Maynard Smith, J., 1972. Game theory and the evolution of fighting. In *On Evolution*, ed. J. Maynard Smith. Edinburgh: Edinburgh University Press.

———, 1974. The theory of games and the evolution of animal conflict. *Journal of Theoretical Biology* 47:209–221.

———, 1976. Evolution and the theory of games. *American Scientist* 64:41–45.

McDonald, J., 1975. *The Game of Business*. New York: Doubleday.

McDonald, J., and J. W. Tukey, 1949. Colonel Blotto: A problem of military strategy. *Fortune* (June): 102.

McKelvey, R. D., P. C. Ordeshook, and M. D. Winer, 1978. The competitive solution for *n*-person games without transferable utility, with an application to committee games. *APSR* 72:599–615.

McKenney, J. L., 1962. An evaluation of a business game in an MBA curriculum. *Journal of Business* 35:278–286.

———, 1967. *Simulation Gaming for Management Development*. Cambridge, MA: Harvard Business School.

McKinsey, J. C. C., 1952a. *Introduction to the Theory of Games*. New York: McGraw-Hill.

———, 1952b. Some notions and problems of game theory. *Bulletin of the AMS* 58:591–611.

Megiddo, N., 1971. The kernel and the nucleolus of a product of simple games. *Israel Journal of Mathematics* 9:210–221.

———, 1974a. Kernels of compound games with simple components. *Pacific Journal of Mathematics* 50:531–555.

———, 1974b. Nucleoluses of compound simple games. *SIAM Journal of Applied Mathematics* 26:607–621.

Mertens, J. F., and S. Zamir, 1971. The value of two-person zero-sum repeated games with lack of information on both sides. *IJGT* 1:39–64.

Meshalkin, L. D., 1963. A generalization of Sperner's theorem on the number of subsets of a finite set. *Theory of Probability and Its Applications* 8:203–204.

Midgaard, K., 1970. Strategy and ethics in international politics. *Cooperation and Conflict* 4:224–240.

Miller, D. R., 1973. A Shapley value analysis of the proposed Canadian constitutional amendment scheme. *Canadian Journal of Political Science* 6:140–143.

Mills, H., 1960. Equilibrium points in finite games. *SIAM Journal* 8:397–402.

———, 1961. A study in promotional competition. In *Mathematical Models and Methods in Marketing*, eds. B. M. Bass et al. Homewood, IL: Irwin, pp. 245–301.

Mills, W. H., 1954. The four person game—edge of the cube. *Annals of Mathematics* 59:367–378.

———, 1959. The four person game—finite solutions on the face of the cube. *AnMS* 40:135–143.

Milnor, J., 1951. Games against nature. RAND Publication RM-679.

———, 1952. Reasonable outcomes for *n*-person games. RAND Publication RM-916.

———, 1953. Sums of positional games. *AnMS* 28:291–301.

———, 1954. Games against nature. In *Decision Processes*, eds. R. M. Thrall, C. H. Coombs, and R. L. Davis. New York: John Wiley, pp. 49–59.

Milnor, J., and L. S. Shapley, 1957. On games of survival. *AnMS* 39:15–45.

———, 1961. Values of large games, II: Oceanic games. *Mathematics of Operations Research* 3:290–307.

Mishan, E. J., 1960. A survey of welfare economics, 1939–59. *Economic Journal* 70:197–265.

Miyasawa, K., 1962. An economic survival game. *Journal of the Operations Research Society of Japan* 4:95–113.

———, 1964. The *n*-person bargaining game. *AnMS* 52:547–575.

Moriguti, S., and S. Suganami, 1959. Notes on auction bidding. *Journal of the Operations Research Society of Japan* 2:43–59.

Morin, R. E., 1960. Strategies in games with saddlepoints. *Psychological Reports* 7:479–485.

Morishima, M., 1965. Should dynamic utility be cardinal? *Econometrica* 33:869–871.

Morrill, J., 1966. One-person games of economic survival. *NRLQ* 13:49–69.

Morse, P. M., and G. E. Kimball, 1951. *Methods of Operations Research.* New York: John Wiley.

Moulin, H., and J. P. Vial, 1978. Strategically zero-sum games: The class of games whose completely M strategies cannot be improved upon. *IJGT* 7:201–221.

Mueller, D. C., 1967. The possibility of a social welfare function: Comment (on a paper by J. S. Coleman). *American Economic Review* 57:1304–1311.

Munier, B., 1972. Contribution de la théorie des jeux à la critique du théorème de Heckscher, Ohlin, Samuelson. *Economie Appliquée* 25:61–90.

Murakami, Y., 1966. Formal structure of majority decisions. *Econometrica* 34:709–718.

Mycielski, J., 1964a. On the axiom of determinateness. *Fundamenta Mathematicae* 53:205–224.

———, 1964b. Continuous games with perfect information. *AnMS* 52:103–112.

———, 1966. On the axiom of determinateness, II. *Fundamenta Mathematicae* 59:203–212.

Mycielski, J., and H. Steinhaus, 1962. A mathematical axiom contradicting the axiom of choice. *Bull. Acad. Polon. Sci. Ser. Math. Astr. Phys.* 10:1–3.

Mycielski, J., and S. Swierczkowski, 1964. On the Lebesgue measurability and the axiom of determinateness. *Fundamenta Mathematicae* 54:67–71.

Mycielski, J., S. Swierczkowski, and A. Zieba, 1956. On infinite positional games. *Bull. Acad. Polon. Sci. Ser. Math. Astr. Phys.* 4:485–488.

Mycielski, J., and A. Zieba. 1955. On infinite games. *Bull. Acad. Polon. Sci. Ser. Math. Astr. Phys.* 3:133–136.

Myerson, R. B., 1977. Values of games in partition function form. *IJGT* 6:23–31.

———, 1978. Refinements of the Nash equilibrium concept. *IJGT* 7:73–80.

Nagel, J. H., 1968. Some questions about the concept of power. *Behavioral Science* 13:129–137.

———, 1975. *The Descriptive Analysis of Power.* New Haven, CT: Yale University Press.

Nakamura, K., 1973. ψ stability of a cooperative game with side payments. *IJGT* 2:129–140.

Nanson, E. J., 1882. Methods of election. *Transactions and Proceedings of the Royal Society of Victoria* 19:197–240.

Nash, J. F., Jr., 1950a. The bargaining problem. *Econometrica* 18:155–162.

———, 1950b. Rational non-linear utility. RAND informal note (reproduced here as Appendix A.2).

———, 1950c. Equilibrium points in n-person games. *PNAS* 36:48–49.

———, 1951. Noncooperative games. *Annals of Mathematics* 54:289–295.

———, 1953. Two-person cooperative games. *Econometrica* 21:128–140.

Nash, J. F., Jr., and L. S. Shapley, 1950. A simple three-person poker game. *AnMS* 24:105–116.

Neave, E. M., and J. C. Wiginton, 1977. Game theory, behavior and the paradox of the prisoner's dilemma: Resolution by linguistic variables. Publication WP-77-24, School of Business, Queens University, Kingston, Ontario.

Nering, E. D., 1959. Symmetric solutions for general-sum symmetric 4-person games. *AnMS* 40:111–123.

Nichol, A. J., 1934. A re-appraisal of Cournot's theory of duopoly price. *Journal of Political Economy* 42:80–105.

Nikaido, H., 1954. On von Neumann's minimax theorem. *Pacific Journal of Mathematics* 4:65–72.

Nikaido, H., and K. Isoda, 1955. Note on noncooperative games. *Pacific Journal of Mathematics* 5:807–815.

Nikolskii, M. S., 1971. Linear differential games of pursuit with delay (in Russian). *Doklady Akademii Nauk SSSR* 197:1018–1021.

Nitz, L. H., and J. L. Phillips, 1969. The effects of divisibility of payoff on confederative behavior. *Journal of Conflict Resolution* 13:381–387.

Nozick, R., 1968. Weighted voting and "one-man, one-vote." In *Representation* (Nomos X), eds. J. R. Pennock and J. W. Chapman. New York: Atherton, pp. 217–225.

Nyblen, G., 1951. *The Problem of Summation in Economic Science.* Lund: Gleerup.

Ordeshook, P. C., ed., 1978. *Game Theory and Political Science.* New York: New York University Press.

Ore, O., 1962. *Theory of Graphs.* Providence, RI: AMS (Colloquium Publications 38).

Otter, R., and J. Dunne, 1953. Games with equilibrium points. *PNAS* 39:310–314.

Owen, G., 1964. The tensor composition of non-negative games. *AnMS* 52:307–326.

———, 1966. Discriminatory solutions of n-person games. *Proceedings of the AMS* 17:653–657.

———, 1968a. *Game Theory.* Philadelphia, PA: W. B. Saunders.

———, 1968b. n-person games with only 1, $n - 1$, and n-person coalitions. *Proceedings of the AMS* 19:1258–1261.

———, 1971. Political games. *NRLQ* 18:345–355.

———, 1972. Multilinear extensions of games. *Management Science* 18:64–79.

———, 1974. A discussion of minimax. *Management Science* 20:1317.

———, 1975. Multilinear extensions and the Banzhaf value. *NRLQ* 22:741–750.

Oxtoby, J. C., 1957. The Banach–Mazur game and Banach category theorem. *AnMS* 39:159–163.

Park, R. E., 1967. The possibility of a social welfare function: Comment (on a paper by J. S. Coleman). *American Economic Review* 57:1300–1304.

Parsons, T., 1957. The distribution of power in American society. *World Politics* 10:123–143.

Parthasarathy, T., 1966. A note on compound simple games. *Proceedings of the AMS* 17:1334–1340.

———, 1967. Minimax theorems in separable compact metric spaces. *Journal of the Indian Statistical Association* 5:182–191.

———, 1969a. Product solutions for simple games, II. *Proceedings of the AMS* 20:107–114.

———, 1969b. Product solutions for simple games, III. *Proceedings of the AMS* 23:412–420.

———, 1971. Discounted and positive stochastic games. *Bulletin of the AMS* 77:134–136.

Parthasarathy, T., and T. E. S. Raghavan, 1971. *Some Topics in Two-Person Games.* New York: American Elsevier.

———, 1974. Structure of equilibrium in n-person noncooperative games. Working paper, University of Illinois, Chicago.

Parthasarathy, T., and M. Stern, 1977. On Markov games. In *Lecture Notes in Pure and Applied Mathematics,* vol. 30. New York: Marcel Dekker, pp. 1–46.

Pazner, E. A., and D. Schmeidler, 1974. A difficulty in the concept of fairness. *Review of Economic Studies* 41:441–443.

Peleg, B., 1959. On the set of solvable *n*-person games. *Bulletin of the AMS* 65:380–383.

———, 1963a. Solutions to cooperative games without side payments. *Transactions of the AMS* 106:280–292.

———, 1963b. Quota games with a continuum of players. *Israel Journal of Mathematics* 1:48–53.

———, 1963c. Bargaining sets of cooperative games without side payments. *Israel Journal of Mathematics* 1:366–369.

———, 1965a. An inductive method for constructing minimal balanced collections of finite sets. *NRLQ* 12:155–162.

———, 1965b. The kernel of the composition of characteristic function games. *Israel Journal of Mathematics* 3:127–138.

———, 1965c. Composition of general sum games. Publication RM-74, Economics Research Program, Princeton University.

———, 1965d. Composition of kernels of characteristic function games. Publication RM-15, Department of Mathematics, Hebrew University, Jerusalem.

———, 1966. The independence of game theory of utility theory. *Bulletin of the AMS* 72:995–999.

———, 1967. Existence theorem for the bargaining set $\mathcal{M}_1^{(i)}$. In *Essays in Mathematical Economics,* ed. M. Shubik. Princeton, NJ: Princeton University Press, pp. 53–56.

———, 1968. On weights of constant-sum majority games. *SIAM Journal of Applied Mathematics* 16:527–532.

———, 1969. The extended bargaining set for cooperative games without side payments. Publication RM-44, Department of Mathematics, Hebrew University, Jerusalem.

———, 1970. Utility functions for partially ordered topological spaces. *Econometrica* 38:93–96.

———, 1978. Consistent voting systems. *Econometrica* 46:153–162.

Pen, J., 1952. A general theory of bargaining. *American Economic Review* 46:24–42.

Petrosjan, L. A., and N. V. Murzov, 1967. Dynamic pursuit games. *Soviet Mathematics-Doklady* 8:272–275.

Petrov, N. N., 1970. A proof of the existence of value of a pursuit game with restricted time (in Russian). *Differencial nye Uravnenija* 6:784–794.

Pfanzagl, J., 1959. A general theory of measurement: Applications to utility. *NRLQ* 6:283–294.

Phillips, J. L., and L. H. Nitz, 1968. Social contacts in a three-person political convention simulation. *Journal of Conflict Resolution* 12:206–214.

Plott, C., 1967. A notion of equilibrium and its possibility under majority rules. *American Economic Review* 57:787–806.

———, 1971. Recent results in the theory of voting. In *Frontiers of Quantitative Economics,* ed. M. D. Intrilligator. Amsterdam: North-Holland, pp. 109–127.

Polanyi, K., C. M. Arensburg, and H. W. Pearson, 1957. *Trade and Market in the Early Empires.* Glencoe, IL: Free Press.

Pollak, R. A., 1965. "Dynamic utility" [by R. Frisch]: A comment. *Econometrica* 33:872–877.

———, 1967. Additive von Neumann–Morgenstern utility functions. *Econometrica* 35:485–494.

Pollatschek, M. A., and B. Avi-Itzhak, 1969. Algorithms for stochastic games. *Management Science* 15:339–415.

Ponssard, J.-P., 1971. Information usage in non-cooperative game theory. Ph.D. dissertation, Stanford University.

Pontryagin, L. S., 1957. Some mathematical problems arising in connection with the theory of optimal automatic control systems (in Russian). In *Proceedings, Conference on Basic Problems in Automatic Control and Regulation,* Academy of Sciences, Moscow.

———, 1966. On the theory of differential games. *Russian Mathematical Surveys* 21:193–246.

———, 1970. Linear differential games of evasion. *Soviet Mathematics-Doklady* 11:359–361.

Postlewaite, A., 1979. Manipulation via endowments. *Review of Economic Studies* 46:255–262.

Postlewaite, A., and R. W. Rosenthal, 1974. Disadvantageous syndicates. *Journal of Economic Theory* 9:324–326.

Postlewaite, A., and D. Schmeidler, 1978. Approximate efficiency of non Walrasian Nash equilibria. *Econometrica* 46:127–135.

Pruitt, D. G., and M. K. Kimmel, 1977. Twenty years of experimental gaming: Critique, synthesis and suggestions for the future. *Annual Review of Psychology* 28:363–392.

Quade, E. S., 1964. Methods and procedures. In *Analysis for Military Decisions,* ed. E. S. Quade. Chicago, IL: Rand McNally, pp. 149–176.

Quine, W. V. O., 1944. *Mathematical Logic.* New York: W. W. Norton.

Rabin, M. O., 1957. Effective computability of winning strategies. *AnMS* 39:147–157.

Rader, T., 1963. The existence of a utility function to represent preferences. *Review of Economic Studies* 31:229–232.

Radner, R., 1968. Competitive equilibrium under uncertainty. *Econometrica* 36:31–58.

Rae, D. W., 1969. Decision rules and individual values in constitutional choice. *APSR* 63:40–56.

Raiffa, H., 1953. Arbitration schemes for generalized two-person games. In *Contributions to the Theory of Games,* eds. H. Kuhn and A. W. Tucker. Princeton, NJ: Princeton University Press, pp. 361–387.

———, 1968. *Decision Analysis: Introductory Lectures on Choices under Uncertainty.* Reading, MA: Addison-Wesley.

Raiffa, H., and R. Schlaiffer, 1961. *Applied Statistical Decision Theory.* Cambridge, MA: Division of Research, Graduate School of Business Administration, Harvard University.

RAND Corporation, 1971, 1977. *A Bibliography of Selected RAND Publications.* RAND Publication SB1039.

Rapoport, A., 1960. *Fights, Games and Debates.* Ann Arbor, MI: University of Michigan Press.

———, 1964. *Strategy and Conscience.* New York: Harper & Row.

———, 1967. Games which simulate deterrence and disarmament. *Peace Research Review* 1:1–76.

Rapoport, A., and A. M. Chammah, 1965. *Prisoner's Dilemma.* Ann Arbor, MI: University of Michigan Press.

Rapoport, A., and M. J. Guyer, 1966. A taxonomy of 2 × 2 games. *General Systems* 11:203–214.

Rapoport, A., M. J. Guyer, and D. G. Gordon, 1976. *The 2 × 2 Game.* Ann Arbor, MI: University of Michigan Press.

Rawls, J., 1971. *A Theory of Justice.* Cambridge, MA: Harvard University Press.

Reichardt, R., 1968. Three-person games with imperfect coalitions: A sociologically relevant concept in game theory. *General Systems* 13:189–204.

Restrepo, R., 1957. Tactical problems involving several actions. *AnMS* 39:313–335.

Richardson, L. F., 1960a. *Arms and Insecurity.* Chicago, IL: Quadrangle Books.

———, 1960b. *Statistics of Deadly Quarrels.* Chicago, IL: Quadrangle Books.

Richardson, M., 1946. On weakly ordered systems. *Bulletin of the AMS* 52:113–116.

———, 1953a. Solutions of irreflexive relations. *Annals of Mathematics* 58:573–590.

———, 1953b. Extension theorems for solutions of irreflexive relations. *PNAS* 39:649–655.

———, 1955. Relativization and extension of solutions of irreflexive relations. *Pacific Journal of Mathematics* 5:551–584.

———, 1956. On finite projective games. *Proceedings of the AMS* 7:458–465.

Richter, M. K., 1971. Coalitions, core, and competition. *Journal of Economic Theory* 3:323–334.

Rigby, F. D., 1959. Introductory note to L. S. Shapley's "Equilibrium points in games with vector payoffs." *NRLQ* 6:57–58.

Riker, W. H., 1959. A test of the adequacy of the power index. *Behavioral Science* 4:120–131.

———, 1961. Voting and the summation of preferences: An interpretative bibliographic review of selected developments during the last decade. *APSR* 55:900–911.

———, 1962. *The Theory of Political Coalitions.* New Haven, CT: Yale University Press.

———, 1964. Some ambiguities in the notion of power. *APSR* 58:341–349.

———, 1966. A new proof of the size principle. In *Mathematical Applications in Political Science, II,* ed. J. Bernd. Dallas, TX: Arnold Foundation, Southern Methodist University.

———, 1967. Bargaining in a three person game. *APSR* 61:642–656.

Riker, W. H., and R. G. Niemi, 1964. Anonymity and rationality in the essential three-person game. *Human Relations* 17:131–141.

Riker, W., and P. Ordeshook, 1968. A theory of the calculus of voting. *APSR* 62:25–42.

———, 1973. *An Introduction to Positive Political Theory.* Englewood Cliffs, NJ: Prentice-Hall.

Riker, W. H., and L. S. Shapley, 1968. Weighted voting: A mathematical analysis for instrumental judgments. In *Representation* (Nomos X), eds. J. R. Pennock and J. W. Chapman. New York: Atherton, pp. 199–216.

Riker, W. H., and J. Zavoina, 1970. Rational behavior in politics: Evidence from a three-person game. *APSR* 64:48–60.

Riley, V., and J. P. Young, 1957. *Bibliography on War Gaming.* Chevy Chase, MD: Johns Hopkins, Operations Research Office.

Roberts, F. S., 1971. Homogeneous families of semiorders and the theory of probabilistic consistency. *Journal of Mathematical Psychology* 8:248–263.

Roberts, J., and H. Sonnenschein, 1977. On the foundations of the theory of monopolistic competition. *Econometrica* 45:101–113.

Robinson, J., 1951. An iterative method of solving a game. *Annals of Mathematics* 54:296–301.

Rogers, P. D., 1969. Nonzero-sum stochastic games. Publication ORC-69-8, Operations Research Center, University of California, Berkeley.

Romanovski, I. V., 1961. Random walks of game-type (in Russian). *Teoriya Veroyotnostei i ee Primeneniya* 6:426–429.

Rosenfeld, J. L., 1964. Adaptive competitive decision. *AnMS* 52:69–83.

Rosenmüller, J., 1968. On some classes of game functions. Mathematics Institute Preprint 20, Aarhus University, Aarhus, Denmark.

———, 1969. On core and value. Mimeographed paper, School of Mathematics, University of Minnesota.

———, 1971a. *Kooperative Spiele und Märkte.* Berlin: Springer-Verlag.

———, 1971b. On core and value. In *Methods of Operations Research,* ed. R. Herr. Meisenhein: Anton Hain.

———, 1971c. On a generalization of the Lemke-Howson algorithm to noncooperative *N*-person games. *SIAM Journal of Applied Mathematics* 21:73–79.

Rosenthal, R. W., 1970. Stability analysis of cooperative games in effectiveness form. Technical Report 70-11, Operations Research Department, Stanford University.

———, 1972. Cooperative games in effectiveness form. *Journal of Economic Theory* 51:88–101.

Roth, A. E., 1979. *Axiomatic Models of Bargaining.* Lecture Notes in Economics and Mathematical Systems 170. Berlin: Springer-Verlag.

———, 1980. Values for games without sidepayments: Some difficulties with current concepts. *Econometrica* 48:457–466.

Roth, B., 1975. Coalition formation in the triad: An integration of game theoretic and social psychological perspectives. Ph.D. dissertation, New School for Social Research, New York.

Rothenberg, J., 1961. *The Measurement of Social Welfare.* Englewood Cliffs, NJ: Prentice-Hall.

Rothkopf, M. A., 1969. A model of rational competitive bidding. *Management Science* 15:362–373.

———, 1977. Bidding in simultaneous auctions with a constraint on exposure. *Operations Research* 25:620–629.

Rozenoer, L. T., 1959. L. S. Pontrjagin's maximum function principle in its application to the theory of optimum systems (in Russian). *Automatica i Telemakhanika* 20:1320–1334, 1441–1458, 1561–1578.

Rubin, H., 1949. Postulates for the existence of measurable utility and psychological probability (abstract). *Bulletin of the AMS* 55:1050–1051.

Sakaguchi, M., 1956. Stochastic games of survival. *Reports of the University of Electro-Communications* (Osaka) 8:55–59.

Salisbury, R. F., 1968. *Anthropological Economics.* Englewood Cliffs, NJ: Prentice-Hall.

Samuelson, P. A., 1948. *Foundations of Economic Analysis.* Cambridge, MA: Harvard University Press.

Sanghvi, A., and M. Sobel, 1976. Bayesian games as stochastic processes. *IJGT* 5:1–22.

Satterthwaite, M., 1975. Strategy-proofness and Arrow's conditions: Existence and correspondence theorems for voting procedures and social welfare functions. *Journal of Economic Theory* 10:187–217.

Sauermann, H., ed., 1967. *Contributions to Experimental Economics,* vol. 1. Tübingen: Mohr.

———, 1970. *Contributions to Experimental Economics,* vol. 2. Tübingen: Mohr.

———, 1972. *Contributions to Experimental Economics,* vol. 3. Tübingen: Mohr.

Savage, L. J., 1954. *The Foundations of Statistics.* New York: John Wiley.

Sawyer, J., and H. Guetzkow, 1965. Bargaining and negotiation in international relations. In *International Behavior, A Social-Psychological Analysis,* ed. H. C. Kelman. New York: Holt, Rinehart & Winston, chapter 13.

Scarf, H. E., 1957. On differential games with survival payoff. *AnMS* 39:393–405.

———, 1967. The core of an *n*-person game. *Econometrica* 35:50–69.

———, 1971. On the existence of a cooperative solution for a general class of *n*-person games. *Journal of Economic Theory* 3:169–181.

Scarf, H. E., and L. S. Shapley, 1957. Games with partial information. *AnMS* 39:213–229.

———, 1973. On cores and indivisibility. In *Studies in Optimization,* eds. G. Dantzig and C. Eaves. Washington, DC: Mathematical Association of America.

Schelling, T. C., 1960a. *The Strategy of Conflict.* Cambridge, MA: Harvard University Press.

———, 1960b. Experimental games and bargaining theory. *World Politics* 14:47–68.

———, 1978. *Micromotives and Macrobehavior.* New York: Norton.

Schleicher, H., 1971. *Staatshaushalt und Strategie.* Berlin: Duncker & Humblot.

———, ed. 1979. *Jeux, Information et Groupes.* Paris: Economica.

Schmeidler, D., 1967. On balanced games with infinitely many players. Publication RM-28, Department of Mathematics, Hebrew University, Jerusalem.

———, 1969a. Competitive equilibria in markets with a continuum of traders and incomplete preferences. *Econometrica* 37:578–585.

———, 1969b. The nucleolus of a characteristic function game. *SIAM Journal of Applied Mathematics* 17:1163–1170.

———, 1971. A condition for the completeness of partial preference relations. *Econometrica* 39:403–404.

———, 1973. An individually rational procedure for planning the provision of public goods. Paper presented to NSD-SSRC Conference on Individual Rationality, University of Massachusetts, Amherst, mimeographed.

Schofield, N., 1978. Generalized bargaining sets for cooperative games. *IJGT* 7:183–199.

Schotter, A., 1981. *The Economic Theory of Social Institutions.* New York: New York University Press.

Schroeder, R. G., 1970. Linear programming solutions to ratio games. *Operations Research* 18:300–305.

Scott, D., 1964. Measurement structures and linear inequalities. *Journal of Mathematical Psychology* 1:233–247.

Selten, R., 1960. Bewertung strategischer Spiele. *Zeitschrift für die gesamte Staatswissenschaft* 116:221–282.

———, 1964. Valuation of *n*-person games. In *Advances in Game Theory,* eds. M. Dresher, L. S. Shapley, and A. W. Tucker. Princeton, NJ: Princeton University Press, pp. 577–626.

———, 1965. Spieltheoretische Behandlung eines oligopolmodells mit Nachfrageträgheit. *Zeitschrift für die gesamte Staatswissenschaft* 121:301–324, 667–689.

———, 1972. Equal share analysis of characteristic function experiments. In *Contributions to Experimental Economics,* vol. 3, ed. H. Sauermann. Tübingen: Mohr.

———, 1973. A simple model of imperfect competition where 4 are few and 6 are many. *IJGT* 2:141–201.

———, 1975. Reexamination of the perfectness concept for equilibrium points in extensive games. *IJGT* 4:25–55.

———, 1978. A note on evolutionary stable strategies in asymmetric animal conflicts. Research Paper 73, Institute of Mathematical Economics, University of Bielefeld.

Sen, A. K., 1964. Preferences, votes and the transitivity of majority decisions. *Review of Economic Studies* 31:163–165.

———, 1970. *Collective Choice and Social Welfare.* San Francisco, CA: Holden-Day.

———, 1973. On ignorance and equal distribution. *American Economic Review* 63:1022–1024.

Shafer, W., and H. Sonnenschein, 1974. The Arrow-Debreu lemma on abstract economies with noncomplete and nontransitive preferences. Discussion Paper 94, Center for Mathematical Studies in Economics and Management Science, Northwestern University.

Shakun, M. F., 1965. Advertising expenditures in coupled markets—A game-theory approach. *Management Science* 11:42–47.

Shapiro, N. Z., and L. S. Shapley, 1978. Values of large games, I: A limit theorem. *Mathematics of Operations Research* 3:1–9.

Shapley, L. S., 1950. Information and the formal solution of many-moved games (abstract). In *Proceedings, International Congress of Mathematicians,* vol. 1, pp. 574–575.

———, 1951a. The noisy duel: Existence of a value in the singular core. RAND Publication RM-641.

———, 1951b. Notes on the *n*-person game, II: The value of an *n*-person game. RAND Publication RM-670.

———, 1952a. Notes on the *n*-person game, III: Some variants of the von Neumann–Morgenstern definition of solution. RAND Publication RM-817.

———, 1952b. [Untitled abstract.] *Econometrica* 20:91.

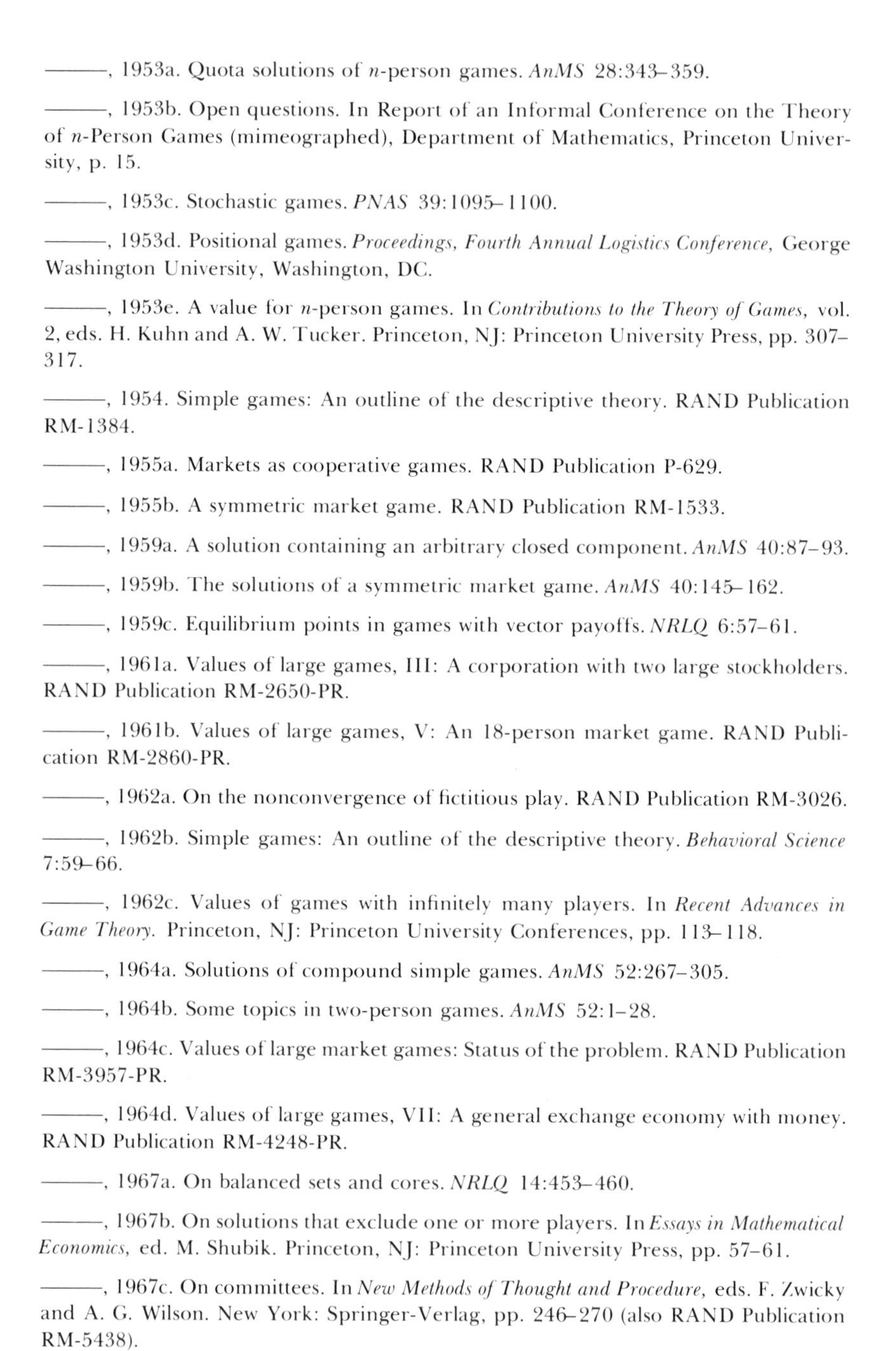

———, 1953a. Quota solutions of *n*-person games. *AnMS* 28:343–359.

———, 1953b. Open questions. In Report of an Informal Conference on the Theory of *n*-Person Games (mimeographed), Department of Mathematics, Princeton University, p. 15.

———, 1953c. Stochastic games. *PNAS* 39:1095–1100.

———, 1953d. Positional games. *Proceedings, Fourth Annual Logistics Conference,* George Washington University, Washington, DC.

———, 1953e. A value for *n*-person games. In *Contributions to the Theory of Games,* vol. 2, eds. H. Kuhn and A. W. Tucker. Princeton, NJ: Princeton University Press, pp. 307–317.

———, 1954. Simple games: An outline of the descriptive theory. RAND Publication RM-1384.

———, 1955a. Markets as cooperative games. RAND Publication P-629.

———, 1955b. A symmetric market game. RAND Publication RM-1533.

———, 1959a. A solution containing an arbitrary closed component. *AnMS* 40:87–93.

———, 1959b. The solutions of a symmetric market game. *AnMS* 40:145–162.

———, 1959c. Equilibrium points in games with vector payoffs. *NRLQ* 6:57–61.

———, 1961a. Values of large games, III: A corporation with two large stockholders. RAND Publication RM-2650-PR.

———, 1961b. Values of large games, V: An 18-person market game. RAND Publication RM-2860-PR.

———, 1962a. On the nonconvergence of fictitious play. RAND Publication RM-3026.

———, 1962b. Simple games: An outline of the descriptive theory. *Behavioral Science* 7:59–66.

———, 1962c. Values of games with infinitely many players. In *Recent Advances in Game Theory.* Princeton, NJ: Princeton University Conferences, pp. 113–118.

———, 1964a. Solutions of compound simple games. *AnMS* 52:267–305.

———, 1964b. Some topics in two-person games. *AnMS* 52:1–28.

———, 1964c. Values of large market games: Status of the problem. RAND Publication RM-3957-PR.

———, 1964d. Values of large games, VII: A general exchange economy with money. RAND Publication RM-4248-PR.

———, 1967a. On balanced sets and cores. *NRLQ* 14:453–460.

———, 1967b. On solutions that exclude one or more players. In *Essays in Mathematical Economics,* ed. M. Shubik. Princeton, NJ: Princeton University Press, pp. 57–61.

———, 1967c. On committees. In *New Methods of Thought and Procedure,* eds. F. Zwicky and A. G. Wilson. New York: Springer-Verlag, pp. 246–270 (also RAND Publication RM-5438).

———, 1969. Utility comparison and the theory of games. In *La Décision: Agrégation et dynamique des ordres de préférence.* Paris: Editions du CNRS, pp. 251–263.

———, 1971a. Cores of convex games. *IJGT* 1:11–26.

———, 1971b. On Milnor's class "L." RAND Publication R-795-RC.

———, 1972. Simple games: Application to organization theory. RAND Publication.

———, 1973a. On balanced games without side payments. In *Mathematical Programming,* eds. T. C. Hu and S. M. Robinson. New York: Academic Press, pp. 261–290 (also RAND Publication P-4910).

———, 1973b. An example of a nonconverging kernel in a replicated market game. RAND Publication IN-22419-NSF.

———, 1975a. Cardinal utility from intensity comparison. RAND Publication R-1683-PR.

———, 1975b. An example of a slow-converging core. *International Economic Review* 16:345–351.

———, 1976. Noncooperative general exchange. In *Theory and Measurement of Economic Externalities,* ed. A. Y. Lin. New York: Academic Press.

———, 1977. A comparison of power indices and a nonsymmetric generalization. RAND Publication P-5872.

Shapley, L. S., and H. E. Scarf, 1974. On cores and indivisibility. *Journal of Mathematical Economics* 1:23–38.

Shapley, L. S., and M. Shubik, 1953. Solutions of n-person games with ordinal utilities (abstract). *Econometrica* 21:348–349.

———, 1954. A method for evaluating the distribution of power in a committee system. *APSR* 48:787–792.

———, 1966. Quasi-cores in a monetary economy with non-convex preferences. *Econometrica* 34:805–827.

———, 1967a. Concepts and theories of pure competition. In *Essays in Mathematical Economics,* ed. M. Shubik. Princeton, NJ: Princeton University Press, pp. 63–79.

———, 1967b. Ownership and the production function. *Quarterly Journal of Economics* 81:88–111.

———, 1969a. On market games. *Journal of Economic Theory* 1:9–25.

———, 1969b. On the core of an economic system with externalities. *American Economic Review* 59:678–684.

———, 1969c. Price strategy oligopoly with product variation. *Kyklos* 22:30–44.

———, 1969d. Pure competition, coalitional power, and fair division. *International Economic Review* 10:337–362.

———, 1972a. The assignment game, I: The core. *IJGT* 2:111–130.

———, 1972b. Convergence of the bargaining set for differentiable market games. Working notes, mimeographed.

———, 1972c. The kernel and bargaining set for market games. Working notes, mimeographed.

———, 1977. Trade using one commodity as a means of payment. *Journal of Political Economy* 85:937–968.

Shapley, L. S., and R. N. Snow, 1950. Basic solutions of discrete games. *AnMS* 24:27–35.

Sherman, R. 1972. *Oligopoly: An Empirical Approach.* Lexington, MA: Lexington Books.

Shiffman, M., 1953. Games of timing. In *Contributions to the Theory of Games, II,* eds. H. W. Kuhn and A. W. Tucker. Princeton, NJ: Princeton University Press, pp. 97–123.

Shishko, R., 1970. A survey of solution concepts for majority rule games. Mimeographed notes, Department of Economics, Yale University.

Shitovitz, B., 1973. Oligopoly in markets with a continuum of traders. *Econometrica* 41:467–501.

Shubik, M., 1952. A business cycle model with organized labor considered. *Econometrica* 20:284–294.

———, 1954a. Does the fittest necessarily survive? In *Readings in Game Theory and Political Behavior,* ed. M. Shubik. New York: Doubleday.

———, ed. 1954b. *Readings in Game Theory and Political Behavior.* New York: Doubleday.

———, 1955a. A comparison of treatments of a duopoly problem, *Econometrica* 23:417–431.

———, 1955b. The uses of game theory in management science. *Management Science* 2:40–54.

———, 1955c. Edgeworth market games. Seminar notes, Center for Advanced Study in the Behavioral Sciences, Palo Alto, CA.

———, 1956. A game theorist looks at the Sherman Antitrust Act and the automobile industry. *Stanford Law Review* 8:594–630.

———, 1957. Market form: Intent of the firm and market behavior. *Zeitschrift fur Nationalökonomie* 17:186–196.

———, 1959a. Edgeworth market games. In *Contributions to the Theory of Games, IV,* eds. A. W. Tucker and R. D. Luce. Princeton, NJ: Princeton University Press, pp. 267–278.

———, 1959b. *Strategy and Market Structure.* New York: John Wiley.

———, 1960. Games, decisions and industrial organization. *Management Science* 6:455–473.

———, 1961. Objective functions and models of corporate optimization. *Quarterly Journal of Economics* 73:345–375.

———, 1962a. Incentives, decentralized control, the assignment of joint costs and internal pricing. *Management Science* 8:325–343.

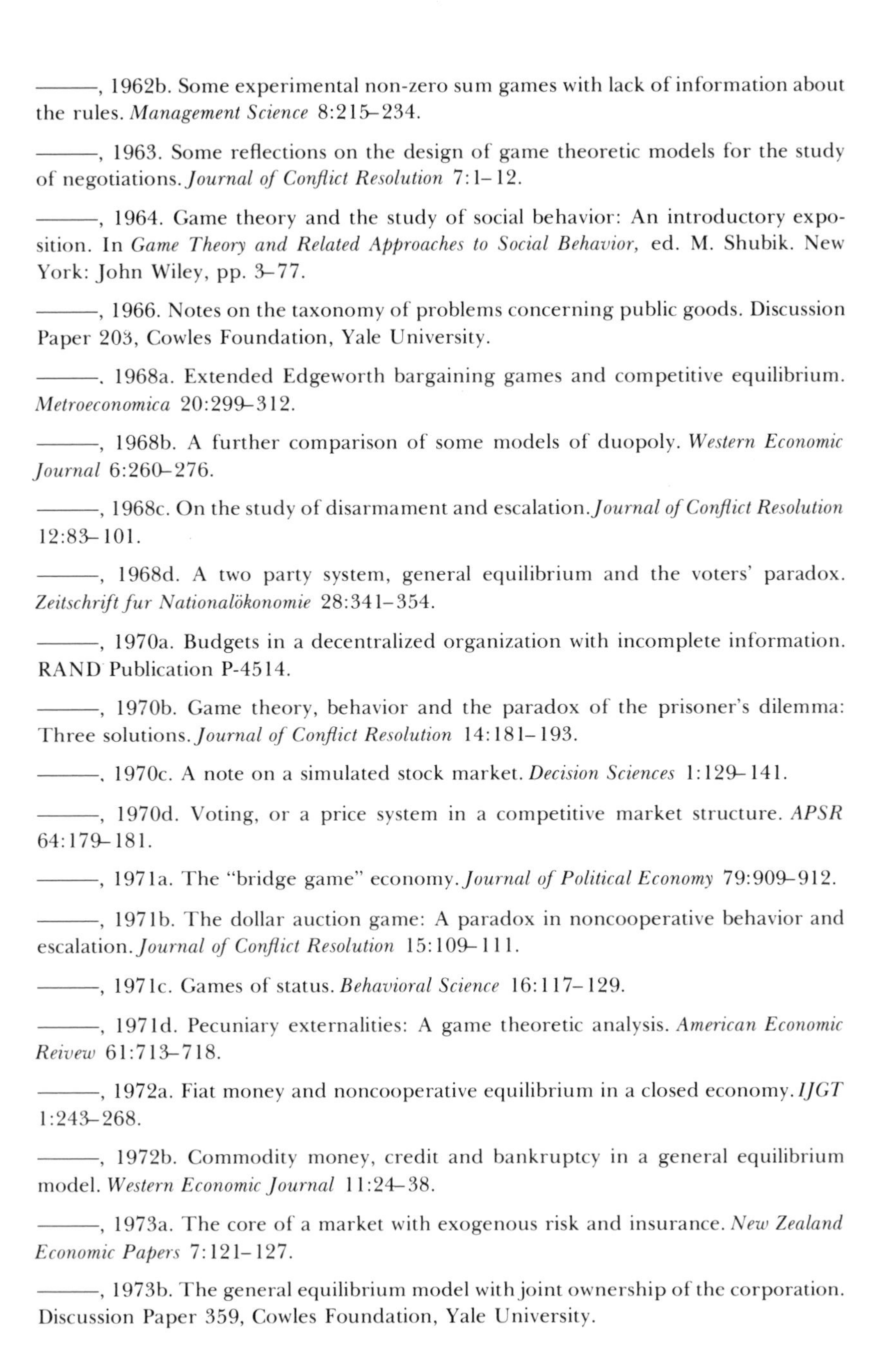

———, 1962b. Some experimental non-zero sum games with lack of information about the rules. *Management Science* 8:215–234.

———, 1963. Some reflections on the design of game theoretic models for the study of negotiations. *Journal of Conflict Resolution* 7:1–12.

———, 1964. Game theory and the study of social behavior: An introductory exposition. In *Game Theory and Related Approaches to Social Behavior*, ed. M. Shubik. New York: John Wiley, pp. 3–77.

———, 1966. Notes on the taxonomy of problems concerning public goods. Discussion Paper 203, Cowles Foundation, Yale University.

———, 1968a. Extended Edgeworth bargaining games and competitive equilibrium. *Metroeconomica* 20:299–312.

———, 1968b. A further comparison of some models of duopoly. *Western Economic Journal* 6:260–276.

———, 1968c. On the study of disarmament and escalation. *Journal of Conflict Resolution* 12:83–101.

———, 1968d. A two party system, general equilibrium and the voters' paradox. *Zeitschrift fur Nationalökonomie* 28:341–354.

———, 1970a. Budgets in a decentralized organization with incomplete information. RAND Publication P-4514.

———, 1970b. Game theory, behavior and the paradox of the prisoner's dilemma: Three solutions. *Journal of Conflict Resolution* 14:181–193.

———, 1970c. A note on a simulated stock market. *Decision Sciences* 1:129–141.

———, 1970d. Voting, or a price system in a competitive market structure. *APSR* 64:179–181.

———, 1971a. The "bridge game" economy. *Journal of Political Economy* 79:909–912.

———, 1971b. The dollar auction game: A paradox in noncooperative behavior and escalation. *Journal of Conflict Resolution* 15:109–111.

———, 1971c. Games of status. *Behavioral Science* 16:117–129.

———, 1971d. Pecuniary externalities: A game theoretic analysis. *American Economic Reivew* 61:713–718.

———, 1972a. Fiat money and noncooperative equilibrium in a closed economy. *IJGT* 1:243–268.

———, 1972b. Commodity money, credit and bankruptcy in a general equilibrium model. *Western Economic Journal* 11:24–38.

———, 1973a. The core of a market with exogenous risk and insurance. *New Zealand Economic Papers* 7:121–127.

———, 1973b. The general equilibrium model with joint ownership of the corporation. Discussion Paper 359, Cowles Foundation, Yale University.

———, 1973c. Information, duopoly and competitive markets: A sensitivity analysis. *Kyklos* 26:736–761.

———, 1975a. Competitive equilibrium, the core, preferences for risk and insurance markets. *Economic Record* 51:78–83.

———, 1975b. *Games for Society, Business and War.* Amsterdam: Elsevier.

———, 1975c. *The Uses and Methods of Gaming.* New York: Elsevier.

———, 1976. A noncooperative model of a closed economy with many traders and two bankers. *Zeitschrift fur Nationalökonomie* 36:49–60.

———, 1977. Competitive and controlled price economies: The Arrow-Debreu model revisited. In *Equilibrium and Disequilibrium in Economic Theory,* ed. G. Schwodiauer. Dordrecht: Reidel.

———, 1978a. Opinions on how one should play a three person nonconstant sum game. *Games and Simulation* 9:302–308.

———, 1978b. A theory of money and financial institutions. *Economie Appliquée* 31:61–84.

———, 1979. Cooperative game solutions: Australian, Indian and U.S. opinions. Discussion Paper 517, Cowles Foundation, Yale University.

———, 1980. The capital stock modified competitive equilibrium. In *Models of Monetary Economies,* eds. J. H. Karaken and N. Wallace. Minneapolis, MN: Federal Reserve Bank of Minneapolis.

———, 1981. Society, land, love or money: A strategic model of how to glue the generations together. Discussion Paper 577, Cowles Foundation, Yale University.

Shubik, M., with R. E. Levitan, 1980. *Market Structure and Behavior.* Cambridge, MA: Harvard University Press.

Shubik, M., and M. Riese, 1972. An experiment with ten duopoly games and beat-the-average behavior. In *Contributions to Experimental Economics,* vol. 3, ed. H. Sauermann. Tübingen: Mohr.

Shubik, M., and M. J. Sobel, 1980. Stochastic games, oligopoly theory and competitive resource allocations. In *Dynamic Optimization and Mathematical Economics,* ed. P.-T. Liu. New York: Plenum.

Shubik, M., and G. L. Thompson, 1959. Games of economic survival. *NRLQ* 6:111–123.

Shubik, M., and L. Van der Heyden, 1978. Logrolling and budget allocation games. *IJGT* 7:151–162.

Shubik, M., and R. J. Weber, 1981a. Competitive valuation of cooperative games. *Mathematics of Operations Research* (forthcoming).

———, 1981b. Systems defense games: Colonel Blotto, command and control. *NRLQ* 28:281–287.

Shubik, M., and W. Whitt, 1973. Fiat money in an economy with one nondurable good and no credit (a noncooperative sequential game). In *Topics on Differential Games*, ed. A. Blaquiere. Amsterdam: North-Holland, pp. 401–448.

Shubik, M., and C. Wilson, 1977. The optimal bankruptcy rule in a trading economy using fiat money. *Zeitschrift fur Nationalökonomie* 37:337–364.

Shubik, M., G. Wolf, and H. Eisenberg, 1972. Some experiences with an experimental oligopoly business game. *General Systems* 13:61–75.

Shubik, M., G. Wolf, and S. Lockhart, 1971. An artificial player for a business market game. *Simulation and Games* 2:27–43.

Shubik, M., G. Wolf, and B. Poon, 1974. Perception of payoff structure and opponent's behavior in related matrix games. *Journal of Conflict Resolution* 18:646–656.

Shubik, M., and H. P. Young, 1978. The nucleolus as a noncooperative game solution. In *Game Theory and Political Science*, ed. P. Ordeshook. New York: New York University Press, pp. 511–528.

Siegel, S., and L. E. Fouraker, 1960. *Bargaining and Group Decision Making*. New York: McGraw-Hill.

Simon, H. A., 1953. Notes on the observation and measurement of political power. *Journal of Politics* 15:500–516.

———, 1956. A comparison of game theory and learning theory. *Psychometrika* 21:267–272.

Simon, R. I., 1967. The effects of different encodings on complex problem solving. Ph.D. dissertation, Yale University.

Sion, M., and P. Wolfe, 1957. On a game without a value. *AnMS* 39:299–306.

Sjoberg, G., 1953. Strategy and social power: Some preliminary foundation. *Southwestern Social Science Quarterly* 33:297–308.

Skala, H. J., 1967. Uber die Existenz von "cores." Workshop on Game Theory, Institute for Advanced Studies, Vienna, mimeographed.

Skérus, S. L., and I. P. Jačiauskas, 1971. Coalitional differential three person game (in Russian with English and Lithuanian summaries). *Lietuvos Mathematikos Rinkiys* 11:887–898.

Sloss, J. L., 1971. Stable points of directional preference relations. Technical Report 71-7, Operations Research House, Stanford University.

Smale, S., 1980. The prisoner's dilemma and dynamical systems associated to noncooperative passes. *Econometrica* 48:1617–1634.

Smith, D. G., 1965. The probability distribution of the number of survivors in a two-sided combat situation. *Operational Research Quarterly* 16:429–437.

———, 1967. A duel with silent-noisy gun versus noisy gun. *Colloquium Mathematicum* 17:131–146.

———, 1969. Stochastic games (abstract). *Notices of the AMS* 16:176.

Smith, V. L., 1965. Experimental auction markets and the Walrasian hypothesis. *Journal of Political Economy* 73:387–393.

———, 1966. Bidding theory and the treasury bill auction: Does price discrimination increase bill prices? *Review of Economics and Statistics* 48:141–146.

———, 1967. Experimental studies of discrimination vs. competition in sealed-bid auction markets. *Journal of Business of the University of Chicago* 40:56–84.

———, 1978. Experimental mechanisms for public choice. In *Game Theory and Political Science*, ed. P. Ordeshook. New York: New York University Press, pp. 323–355.

———, 1979. *Research in Experimental Economics*, vol. 1. Greenwich, CT: JAI Press.

Smithies, A., and L. J. Savage, 1940. A dynamic problem in duopoly. *Econometrica* 8:130–148.

Smorodinsky, M., and E. Kalai, 1975. Other solutions to Nash's bargaining problem. *Econometrica* 43:513–518.

Snell, J. L., 1952. Applications of martingale system theorems. *Transactions of the AMS* 73:293–312.

Sobel, M. J., 1970. An algorithm for a game equilibrium. Discussion Paper 7035, Center for Operations Research and Econometrics, Louvain, Belgium.

———, 1971. Noncooperative stochastic games. *Annals of Mathematical Statistics* 42:1930–1935.

———, 1978. Stochastic games. Unpublished manuscript.

Sobolev, A. I., 1975. The characterization of optimality principles in cooperative games by functional equations (in Russian with English summary). *Mathematical Methods in the Social Sciences* 6:150–165.

Skolina, N. A., and O. N. Bondareva, 1971. Construction of the solution for a class of games with nonempty core by a decomposition method (in Russian with English summary). *Vestnik Leningradskii Universitet* 7:40–46.

Sonnenschein, H., 1965. The relationship between transitive preference and the structure of the choice space. *Econometrica* 33:624–634.

———, 1967. Reply to "A note on orderings" (by P. Lorimer). *Econometrica* 35:540–541.

Sperner, E., 1928. Ein Satz uber Untermengen einer Endlichen Menge. *Mathematische Zeitschrift* 27:544–548.

Stackelberg, H. F. von, 1952. *The Theory of the Market Economy* (translated from German; original, 1934). London: W. Hodge.

Stahl, I., 1972. *Bargaining Theory*. Stockholm: Economics Research Institute.

Stark, R., and R. Mayer, Jr., 1971. Some multi-contract decision theoretic competitive bidding models. *Operations Research* 19:469–483.

Stark, R., and M. H. Rothkopf, 1979. Competitive bidding: A comprehensive bibliography. *Operations Research* 27:364–390.

Starr, A. W., and Y. C. Ho, 1969. Non zero sum differential games. *Journal of Optimization Theory and Applications* 3:184–206.

Starr, R. M., 1966. A probability weighted value for n-person games. RAND Publication D-15049-PR.

———, 1969. Quasi-equilibria in markets with non-convex preferences. *Econometrica* 37:25–38.

———, 1974. The price of money in a pure exchange monetary economy. *Econometrica* 42:45–54.

Stearns, R. E., 1964a. Three-person cooperative games without side payments. *AnMS* 52:377–406.

———, 1964b. On the axioms for a cooperative game without side payments. *Proceedings of the AMS* 15:82–86.

———, 1965. A game without side payments that has no solution. In Report of the Fifth Conference on Game Theory (mimeographed), Department of Mathematics, Princeton University.

———, 1967. A formal information concept for games with incomplete information. In *Report to the U.S. Arms Control and Disarmament Agency.* Mathematica, Policy Research, Inc., Princeton, NJ, pp. 405–433.

———, 1968. Convergent transfer schemes for N-person games. *Transactions of the AMS* 134:449–459.

Steinhaus, H., 1948. The problem of fair division. *Econometrica* 16:101–104.

———, 1949. Sur la division pragmatique. *Econometrica* (supplement) 17:315–319.

Stern, D., 1966. Some notes on oligopoly theory and experiments. In *Essays in Mathematical Economics,* ed. M. Shubik. Princeton, NJ: Princeton University Press, pp. 255–281.

Stevens, S. S., 1957. On the psychophysical law. *Psychological Review* 64:153–181.

———, 1961. The psychophysics of sensory function. *American Scientist* 48:226–253.

Stigler, G. J., 1940. Notes on the theory of duopoly. *Journal of Political Economy* 48:521–541.

———, 1947. The kinky oligopoly demand curve and prices. *Journal of Political Economy* 55:432–449.

———, 1964. A theory of oligopoly. *Journal of Political Economy* 72:44–61.

Stone, J. J., 1958. An experiment in bargaining games. *Econometrica* 26:286–296.

Straffin, P. D., 1976. Probability models for measuring voting power. Technical Report 320, College of Engineering, Cornell University.

———, 1977. Homogeneity, independence and power indices. *Public Choice* 30:107–118.

Suppes, P., 1961. Behavioristic foundations of utility. *Econometrica* 29:186–202.

Suppes, P., and J. Carlsmith, 1962. Experimental analysis of a duopoly situation from the standpoint of mathematical learning theory. *International Economic Review* 3:60–78.

Suppes, P., and M. Winet, 1955. An axiomatization of utility based on the notion of utility differences. *Management Science* 1:259–270.

Suzdal, V., 1976. *Game Theory for the Navy.* Moscow: Voyenizdat.

Suzuki, M., and M. Nakayama, 1976. The cost assignment of cooperative water resource development—A game theoretic approach. *Management Science* 22:1081–1086.

Sweat, C. W., 1968. Adaptive competitive decision in repeated play of a matrix game with uncertain entries. *NRLQ* 15:424–448.

———, 1969. A duel involving false targets. *Operations Research* 17:478–488.

———, 1971. A single-shot noisy duel with detection uncertainty. *Operations Research* 19:170–181.

Sweezy, P., 1939. Demand under conditions of oligopoly. *Journal of Political Economy* 47:568–573.

Takahashi, M., 1962. Stochastic games with infinitely many strategies. *Journal of Science* (Hiroshima University) 26:123–134.

Talamanca, M., 1954. Contributi allo studio delle vendite all'asta nel mondo classico. Serie VIII, Vol. VI, Fascicolo 2, Accademia Nationale dei Lincei, Roma.

Taylor, J. G., 1974. Solving Lanchester-type equations for "modern warfare" with variable coefficients. *Operations Research* 22:756–770.

———, 1979. Recent developments in the Lanchester theory on combat. In *IFORS Proceedings,* ed. K. B. Haley. Amsterdam: North-Holland, pp. 773–806.

Telser, L. G., 1972. *Competition, Collusion and Game Theory.* Chicago, IL: University of Chicago Press.

———, 1979. *Economic Theory and the CORE.* Chicago: University of Chicago Press.

Thompson, F. B., 1952a. Equivalence of games in extensive form. RAND Publication RM-759.

———, 1952b. Behavior strategies in finite games. RAND Publication RM-769.

Thompson, G. L., 1953a. Signaling strategies in n-person games. *AnMS* 28:267–277.

———, 1953b. Bridge and signaling. *AnMS* 28:279–289.

Thrall, R. M., 1954. Applications of multidimensional utility theory. In *Decision Processes,* eds. R. M. Thrall, C. H. Coombs, and R. L. Davis. New York: John Wiley, pp. 181–186.

———, 1962. Generalized characteristic functions for n-person games. In *Recent Advances in Game Theory.* Princeton, NJ: Princeton University Conferences, pp. 157–160.

Thrall, R. M., C. H. Coombs, and R. L. Davis, eds., 1954. *Decision Processes.* New York: John Wiley.

Thrall, R. M., and W. F. Lucas, 1963. n-person games in partition function form. *NRLQ* 10:281–298.

Trivers, R. L., 1971. The evolution of reciprocal altruism. *Quarterly Review of Biology* 46:35–57.

Tucker, A. W., 1950. A two-person dilemma. Mimeographed paper, Stanford University.

Tullock, G., 1967. The general irrelevance of the general possibility theorem. *Quarterly Journal of Economics* 81:256–270.

Turner, J. E., and D. J. Rapoport, 1971. Concepts of preference in economics and biology. In *Differential Games and Related Topics,* eds. H. Kuhn and G. Szegö. Amsterdam: North-Holland.

Tversky, A., 1969. Intransitivity of preferences. *Psychological Review* 76:31–48.

Vajda, S., 1956. *Theory of Games and Linear Programming.* New York: John Wiley.

Vakriniene, S. P., 1968. Antagonistic dynamic games based on repeated bimatrix games (in Russian with English summary). *Lietuvos Matematikos Rinkinys* 7:397–402.

———, 1970. The dynamic game when the interests of the players coincide (in Russian with English summary). *Lietuvos Mathematikos Rinkinys* 10:229–234.

van Heijenoort, J., 1967. *From Frege to Gödel.* Cambridge, MA: Harvard University Press.

Vickrey, W., 1959. Self-policing properties of certain imputation sets. *AnMS* 40:213–246.

———, 1960. Utility strategy and social decision rules. *Quarterly Journal of Economics* 74:507–535.

———, 1961. Counterspeculation, auctions, and competitive sealed tenders. *Journal of Finance* 16:8–37.

Vilkas, E., 1963. Axiomatic definitions of value of a matrix game (in Russian). *Teoriya Veroyatnostei i ee Premeneniya* 8:324–327.

Vilkov, V. B., 1972. Some theorems about core-solutions for games without side payments (in Russian). *Vestnik Leningradskii Universitet* 19:5–8.

Ville, J. A., 1938. Sur la théorie générale des jeux où intervient l'habilité des jouers. In *Traité du calcul des probabilités et de ses applications,* vol. 4, ed. E. Borel. Paris: Gauthier-Villars, pp. 105–113.

Vind, K., 1964. Edgeworth allocation in an exchange economy with many traders. *International Economic Review* 15:165–177.

———, 1965. A theorem on the core of an economy. *Review of Economic Studies* 32:47–48.

Vogel, W., 1956. Die Annaherung guter Strategien bei einer gewissen Klasse von Spielen. *Mathematische Zeitschrift* 65:283–308.

von Neumann, J., 1928. Zur theorie der Gesellschaftespiele. *Mathematische Annalen* 100:295–320.

———, 1937. Uber ein ökonomisches Gleichungssystem und eine Verallgemeinerung des Brouwerschen Fixpunktsatzes. In *Ergebnisse eines Math. Coll.* (Vienna), ed. K. Menger, vol. 8, pp. 73–83.

———, 1945. A model of general economic equilibrium. *Review of Economic Studies* 13:1–9.

———, 1953a. A certain zero-sum two-person game equivalent to an optimal assignment problem. *AnMS* 28:5–12.

———, 1953b. Communication on the Borel notes. *Econometrica* 21:123–125.

von Neumann, J., and O. Morgenstern, 1944. *Theory of Games and Economic Behavior.* Princeton, NJ: Princeton University Press (2nd ed., 1947).

———, 1961. Symmetric solutions of some general *N*-person games. RAND Publication P-2169 (based on an unpublished 1946 manuscript).

Vorobyev, N. N., 1958. Equilibrium points in bimatrix games. *Theory of Probability and Its Applications* 3:297–309.

———, 1963. Partitioning of strategies in extensive games (in Russian). *Problemy Kibernetiki* 7:5–20.

———, 1966. Games with imperfectly known rules (in Russian). *Tezisi Kratkych Nauсznych Soobshchenij,* Sekcia 13, str. 15.

———, 1968. The "attack-défense" game (in Russian). *Lietuvos Mathematikos Rinkinys* 8:437–444.

———, 1970a. Positzionniye igri i prinyatiye statisticheskix reshenii. *Mathematische Operations forschung und Statistik* 1:45–53.

———, 1970b. The development of game theory. Working Paper 2 (1970), Department of Economics, New York University (translated from Russian); German translation, *Entwicklung der Spieltheorie* (Berlin: VEB Deutscher Verlag der Wissenschaften, 1975).

———, 1970c. The present state of the theory of games. *Russian Mathematical Surveys* 25:77–136.

———, 1976. *Game Theory: Bibliography* (in Russian). Leningrad: Soviet Academy of Sciences.

———, 1981. *Game Theory: Bibliography 1969–1974.* (in Russian). Leningrad: Soviet Academy of Sciences (forthcoming).

Vorobyev, N. N., and I. N. Vrublevskaya, eds., 1967. *Positzionnie Igre* [Positional games]. Moscow: Nauka.

Vrublevskaya, I. N., 1968. Ob igre odnogo napadaioshchego protiv neskol'kix zashchitnikov. *Lietuvos Mathematikos Rinkinys* 8:447–459.

———, 1970. Svoistva resheniya igri odnogo napadaioshchego protiv neskol'kix zashchitnikov. *Lietuvos Mathematikos Rinkinys* 10:235–251.

Waggener, H., and G. Suzuki, 1967. Bid evaluation for procurement of aviation fuel at DFSC: A case study. *NRLQ* 14:115–130.

Wagner, H., 1969. *Principles of Operations Research.* Englewood Cliffs, NJ: Prentice-Hall.

Wagner, R., 1969. The concept of power and the study of politics. In *Political Power,* eds. R. Bell et al. New York: Free Press, pp. 3–12.

Wald, A., 1945a. Statistical decision functions which minimize the maximum risk. *Annals of Mathematics* 46:265–280.

———, 1945b. Generalization of a theorem by von Neumann concerning zero sum two person games. *Annals of Mathematics* 46:281–286.

———, 1947. Review of *Theory of Games and Economic Behavior. Review of Economic Statistics* 29:47–52.

———, 1951. On some systems of equations of mathematical economics. *Econometrica* 19:368–403.

Walras, L., 1874. *Elements l'économie politique pure.* Lausanne: L. Corbaz. English translation, *Elements of Pure Economics* (London: Allen and Unwin, 1954).

Washburn, A. R., 1971. *An Introduction to Evasion Games.* Monterey, CA: U.S. Naval Postgraduate School.

Weber, R. J., 1973a. A generalized discriminatory solution for a class of n-person games. Technical Report 174, Department of Operations Research, Cornell University.

———, 1973b. Discriminatory solutions for $[n, n - 2]$-games. Technical Report 175, Department of Operations Research, Cornell University.

———, 1979. Subjectivity in the valuation of games. Discussion Paper 515, Cowles Foundation, Yale University.

———, 1981. Probabilistic values for games. *Mathematics of Operations Research* (forthcoming).

Wesley, E., 1971. An application of nonstandard analysis to game theory. *Journal of Symbolic Logic* 36:385–394.

Wieczorek, A., 1976. Coalition games without players. Publication 237, Computation Centre, Polish Academy of Sciences.

Willis, R. H., and N. J. Long, 1967. An experimental simulation of an international truel. *Behavioral Science* 12:24–32.

Wilson, R. B., 1967. Competitive bidding with asymmetrical information. *Management Science* 13:816–820.

———, 1968. A class of solutions for voting games. Working Paper 156, Graduate School of Business, Stanford University.

———, 1969. An axiomatic model of logrolling. *American Economic Review* 3:331–341.

———, 1970. The finer structure of revealed preference. *Journal of Economic Theory* 2:348–353.

———, 1971a. A game theoretic analysis of social choice. In *Social Choice,* ed. B. Lieberman. New York: Gordon & Breach, pp. 393–407.

———, 1971b. Computing equilibria of *N*-person games. *SIAM Journal of Applied Mathematics* 21:80–87.

———, 1971c. Stable coalition proposals in majority-rule voting. *Journal of Economic Theory* 3:254–271.

———, 1972a. Consistent modes of behavior. Technical Report 55, Institute for Mathematical Studies in the Social Sciences, Stanford University.

———, 1972b. The game-theoretic structure of Arrow's general possibility theorem. *Journal of Economic Theory* 5:14–20.

———, 1972c. Social choice theory without the Pareto principle. *Journal of Economic Theory* 5:478–486.

Winder, R. O., 1968. Fundamentals of threshold logic. In *Applied Automata Theory,* ed. J. T. Tau. New York: Academic Press, pp. 235–318.

———, 1969. The status of threshold logic. *RCA Review* 30:62–84.

Wohlstetter, A., 1964. Sin and games in America. In *Game Theory and Related Approaches to Social Behavior,* ed. M. Shubik. New York: John Wiley, pp. 209–225.

Wolf, G., and M. Shubik, 1974. Concepts, theories and techniques: Solution concepts and psychological motivation in prisoner's dilemma games. *Decision Sciences* 5:153–163.

Wolfe, P., 1955. The strict determinateness of certain infinite games. *Pacific Journal of Mathematics* 5:891–897.

Yasuda, Y., 1970. A note on the core of a cooperative game without side payment. *NRLQ* 17:143–149.

———, 1971. The structure of an Edgeworth market. Working Paper, Department of Social Engineering, Tokyo Institute of Technology.

Yoshikawa, T., 1970. An example of stochastic multistage games with noisy state observation. *IEEE Transactions on Automatic Control* 15:455–458.

Young, H. P., 1978a. A tactical lobbying game. In *Game Theory and Political Science,* ed. P. C. Ordeshook. New York: New York University Press, pp. 391–404.

———, 1978b. Power prices and incomes in voting systems. *Mathematics of Programming* 14:129–148.

Zabel, E., 1967. A dynamic model of the competitive firm. *International Economic Review* 8:199–208.

———, 1969. The competitive firm and price expectations. *International Economic Review* 10:467–478.

———, 1970. Monopoly and uncertainty. *Review of Economic Studies* 37:205–219.

Zachrisson, L. E., 1964. Markov games. *AnMS* 52:211–253.

Zadeh, L., 1965. Fuzzy sets. *Information and Control* 8:338–353.

Zamir, S., 1969a. On the relation between the values of finitely and infinitely repeated games of incomplete information. Publication RM-43, Department of Mathematics, Hebrew University, Jerusalem.

———, 1969b. On the value of a finitely repeated game of incomplete information with a general information function. Publication RM-46, Department of Mathematics, Hebrew University, Jerusalem.

———, 1970. The inequivalence between the two approaches to repeated games with incomplete information. Publication RM-55, Department of Mathematics, Hebrew University, Jerusalem.

Zauberman, A., ed., 1975. *Differential Games and Other Game Theoretic Topics in Soviet Literature.* New York: New York University Press.

Zeckhauser, R., 1973. Determining the qualities of a public good. *Western Economic Journal* 11:39–60.

Zermelo, E., 1913. Uber eine Anwendung der Mengenlehre auf die Theorie des Schachspiels. *Proceedings, Fifth International Congress of Mathematicians* 2:501–504.

Zeuthen, F., 1930. *Problems of Monopoly and Economic Warfare.* London: G. Routledge and Sons.

Index